Ships and Shipping in the North Sea and Atlantic, 1400–1800

Professor Richard W. Unger

Richard W. Unger

Ships and Shipping in the North Sea and Atlantic, 1400–1800

Routledge
Taylor & Francis Group
LONDON AND NEW YORK

First published 1997 in the Variorum Collected Studies Series by Ashgate Publishing

Reissued 2018 by Routledge
2 Park Square, Milton Park, Abingdon, Oxon, OX14 4RN
52 Vanderbilt Avenue, New York, NY 10017

Routledge is an imprint of the Taylor & Francis Group, an informa business

Publisher's Note
The publisher has gone to great lengths to ensure the quality of this reprint but points out that some imperfections in the original copies may be apparent.

Disclaimer
The publisher has made every effort to trace copyright holders and welcomes correspondence from those they have been unable to contact.

A Library of Congress record exists under LC control number:

ISBN 13: 978-1-138-38624-2 (hbk)
ISBN 13: 978-0-429-42690-2 (ebk)

CONTENTS

SHIPBUILDING, GOVERNMENTS AND SHIPPING

This volume contains xii + 316 pages

PUBLISHER'S NOTE

The articles in this volume, as in all others in the Collected Studies Series, have not been given a new, continuous pagination. In order to avoid confusion, and to facilitate their use where these same studies have been referred to elsewhere, the original pagination has been maintained wherever possible.

Each article has been given a Roman number in order of appearance, as listed in the Contents. This number is repeated on each page and quoted in the index entries.

PREFACE

Dutch trade and shipping have long been recognized as the source of the short period of prominence which those seven provinces enjoyed in European international politics in the seventeenth century. What made the success even more dramatic were the simultaneous political and economic difficulties of virtually all other European states. Dutch practices were looked on at the time as the ultimate key to prosperity and power. Much contemporary discussion turned on how to emulate the Dutch while destroying the basis for their power and prosperity. The critical role of exchange between East and West in the shifting balance of power in Renaissance northern Europe was the starting point for the entire programme of study covered in this volume. The search for answers led to a concern for the sources of Dutch success. Questions of how the Dutch did it date from the seventeenth century as do some of the answers proposed in the writing which follows in this book.

The capital goods of Dutch shipping, the ships, were known to be better, at least in business terms, than those of their competitors. Why that was so, how that was so and what caused that to be so led to an examination of a range of aspects of the Dutch shipbuilding industry. The general study in the book, *Dutch Shipbuilding Before 1800: Ships and Guilds*, Assen, 1978, has been supplemented and expanded with specific work on certain regions and towns, on the intricacies of technical advance and the representation of the developments made in Dutch ships. Those more particular inquiries of the sources are gathered here.

Dutch shipbuilding and ship design did not occur in a vacuum. Changes were very much a part of larger developments in Europe and a series of studies on the actions of governments to promote and direct shipbuilding has been complemented by studies on the effects of developments in shipbuilding on the actions of government. Diffusion of new methods in shipbuilding should have proven historically to have been easier than with virtually any other technology since the products embodying improvements travelled widely. Still the pace was slow and there are some efforts, when talking about governments, to test the influence of geography, institutions and public authorities on shipbuilders' borrowing of techniques and methods from elsewhere.

The results of improvements in ships and shipbuilding methods did, it would appear, give Dutch merchants and fishermen real competitive advantages. One of the greatest successes of Holland was in the herring fishery and the origins of that success can be found in the fifteenth century. The complex interconnection of the Dutch trading network and the ways the Dutch held and improved their competitive position is well illustrated by the herring fishery and the trade in cured herring. It is clear in other trades as well from coal to beer.

Two of the items included in this volume appeared originally in Dutch. Rather than reproducing the original English versions the published texts have been translated back into English. This has allowed keeping the sense of what did appear in print and also the addition of some new information. In one case a list of suggestions for further reading has been included and in the other the sources and notes have been supplemented by works that have come out since the original Dutch version appeared. The goal was to retain as much of the earlier sense and conclusions as possible while making the two pieces more useful. The illustrations from the originals have, however, been omitted.

The individual studies are brought together first with an examination of the foundation for Dutch competitiveness in the superiority of shipbuilding, then an examination of the relationship in general between shipbuilding and the larger economy and society, and finally with some examination of the specific cases of trades in northern Europe, of exchanges between East and West from the later Middle Ages down to the end of the eighteenth century. The influence of technological change on the development of human society is an issue which remains unresolved. Though for a while considered worthy of vigourous research the topic and the whole range of issues connected with it has fallen from the forefront of historical study. It belongs to a long list of discarded questions and problems, identified but never resolved and overwhelmed by the novelty of fascination with a different set of problems. It is always valuable to revisit the discarded and forgotten issues in the historical storehouse of ideas now and again if for no other reason than to be reminded of the variety of experience and variety of influences that have existed and have formed the lives of people in the past and, by implication, the present.

The original publishers of the articles and essays offered here have been kind enough to give their permission for reproduction of the items: the editors of the *Tijdschrift voor Zeegeschiedenis* and the officers of the Nederlandse Vereniging voor Zeegeschiedenis; David & Charles PLC, the former publishers of *Maritime History*; the honourable editor of *The Mariner's Mirror*; the editor of the *Zeeuws Tijdschrift*; the Publications and Exhibitions Department of The Getty Research Institute for the History of Art and the Humanities; Unieboek B. V.; the University of Chicago Press, publishers of *Technology and Culture*; Robert William Love, the editor of *Changing Interpretations and New Sources of Naval History*; the officers of the International Commission of Maritime History; the president of the Instituto de Investigaçao Científica e Tropica and the president of the Academia Marinha, both of Lisbon; Barry M. Gough, editor of *The American Neptune*, Peabody Museum of Salem - Essex Institute, East India Square, Salem, Massachusetts 01970, U. S. A.; Dott.ssa M. C. Barabino, director of the Direzione Communicazione of the Commune of Genoa; the Regents of the University of California and the University of California Press, publishers of *Viator*; the Cambridge University Press, publishers of the *Journal of Economic History*; Christiane Villain-Gandossi, the editor of *Medieval Ships and the Birth of Technological Societies*.

The research and writing that is brought together in this volume carries only my name but it is the product of the efforts of many archivists, librarians, research assistants, colleagues, friends and members of my close family. Many of them played a critical role in the preparation and completion of specific studies. Though they deserve to be mentioned it is one individual whose contribution is and has been truly outstanding. It is to the individual who gave me early direction, who has supported my work unfailingly over the past thirty years, and who has been the leader in the promotion of the study of Dutch maritime history at home and internationally that I would like to dedicate this volume. My deepest gratitude and that of many of us who investigate the past of ships and shipping of the Netherlands goes to Jaap Bruijn.

RICHARD W. UNGER

Vancouver
May 1997

I

DUTCH SHIPCARPENTERS' GUILDS, C. 1400 TO C. 1600*

Though less commonly examined historically than ship design or shipping, Dutch shipbuilding was and remains an integral part of the maritime history of the Netherlands. Economically, shipbuilding supplied capital goods to the shipping industry and therefore made an important contribution to this trade sector of the economy. More than that it was the shipbuilder who created novel designs. Technical improvements in ship design merely reflected progress in the shipbuilding industry. The importance of high quality capital goods, goods which included properly suited technical features, is most often recognized by writers on the herring fishery. It is said that the new type of herring net introduced in the early fifteenth century could not be effectively employed until the herring buss came into use.

Clearly the sources of that new type or any new type of vessel, the way the design evolved and the methods of building such a type are important to the history of the economy. Though the need to explore the history of shipbuilding may be widely appreciated, such study has proven to be rather difficult. To all those problems of sources and of technical information common to any examination of maritime history is added a general lack of information about shipbuilding itself. Until the appearance of modern corporate organizations with public responsibility shipbuilding was in the hands of families noted for their conservatism and the haphazard nature of their record keeping. Since historically the tendency has been toward greater formalization, toward more complete records, then of course the study of shipbuilding is most difficult and most problematic for the earliest years, indeed for the years before the Golden Century. The development of the "fluit" in 1595 and its refinement over the following two decades combined with the appearance of a highly centralized and efficient shipbuilding industry in the Zaanstreek marked the establishment of Dutch shipbuilding as the best in Europe. That combined with the development of the Admiralties and a more scientific approach to ship design, another product of the seventeenth century, all combined to generate more extensive records dealing with ship design and with the shipbuilders implementing those designs. Though records may be less extensive still the history of Dutch shipbuilding before 1595 deserves equal or greater attention.

Craft were probably built in prehistoric times along the coast that now forms part of the two maritime provinces of Holland and Zeeland. The extensive commerce of Frisians in the eight and ninth centuries was in part shipped in their own vessels built in the two provinces. The industry, if it can be called an industry in those years, was unstable. A builder would temporarily appropriate a small stretch of beach and there piece together the vessel from a small quantity of wood, wood which might even be found locally. A stable, urban-based and Dutch, that is non-Frisian, industry with fixed and continuing shipbuilding yards did not appear until the time of Count Floris the Fifth (d. 1296).1) From those beginnings in the late thirteenth century the history of Dutch shipbuilding to the late sixteenth century is both complex and elusive. In those years expecially after about 1450, there was rapid technical progress in ship design throughout Europe and in Holland and Zeeland as well. The design and construction of

* Paper prepared with the help of a Fullbright-Hays Related Fellowship from the U.S. Government.
Footnotes see p. 51

the "fluit" was the final step in that progression. It was also a time of the expansion of Dutch commerce into many new trades, some of which Dutch shippers came to dominate. It was also a time of growth in the Dutch fishery and of significant design improvement in Dutch fishing vessels. But evidence from the shipbuilding industry is sparse. The progression in ship design is obscure even if the principle features of the advance are known. No continuing government agency existed over the three centuries to regulate shipbuilding and thus to generate documentary evidence on the industry. Contracts between shippers and builders, for the construction of vessels, an excellent source for information about the design of ships as well as about the shipbuilders and their ways of doing business, do not appear in the two provinces until the late sixteenth century. The parties to such purchases did not need to have such agreements. The usually, at least before 1550, lived near each other and could regularly consult on the design and construction of the ship. What is more, either or both parties might well be illiterate. Though contracts may be few and government documents extremely scarce, one body of evidence does exist. The information comes not from consumers but from producers, from the shipbuilders' organization, their guilds.

Craft guilds were a common feature of European industrial organization in the fifteenth and sixteenth centuries. The demographic disaster of the mid-fourteenth century associated with the Black Death and the economic consequences generated by that disaster caused an increased demand for skilled labourers. The guilds were established or expanded to train those skilled artisans. Shipbuilders did not escape the general European expansion but shipcarpenters' guilds tended to appear well after the Black Death, that is, not until the early fifteenth century. The delay may have come from a slack demand for tonnage associated with the sharp fall in the volume of international trade in the second half of the fourteenth century. Dutch shipbuilders were not alone in choosing to establish guilds. Such groups appeared in the Hanse towns, in England and in France in the fifteenth and sixteenth centuries. Dutch shipcarpenters' guilds were little different from those in other regions or from other craft guilds in the Netherlands. Loosely defined guilds were groups with an economic basis and a legal or quasi-legal position as a public authority which placed their members under some kind of collective discipline in the practice of their common craft.2) The guild was to train workers for the specific skill in question and also to control or be the agency of control over production and methods of production. The guild received its authority either tacitly or in a charter from some governmental body. Thus it is difficult to say whether a specific guild was an agency of government or of the trade itself, wheter the guild was acting in the common interest of all persons in the town or in the specific interest of the guild membership. Of course the purpose of guild action varied with circumstances, with time and with the nature of the government body authorizing the existence of the guild. Many duties, economic, political and social, devolved on the guilds and the benefits or ill-effects of guild organization varied with the actions taken by the guilds, with how the guilds used the authority given them. The guild could encompass more than one professional group. By the fourteenth century shipcarpenters were already seen as a group apart, as men with a specific skill which required special training. Yet, shipcarpenters often found themselves subsumed in general guilds of wood-workers. Usually called carpenters' guilds, these groups included chair and window-makers. All building trades might be added, even bricklayers. It was still possible in the fifteenth and sixteenth centuries for a shipbuilder to carry on some other trade or trades such as general carpentry or even farming. But the inverse was not true. Shipbuilding was closed and increasingly closed to all except those with precise knowledge of new ship design and building methods. Since shipcarpenters fell into a unique category as early as the fourteenth century, the tendency for those shipcarpenters was to form a separate guild apart from other carpenters. But breaking away from parent guilds was slow in coming. Among other considerations, shipcarpenters apparently decided to wait until the market for their services was large enough to support a reasonable number of men with their specific skill and thus large enough to make such a guild viable. That reasonable number was probably very low. Perhaps as few as ten shipcarpenters would feel able to establish an independent guild in their town.

I

Despite their quasi-public character, usually a guarantee of great quantities of paper work and therefore surviving evidence, guilds have left few records. What has survived does not, for example, cover all those towns in which shipbuilding took place. Guilds were, by their nature, urban so that building and, more important, what repair work was done on the beaches outside the towns is completely beyond the records of guilds. Few of the guild activities were documented in detail. The result is a tantalizingly scarce body of information, even more tantalizing since it is the principal source of information for the Dutch shipbuilding industry before 1600.
General documents have sporadically survived for some of the Dutch shipcarpenters' guilds. The general items give clues to the social activities of the organizations, to the extent of guild activity in town life and in the lives of the members. Such scraps of evidence also offer a test of the involvement of the guilds in the economy of the town and province. No specific body of information in guild records yields consistent information on those important issues. What is needed is an extensive series of account books like that of the Hoorn guild. Unfortunately, that detailed compilation of income and expenditure starts only in 1593. There are no other accounts from before the introduction of the "fluit". Town ordinances dealing with guilds like those of Amsterdam on guildsmen's caulking (1539) are informative but limited, sporadic and few indeed. The enrollment book of the Goes carpenters' guild shows the number of shipcarpenters and all woodworkers in the guild and presumably in the town. It is the only such book surviving and only covers the later sixteenth century. 3) Of these general pieces the best body of evidence is records of litigation. Chartered guilds were by definition juristic personalities, could sue and be sued in court. If the guild ran smoothly then there was little need for the paperwork and documents which came out of disputes. At Dordrecht not all went well, so records from court cases are rather extensive. Since they are records of the guild they are largely the opinion of the shipcarpenters' organizations. Similar views on court actions have survived at Veere from the late sixteenth century along with the results of petitions to the town government. Dordrecht is the only guild to leave record of a petition to the provincial government. The entire haphazard collection, though informative, is severely limited. The greatest problem, beyond the lack of information and the reliability of some of it, is that documents of one town are not strictly comparable with those of others. Fortunately, though, there is one of documents which to a large degree eliminates that difficulty.
Guild letters, "gildebrieven", are statement of regulations which are to apply to all guild members for the life of the guild. The surviving letters of Dutch shipcarpenters' guilds of the fifteenth and sixteenth centuries all deal with similar topics. Further, it is clear that with perhaps one exception the letters were not copied one from the other, but rather composed for each specific town to meet prevailing local circumstances. The similarity of topics covered in the letters is striking. The guild letter was first at most importantly a formal announcement of "incorporation", of the establishment of a legal personality. A charter from some political authority might often begin the guild letter to announce the legal establishment of the guild. A statement of charter was not necessary: the promulgation of the letter was enough. The letter was a formal statement of specific regulations, of bye-laws which covered many aspects of the professional and personal lives of shipcarpenters. The letters invariably dealt with requirements for entrance. Each guild has a policy on the citizenship of applicants. Fees on entrance were fixed and training requirements were often laid down, including in a few late sixteenth century instances, the completion of some specific task overseen by guild inspectors as proof of ability. The guild letter set out dates for general meetings of members but there was in general little said specifically about procedure in those meetings. The connection of the guild with the church, at least in letters which appeared before 1572, and general public services to be carried out by the corporation were outlined. More specific were the benefits accruing to members in case of sickness or death. Within the letters is information about the sources of corporate income and the expenditure of that income. Much of each letter is devoted to the constitutional organization of the guild, how it was to be governed, and the protection to be guaranteed members and the officers who took on specified duties for the group.

I

All the regulations in the guild letter are merely a statement of intent. That statement may or may not reflect fact. Indeed with such limited supplemental information from the general evidence there is no way to demonstrate the effectiveness of the bye-laws. With no source for comparison with the statement of the letters then they must be used very carefully. Apparently the choice is to either make the heroic assumption that the letters were valid, that indeed the bye-laws were adhered to by members, or to simply give up trying to ascertain the nature of organization of the shipbuilding industry during those important years before the Golden Century. There is one way however to avoid the pitfall of questionable enforcement. Some conclusions are possible taken not from the content of the letters but rather from their general nature and from the fact that the guild letters appeared when and where they did.

From the fourteenth through the sixteenth centuries fourteen guild letters have survived which apply specifically to shipcarpenters. Eight of the fourteen come from Holland while the remaining six were drawn up in the much less populous province of Zeeland. Shipcarpenters might fall under sets of bye-laws not specifically for their trade. Such bye-laws are not included such as the 1491 Reimerswaal regulation. There the shipcarpenters were subsumed in the rest of the wood working trades so they may have formed a very small group indeed. Moreover by the late fifteenth century, the town was already in an advanced state of decline because of flooding which had covered almost all the land around the town. 4) The fourteen were not the only guild letters drawn up over the three hundred years. Some regulations may have been destroyed with the Reformation. Guilds had church responsibilities and one of their principal concerns was the maintenance of a church altar. With the change in religious practice it was necessary not only to destroy the altar but also to change guild bye-laws dealing with all aspects of religion. The conversion to a secular organization is clearly documented in the Veere guild letter of 1585. The changes, deletions and corrections made in the letter of 1520 were almost exclusively to accomodate the new approach to religion and the new character of the guild. 5) Indeed the prime reason the guild drew up a new set of regulations at all may well have been the change in religious practice. It is possible that some guilds merely destroyed the old set of bye-laws and did not bother to draw up a new letter, but there is no evidence that guilds destroyed such letters. If what has survived is not the full body of guild letters promulgated before the Reformation, then the set may well be complete for after the Reformation.

The extant documents from Holland include four guild letters from Dordrecht (1397, 1437, 1548, 1587). 6) The earlier two are rather sketchy in the regulation of shipcarpenters. The Alkmaar shipcarpenters received a guild letter in 1521. 7) The three letters of the Amsterdam guild were issued between 1578 and 1589, 1589 and 1591. Some less complete ordinances for Amsterdam shipcarpenters exist for the early sixteenth century. 8) For Zeeland the earliest document of interest is the pre-Reformation letter of Veere (1520) which was followed by the revised version (1585). A guild of shipcarpenters at Brouwershaven received a charter and guild letter in 1545 and that is apparently the only record of such an organization. 9) The guild letters of Vlissingen (1575) and Arnemuiden (1576) were drawn up soon after the Revolt and the change in religion. 10) The Zierikzee letter (1597) was the last to appear in the two provinces in the sixteenth century. 11) Though those are the only guild letters which have survived that is not to say that these were the only guilds of shipcarpenters. Since shipcarpenters could be subsumed in other guilds, the guild letters do not show conclusively the establishment of a shipcarpenters' guild; only the establishment of a separate or independent shipcarpenters' guild. The Amsterdam craftsmen belonged to at least three other guilds before striking out on their own between 1578 and 1589. 12) Further, some guilds though independent did not bother with a formal guild letter. The Haarlem shipcarpenters were fully organized. The earliest extent bye-law is of 1538. 13) No formal letter or charter for the guild has survived however. Rotterdam shipcarpenters were in with the general woodworkers until 1554 when they established an independent organization. The earliest surviving guild letter is from the mid-eighteenth century. 14) So the eight Holland and Zeeland towns which show surviving guild letters are, with the exception of Rotterdam, those which had formally organised and chartered shipcarpen-

ters' guilds. The nine were by no means the only towns with shipcarpenters' guilds nor for that matter were they the only towns with a shipbuilding industry. The problem then is not why shipcarpenters formed guilds - for that there were usually sound economic reasons - but rather why such groups sought formal regulation.
First, any change in town government, in the personnel ruling the town was the inspiration for guilds to seek such formal recognition of their groups. Acceptance of the Revolt or of the Reformation in the 1570's commonly meant such a change. At Dordrecht the establishment of all craft guilds, shipcarpenters included, was part of a move to more democratic government. The monopoly of political power held by merchant oligarchs was broken in 1367 when Duke Albrecht, count of Holland, granted a charter for the establishment of craft guilds and gave the craftsmen some political power, power to be exercised through the guilds.15) Such constitutional arrangements common in Flanders and South Germany were known only at Dordrecht in Holland and Zeeland. The chartering of guilds at Arnemuiden and Vlissingen came soon after those towns joined the Revolt. At Amsterdam the shipcarpenters chose independance soon after the "Alteratie". The guild letter is not dated so it may well have been drawn up immediately after the change in government.
Second, it appears that the desire to regulate the trade, the desire to have formal legal recognition of the regulatory body was for the shipcarpenters a defensive measure and defensive in an economic as well as a political sense. The formal arrangement, the chartering of a guild in most cases apparently came out of a fear of loss of relative economic strength, of loss of tonnage sales to the shipping industry. Guilds as informal associations of men with common economic interests plying the same trade could and did exist. Such associations of shipcarpenters may have even been larger than the chartered groups. With no data on membership in either type of group comparison is impossible. Still what is important is not the size of the group but rather how the competitive position of the shipcarpenters changed within the memory of members and how those tradesmen assessed their future. The Veere shipcarpenters must have perceived a decline in their prospects after the shift of the Schelde and the concomitant move of shipping from the eastern to the western arm of that river. A guild letter was issued at Veere before 1520 and though the date is unknown it probably appeared after the change in delta sandbars. The letter of 1520 included revision and greater formalization of the earlier bye-laws. The decision of 1520 suggest another step to combat the deterioration. The Vlissingen guild letter was not only the product of a change in government but was also an emergency measure designed to meet the crisis in shipping generated by the Revolt and the confusion about the future of trade through Antwerp. The emergency nature of the decision to regulate the guild is all but stated in the letter. The Zierikzee shipcarpenters guild existed before the earliest surviving guild letter of 1597. The change in trade patterns in the Schelde delta had an adverse effect on the trade of Zierikzee and therefore presumably on demand for shipbuilding. An earlier guild letter of unknown date was apparently designed to overcome the associated economic difficulties. The 1597 letter appeared in the wake of the Revolt which was most damaging to the town. But more than that Zierikzee was fighting a losing battle with silting in the harbour. Shippers were deterred from choosing the town as a transshipment point and thus from employing local carpenters for repair or new construction work. In both Holland and Zeeland the shipcarpenter's guild most conscious of privilege was that of Dordrecht. The group regularly sought reaffirmation of its position. Guildsmen made a point of maintaining records of their privileges. They went to court frequently to prevent infringement of their privileges and at times they sued to extend those privileges. Dordrecht, the home of the most vigorous of organized shipcarpenters' guilds in the fifteenth and sixteenth centuries, was also a town that experienced stagnation in shipping volume, in population and indeed in almost every indicator of economic well-being. Dordrecht throughout the two centuries was attempting to hold onto the prosperity the town had known in the thirteenth century. The vigorous defensive posture struck by the shipcarpenters' guild was apparently just a reflection of a general attitude held by all the people of Dordrecht. Despite their efforts though business, commerce and shipbuilding were lost to towns further down the

Maas, especially to Rotterdam which had an independent shipcarpenters' guild after 1554. The shipcarpenters of many rapidly growing commercial centers of the fifteenth and sixteenth centuries did not seek the establishment of a formal chartered and regulated organization. Amsterdam and Rotterdam are the exceptions. For Amsterdam shipcarpenters, in a rapidly growing town with a well established guild system and a government deeply involved in regulating all aspects of shipping, the choices were limited. They could let the town government legislate for them. That had obvious disadvantages. Guild organization was common so the shipcarpenters turned first to larger guilds which might protect them and then finally to self-regulation through the chartering of their own independent guild. At Rotterdam the situation was very much the same. At Hoorn a shipcarpenters' guild existed but it did not have a guild letter. Brielle, the port of the herring fishery in the fifteenth century, Enkhuizen, the rapidly growing center of the fishery in the sixteenth century, and Edam, reputated to be the greatest shipbuilding center in sixteenth century Holland, had no shipcarpenters' guilds at all. If the shipcarpenters were organized in these three towns it was on a personal and informal basis. These men apparently did not feel the need for a set of regulations to cover their actions. Shipcarpenters in stagnate or declining towns did. And though they may have been taking a defensive measure that is not to say that the shipcarpenters took restrictive action. To look at all guilds as impediments to economic growth would be ignoring the facts. The regulations laid out in the guild letter may have in fact been relatively open and designed instead to promote growth in the industry. Indeed a regulated industry might prove less restrictive of competition than one informally organized. Both Amsterdam and Rotterdam shipcarpenters had a guild and there is no indication that growth of the industry was stunted there by the method of organization. Indeed, in both towns in the late sixteenth and early seventeenth centuries extensive public investment was made in shipbuilding facilities. To assess the restrictiveness of the regulations not only is it necessary to know the effectiveness of the regulations but also to know about the nature of the market for shipping tonnage in Holland and Zeeland. Throughout the fifteenth and sixteenth centuries that market was growing and it seems that some towns or more precisely some shipcarpenters did not feel that their sales were growing fast enough. So these men sought solace in better or rather in more formal internal organization of their trade.

The guild letters of shipcarpenters' organizations can be used effectively in at least one sense. The simple fact of issuing a guild letter demonstrates the intent or aspiration of the tradesmen involved. Wheter or not the regulations were enforced or how they effected the trade is not crucial. The great fear for the historian who uses guild bye-laws is always that they may give a false picture of the true situation. But here only the intention and not the action is important. The pitfall has been avoided. Conclusions about the desires of shipbuilders derived from these documents are valid but of course conclusions about the shipbuilding industry itself must be treated with a great deal more care. The maritime history of Holland and Zeeland before 1595 must in part be based on an understanding of shipbuilding so this evidence, the guild letters and related guild material, is valuable as a first step in examining the origins of superior Dutch ships. Of course, a great deal of other information must be brought to bear on the shipbuilding industry, its structure, personnel and methods of operation before it is possible to assess the contribution of Dutch shipbuilders to the shipping industry and indeed to Dutch maritime expansion of the fifteenth and sixteenth centuries. The men who built the ship though were conscious of their position in the economy and the legitimate legal structure of their towns. Faced with internal or external difficulties they deliberately chose a course of action to meet those difficulties. Shipbuilders apparently were aware of their options and rather than passively accept the situation they picked a positive course of action from among those options.

NOTES

Dutch shipcarpenters'guilds, c. 1400 to c. 1600

1) J.M. Kerling, Commercial Relations of Holland and Zeeland with England from the late 13th Century to the Close of the Middle Ages. Leiden, E.J. Brill, 1954, p. 3.
2) E. Coornaert, Les Corporations en France avant 1789. Paris, Les Editions Ouvrières, 1968, p. 32.
3) Gemeente Archief Hoorn, 1851, Scheeptimmermansgild Rekeningboek, 1593-1601.
J.G. van Dillen (ed), Bronnen tot de Geschiedenis van het Bedrijfsleven en Gildewezen van Amsterdam. Vol. 1, 1312-1611 (Rijks Geschiedkundige Publicatiën, 69) no. 235.
Gemeente Archief Goes, 423 (V nr. 1750), Gildeboeken van het Timmerliedengilde.
4) R. Fruin (ed), Het Recht der Stad Reimerswaal. The Hague, Martinus Nijhoff, 1905, pp. 54-57.
5) Gemeente Archief Veere (Middelburg), Gildebrief van 1 Jan., 1520, Gildebrief van 10 Jan., 1585.
6) A. van Vollenhoven, Ambachten en Neringen in Dordrecht. The Hague, Martinus Nijhoff, 1923, Appendix 3. A printing of the guild bye-law of 1397, Gemeente Archief, Dordrecht, Afdeling Gilden 254.
Gemeente Archief Dordrecht, Afdeling Gilden 232 (1437), 261 (1549), 228 (1587).
7) Gemeente Archief Alkmaar, Archieven van Gilden 92.
8) J.G. van Dillen (ed), op. cit., 1349 (between 1578 and 1589), 772 (1589).
Gemeente Archief Amsterdam, Gilden Archief 1376 (1591).
For earlier regulation of Amsterdam shipcarpenters, caulking, their payment of the beer excise, etc. see J.G. van Dillen, ed., op. cit., 29 (1517), 228 (1538), 235 (1539), 240 (1539), 547 (1567).
For a survey of the regulations pertaining to Amsterdam shipcarpenters issued before an independent guild was formed see, Leonie van Nierop, Bijdragen tot de geschiedenis van de Amsterdamse scheepsbouw, in: Amstelodamum, Maandblad voor de Kennis van Amsterdam, 42ste jrg. (1955), pp. 39-40, 54-55.
9) Gemeente Archief Brouwershaven (Middelburg), Scheeptimmerlieden Gildebrief (1545).
10) Gemeente Archief Vlissingen, Ordonnanties Vlissingen 5161, 23 (1575).
Gemeente Archief Arnemuiden, 287 (1576).
11) Gemeente Archief Zierikzee, Resolutie Boek van St. Toeloopers of Scheepstimmermansgilde.
12) Van Nierop, loc. cit.
13) J. Huizinga (ed.), Rechtsbronnen der Stad Haarlem. The Hague, Martinus Nijhoff, 1911, pp. 253-254.
14) S.C. van Kampen, De Rotterdamse Particuliere Scheepsbouw in de Tijd van de Republiek, Assen, Born N.V., 1953, p. 163.
15) Matthys Balen Janszoon, Beschryvinge der stad Dordrecht. 2 vols. Dordrecht, Symon onder de Linde, 1677, p. 311. The avowed purpose of the ordinance was to give greater political stability. Janszoon reprinted the 1367 ordinance.

II

Selling Dutch Ships in the Sixteenth Century[1]

I

THE shipbuilders of Holland and Zeeland experimented extensively with different and novel designs in the sixteenth century. The result was rapid change for the better in the quality of Dutch ships.[2] The shipbuilders and the shippers, probably not purposefully, developed methods of doing business which allowed for careful and precise control over the cost of that principal capital asset in the shipping industry, the ship. More than that, though, these ways of doing business actively promoted or at the very least did not inhibit the great technical improvements in ships. The methods were a result of the overall expansion of trade, of rising literacy among all parties involved, of greater experience in dealing with far-flung operations and, probably most important, of technical change in ships. While business methods promoted design improvement at the same time methods were constrained by the improvements.

Novel and unique adjustments were required on both sides of the bargain for a ship, by the shipper, in financing the purchase of his capital good, and by the shipbuilder, in producing the improved capital good. In both cases there was in the sixteenth century increased specialisation, always a source of efficiency. Specialisation by sector such as in shipping continued and increased through the following centuries. But there was also specialisation in task within the shipping industry and especially in the buying of a ship. As the speed of technical change slowed in the seventeenth century and as men came to be more at home within the expanded shipping industry, the trend toward individual specialisation tended on the whole to stop and in some cases even to reverse.

From *Maritime History* volume 3 (David & Charles, 1973) by kind permission of the publishers.

The market for new ships as with any other was the meeting place of buyer and seller. It was in this interchange that decisions not only about price but also about design, financing and the speed of work were determined. The relative cost of the product, its 'lumpiness', the nature of its use and its life expectancy forced arrangements to be made different from those common to other contemporary markets. The ship was the largest single good bought and sold in sixteenth-century Europe. Buying a ship meant a relatively sizeable investment, made even greater by the increasing size and complexity of ships.

II

The ship, unlike that other great investment, land, was not divisible. This indivisibility usually meant that ownership had to be divided. So on one side of the bargain there was rarely a single buyer but rather a group of buyers, a partnership. Joint ownership was the common method for investment in large vessels from antiquity. Not only did single owners not have the capital at their disposal to buy such a large good but more than that partnership distributed risk. Risk was not insignificant given the unreliability of the vessels. There was one ownership syndicate per ship and there were no limits on the different possible ways of dividing the equity.[3] Even with small vessels for use on internal waterways there might be joint ownership. For large ocean-going ships it was almost a necessity. Shares in ships were negotiable, highly liquid assets compared with other sixteenth-century investments and offered those mythical widows and orphans an opportunity to invest without great responsibility often with a relatively low cost of entry.

There were of course other possible investments in the sixteenth century, many of them the same type of partnership share arrangement. So those investors who chose ships tended to be connected with trade or relatively large-scale commerce. Still, men of varying trades did buy shares.[4] To shipowners participation was generally a sideline and rarely their principal occupation. The same was true in the fishing industry where part ownership of herring busses dates from the appearance of that specialised ship type in the early fifteenth century.

The best sixteenth-century example and the source of much of the information on contemporary Dutch business operations is the extensive extant records of the Van Adrichem family of Delft.[5] They developed an abiding interest in the Baltic grain trade after they found that they could not supply customers from local supplies. Though they were involved in ownership of cargo vessels members of the family still held shares in a number of herring busses. The records of the family, covering approximately the last forty years of the sixteenth century, show increased wealth and with that decreased investment in herring busses. More and more they invested in shares of large cargo vessels. Also from buying eighths and fifths of small ships they expanded to owning quarters and halves of larger sea-going vessels. It was a pattern followed by many successful merchants throughout the sixteenth century.[6]

Part ownership of sea-going vessels increased in sixteenth-century Holland and Zeeland. In the fourteenth century, despite the popularity of partner arrangements in the contemporary Mediterranean, Dutch seamen could still expect that by careful saving and with some inheritance they would come to own vessels of their own. They would then act as owner, master, pilot, skipper and merchant. But in the fifteenth century and increasingly in the sixteenth these functions were separated among an ever increasing number of specialists. Most obviously the skipper or master became less and less important in decisions affecting the business operation. It was still common in the sixteenth century for a master to own at least part of the ship but pressure was against even such a reduced share holding. The reason the master retained as much equity as he did was the well-founded belief on the part of the shareholders that a skipper who was part owner would be more prone to protect their investment, that is by navigating safely, and would act when necessary as their spokesman.

Ships from the fourteenth century grew in size and became technically more complex. So ships became more expensive. That meant that the original capital outlay grew as well. Returns to investment rose but so did the threshold of investment to enter long-distance ocean-going commerce. The degree of specialisation in ship design and the nature of the specific trade of course dictated the level of the threshold. In the sixteenth century few masters had the funds or could earn them to purchase or order a new sea-going vessel.[7] The profit situation dictated the introduction of vessels of the latest design. But without the capital masters turned to sleeping partners. As their share of the equity increased relative to that of masters the sleeping partners became active owners. The master turned into a minor shareholder and in some instances just an employee with no capital share at all.[8]

The tendency to land-based ownership affected the fishing industry as much or more than long-distance commerce. Masters who in the fifteenth century had been owners of busses in whole or in part saw their share on the average decline. These masters remained commanders of the crew and in charge of navigation, an important task when it was a matter of finding fish. From 1434 there is an indication that the master or pilot of the herring buss also acted as the bookkeeper and was the initiator and organiser of investment in the ship. The master apparently spoke for all the owners. But around 1600 all that changed. Perhaps because of the increasing wealth of merchants dealing in herring those experts in selling began to take an interest in production. The merchants thus bettered the security of their supply. Like grain merchants who bought shares in Baltic-bound cargo vessels so herring merchants bought herring vessels. In some cases, they owned the busses outright without any partners. Overall the master came to hold a less pivotal place in the operation of the herring boat and therefore in the process of buying the vessel. The centralisation of control was possible in part because of the greater reliability which shipbuilders included in the busses and also the relative standardisation of design. And it was not just the herring buss which exhibited that feature in the late sixteenth century.[9]

Geographical discoveries and expanding interregional commerce in Europe opened extensive new possibilities for ocean-going commerce in the sixteenth century. At the same time the system of part ownership was taxed to such an extent that it was unable to supply the needed ready money. The Portuguese government, and to some extent the Spanish, simply stepped into the gap and supplied the capital for trade to India and the New World. There were various other ways in which governments chose to support shipping and shipbuilding and help overcome the failings of the simple partnership arrangement. Direct subsidies to shipbuilders, the reserving of long-distance trade to natives who in turn would probably buy ships locally, or tariffs to protect domestic shipping all in one way or another made raising capital easier for undertakers of shipping ventures.[10] Either government cash grants lowered the entry threshold or protection of the business increased expected return. Those practices dating from the early thirteenth century proved adequate to the promotion of trade and shipping in Europe. But when it came to matters of exploration, voyages through unfamiliar waters to largely unknown lands all at high risk with questionable profit, then the Iberian governments found that they could not allow the usual system of part ownership to prevail both because of the incapacity of that system to meet the massive capital need and, what may have been as important, there was an important public function involved in these voyages. Competition between European states and also between religions or sects made extra-European trade and the establishment of colonies and connections throughout the world too important to be left to chance. Northern Europeans, following in the wake of the Iberians, were not confronted with quite the same problems. Yet for England and for the German Hanse towns along the North and Baltic Sea coasts subsidies to shipping and to shipbuilding were quite common. And that proved adequate to the task of mobilising capital for trading ventures, at least in the sixteenth century. As the projects became more ambitious, a novel institution, the joint stock company appeared in England in embryonic form. In the first years of the seventeenth century it came forward full-blown in the form of the East India Company. Though often expressed potential royal pleasure with the success of that venture certainly did not damage its chances for survival, in general the English Crown was satisfied with the usual subsidies to commercial investment for fostering that and other efforts in extra-European trade.

In Holland and Zeeland on the other hand the policy of government and the form of business relationships were much more traditional. The unique character of Low Countries politics prevented governments from giving the common direct and indirect subsidies to shipbuilding and shipping. With no special conditions established by government there was no prod to men of affairs to change their ways. Rather shipowning partnership was the model for all kinds of investment in trade. Trade was opened between the new Dutch Republic and Italy in the late sixteenth century by partnerships which not only bought ships but also handled the shipping and hired agents overseas.[11] The extreme example occurred with the most extreme of trades, that to the Spice Islands. Trade to the East Indies at the end of the sixteenth century was in the hands of companies (*voorcompagniëen*) which were large terminal partnerships but with an inner circle of directors who in turn were responsible to their partners, the investors. This division of function was merely an extension of a division which had grown up in larger shipowning partnerships through the sixteenth century. In the papers of the Van Adrichem family, it is clear that one or more of the partners took an active interest in the operation while the others relied on that individual for continuing vigilance. The *voorcompagniëen* were joined as the United East India Company in 1602 as a full-fledged joint stock company. It was the logical outcome of the increasing division between the functions of manager and investor, a division begun in traditional shipowning partnerships. It was also the final step in the separation of business and navigation functions. For the United East India Company financing, ordering and oversight of the building of needed vessels was in no way connected to the men who worked on board. The company in time even built its own wharves and had its ships built by its own employees.

The potential shipowner needed essentially little other than money. The joint stock company made that explicit. Within the standard partnership system it was also essentially true. The ship was a sound investment and increasingly so in the sixteenth century. The investor needed a sizeable sum to begin with but that could be reduced by taking an interest in smaller vessels, or by the simple expedient of increasing the number of partners. Once the money had been put forward the wisdom of the investment choice became more and more clear. There was a sharp initial fall in value after launching. From then on while the value of the shares might slowly depreciate the investment would bring income more regularly as technical improvements in ships and the development of various standard trade routes gave the shipping industry more stability. There was only the income and no more outlays. Wages, repairs and refitting were paid for out of freight income by the business representative of the owners. Dividends would be distributed generally at the end of a voyage. If by chance a shipping venture made a loss then owners were called on to make up the difference and to finance the next venture.[12]

Also in the sixteenth century there was an increase in the chartering of ships to anyone interested in moving goods. The rising volume of commerce contributed to this greater tramping. Instead of owners supplying cargoes, goods were picked up at various ports visited by the vessel.[13] Dutch partnerships were quick to realise the potential and preceded all other Europeans in developing such shipping services for hire. But that along with other improvements in Dutch shipping and shipbuilding were aided by the rise to prominence of a specialist who had a unique combination of skills.

Partners did not need to have any knowledge of ships or shipping but still they found it an advantage for one or two of the syndicate members to have some expertise. The result was the owners' agent who was often the book-keeper of the partnership. He probably started as a merchant so knew about trading relationships. It was from this man that, by a lengthy but direct route, the seventeenth-century joint stock company arose. The agent replaced the skipper in many facets of the operation of trading ventures. He could make tramping a worthwhile alternative for a shipping combine. Above all he decreased the skill required of his fellow partners. The reliance on the specialist made it easier than ever for anyone with cash in hand to invest in shipping. Among the tasks the agent took on for the skipper and for the partners was buying the ship.

III

On the other side of the bargain for a new vessel was the shipbuilder. Usually an individual, sometimes partnerships of two or three men were formed for the construction of big ships.[14] The builder was above all a skilled artisan, a craftsman in wood. The shipmaker employed assistants both skilled and unskilled. There was no great difference between the shipbuilder, the man in charge of the operation, and the skilled shipcarpenter, a man who might be as skilled or more skilled than the master. The master also employed apprentices, men or boys learning the trade. Once the training period was over the distinguishing feature among shipcarpenters was the holding of the contract for construction. The unique supervisory function which came with the contract meant that the master or contractor was at once skilled worker and small businessman. He had the responsibilities of employer, salesman, merchant and also financier, all in a small way. The changes in ship design in the sixteenth century made the business functions of the contractor increasingly important relative to those of the labourer. That led to greater specialisation. As ships became bigger and involved more work per ton, the number of workers per contractor increased. The undertaker did not become a full-time professional contractor, however. The job of design and direction of work required too much skill for the business aspects of his work to dominate completely.[15]

Bigger ships meant that more land was needed for the job of building. The contractor would own or rent a small piece of waterfront property where he often lived. That not only gave him a place to launch the completed vessel but also easy access to supplies. He might well store the principal raw material, the wood, himself or he might buy it from wood merchants. Sheds for wood were in some cases just next to the house of the shipbuilder.[16] The ownership or control over a place to build the ship, along with some tools and the necessary supplies of wood, may indeed have been what separated contractors from all other shipcarpenters, and made it possiple for them to get contracts. The shipbuilding yard was a small and informal place. It was still possible, in the fifteenth century at least, to just find a piece of waterfront property and put a ship together. There was enough preparation of the land involved to make moving undesirable and to make renegotiation of rental agreements advisable in place of closing down and going elsewhere.[17] Part ownership was possible in wharves as in ships and the contractor might take on one or two silent partners. As capital requirements increased in the sixteenth century—both fixed and turnover capital since wood prices went up—so too did the frequency of such arrangements.[18]

Shipbuilding was not the only business of the contractor. Orders for new ships could be erratic so it was wise to have some other operation to hedge against unemployment of both capital and labour. Repair work, caulking, refitting could all fill in slack periods.[19] Another sideline in the fifteenth century was farming. Agriculture dominated the economy of all of Europe and it appears that shipbuilders like many other tradesmen supplemented their incomes by working some land.[20] By the end of the sixteenth century having a farm and shipbuilding had become inconsistent. The economic sense of and the opportunities for specialisation has become too obvious. Contractors became more involved in the shipping industry and, for example, invested in parts of ships rather than in land.[21] Becoming a wood merchant seems to have been the most common supplementary occupation as interest in agriculture disappeared. Shipcarpenters had to have a sound knowledge of wood types and wood markets. They had to buy wood in sizeable quantities for their primary enterprise and on completion of a ship there would be wood left over, the sale of which could prove an important addition to the builders' income.[22] Wood merchants rightly feared the competition of part-time dealers who might dump surplus wood on the local market, wood already paid for under a shipbuilding contract. At Dordrecht, the oldest commercial centre in Holland, as early as 1383 the wood merchants insisted that shipcarpenters refrain from selling any wood. Shipcarpenters could buy from anyone in or out of Dordrecht, saw the wood, use it in any way they wanted but could not sell wood.[23] Only if they joined the wood dealers' guild was that allowed.[24] Elsewhere wood merchants were not as well organised to put pressure on shipcarpenters and even at Dordrecht the wood merchants were not too successful in keeping shipbuilders from selling wood. The court cases and petitions attest to that failure.[25]

The social and economic status of these shipbuilders or contractors is difficult to ascertain. There are many problems with the use of almost any records which might be used for that purpose but still some facts are clear. Land tax registers, for example, reveal that it was at least possible for shipbuilders to rise relatively high in the social and political hierarchy of towns in Holland and Zeeland. Overall they never attained the status or income of the men for whom they built ships. In the late sixteenth and early seventeenth centuries Haarlem and Amsterdam each had mayors who were shipcarpenters by trade.[26] An Amsterdam shipcarpenter had the capital to buy into the first stock offering of the United East India Company in 1602.[27] But these men were exceptional. In the late fifteenth century shipmakers at Haarlem made tax payments well below the average and many payments were very near the minimum. A similar low figure appears in the fifteenth-century records of the town of Hoorn in Holland as well as in sixteenth-century records of nearby Edam.[28] The number of shipcarpenters was apparently small and not many of them were wealthy enough to rise to the level of taxpayers.

If the social status and income of the skilled shipcarpenter and his sometime superior, the contractor, were not high, then that of the semi-skilled or unskilled worker on the wharf must have been among the lowest in Dutch towns. Such men never appear in tax records or in records of ship construction. They may have, and this was decreasingly the case, formed a pool of unskilled labour which could be called on when needed. The sporadic nature of orders and seasonal nature of work made such a group offering muscle power on demand a necessity.

The income and perhaps status as well of Dutch shipbuilders was enhanced by their ability to produce technically superior ships. Shipowners appreciated the advantages of having vessels of the latest design. As the pace of technical change in ship design quickened and reached a peak in the last years of the sixteenth century so too did the interest on the part of builders and owners in having the newest and best design. And it was the relative income of the contractor, the shipmaker, that rose most. The technical advances were his. And though it may have taken some prodding, prodding which came from the method of arriving at the design, still there were direct economic benefits to him from innovating. That fact certainly made his decision about design change easier.

IV

So the buyer and the seller of a ship differed in social position and income and skills. Despite these differences shipbuilders were responsive to the demands of the buyers and that technical responsiveness came in part from an effective and efficient method of carrying out the selling process. Methods were not highly formalised but they did place supplier and demander face to face at frequent intervals. Owners were expected to take an active part in the design and construction of their ship, that complementing the shipcarpenters' experience and knowledge of new developments in ship design.

After the formation of a partnership for owning a vessel, the syndicate would choose an agent. Often the process was self-fulfilling since the man who brought the investors together wanted to become a part owner in a ship and oversee the construction and operation of a vessel. In the sixteenth century the agent was rarely the future master, and in German Hanse ports after 1591 it was illegal for the skipper to contract for the building of a ship without the clear support of at least one other partner.[29] The agent was responsible to his fellow owners for every aspect of the construction of the ship. It was he who placed the contract for construction, he who worked out the dimensions of the future vessel, he who visited the work site regularly to examine progress, he who handled all payments and he who reported to the owners any and all developments. The agent's special skills among other advantages improved control over the business operation of shipbuilding and shipping ventures and seems to have contributed to greater risk-taking in ship design.

The innovation was temporary, however. In the seventeenth century the combined forces of vertical integration in fishing and in extra-European trade which ultimately meant the separation of management and operational functions made the agent, at most, an employee if he survived at all. At the other extreme skippers gained management skills which had been the province of the agent and so made the latter redundant. Not coincidentally in the seventeenth century the pace of technical improvement in ship design slowed markedly. It was only with strenuous government support and advances in science that ships again experienced design improvements in the last years of the seventeenth century.

The agent, as the owners' representative, took on many tasks which had formerly been separated, transactions made more difficult by the increasing complexity of the ship and of trade. In the sixteenth century buyers tended to look beyond their own town and go further afield to find a shipbuilder. There were no restrictions on buyers choosing a builder in another town. Distance within Holland and Zeeland was not great and transport was easy over the flat land or along the many waterways. That presented new problems of oversight of the construction process, problems solved by having an active agent. The lower cost of building elsewhere or of getting a specific type of vessel apparently more than compensated for travel expenses. Such expenses were small relative to the total outlay for construction.[30]

The agent handled all financial arrangements. Payments were usually made to the contractor in instalments. The contractor could thus finance construction without recourse to any other institution. This also spread the payments by investors. The first instalment usually came just after the keel was laid so the shipbuilder had a sizeable amount of cash to get through the most expensive work, building the hull. Three progress payments were common with the second often made at launching and the third payable on delivery. Other arrangements were possible, for example four instalments. The shipping industry was thus able to finance the purchase of capital goods internally.

The agent chose a shipbuilder for the owners and would, together with the builder, work out a design for the ship. They agreed to the size and major proportions of the vessel. Ships were almost without exception custom-built. The type might be mentioned so many individuals presumably knew what such terms meant. The contract drawn up by the two men, the *bestek*, was impressive if for nothing other than its detail. For the fifteenth century such contracts are rare or, for Holland and Zeeland, non-existent. Agreements, even in the sixteenth century, might well be oral, especially for smaller vessels or where the buyer was going to be close by. In the sixteenth century the geographical separation of owners from the construction site, the spread of the ability to write, the general use of written contracts and better ship design all combined to make *bestekken* useful. But they had to be binding guarantees to build, violation of which would require compensation to the aggrieved.[31] The contract therefore had to have status in a court of law. *Bestekken* appeared in the late sixteenth century in Holland. The analogous *Baucerter* appeared in Hanse towns about the same time.[32]

Though not standardised, some features were common to all *bestekken*. Generally they included a precise statement of the principal dimensions of the vessel; length, either of the keel or between the posts, the length and angle of both stem and sternposts as well as the thickness of each, the width usually measured internally, and the depth of the hold. Specific design features might be outlined such as the nature of the stern and the length and thickness of planking to be used in the transom. The internal and external supports were usually described and the extent of detail on these points varied considerably. Exact placement of supports, distance from each other, thickness and form might be laid out. Descriptions of deck planking and any upperworks would finish the description unless there was mention of caulking work. Requirements of specific types of wood might well be inserted throughout. The last paragraphs dealt with price and arrangements for payment. Hull design was confined within rather narrow limits especially given the accuracy of measurement agreed to by both parties and included in the contract. Rigging was rarely mentioned in the *bestekken*. The location of masts might be explained in passing but that was all. The rigging was left entirely to the owners' agent, the shipbuilder having nothing to do with that important part of the design and construction of the ship. The form of the contract came out of the thinking of the shipbuilders themselves about ships. The earliest tracts on ship design describe vessels in exactly the same way and take up questions in much the same order.[33] Even if the agent was laying down exactly what the builder had to do it was within terms of reference set by the latter. The buyers in the person of their agent did not get to set requirements on capacity or sailing quality, the items which were most important to the financial success of their venture. In fact the agent could hope for little more than a stipulation about the height of the deck, ie a guarantee that he could fit cargo of certain dimensions on board.[34]

The owners' agent did not stop with the *bestek*. He was responsible for supplying many different items. A good deal of the work was done not by the contracting master shipcarpenter but by other contractors who dealt, as did the shipcarpenter, directly with the owners' agent. The agent bought the wood with the advice of the contractor. The agent arranged and paid for the proper sawing of the wood, a job done by specialists.[35] Ironwork was always distinct. All iron, iron fittings and nails were purchased by the agent and he took care of their delivery to the shipyard. The cost of ironwork was not included in the shipcarpenter's contract. The agent controlled that and many other types of expenditure. The same arrangement prevailed for the supply of sails and ropes.[36] Though the exact separation of functions varied from one job to the next, still the extent of the agent's active participation in and control over the building of the ship is impressive. A large number of jobs, many of them not insignificant, had to be turned over by the agent to different skilled craftsmen. The agent then also had to have the expertise to judge the quality of work of men as diverse as sailmaker, blockmaker, anchorsmith and compassmaker. The shipmaker probably knew men who could do such subcontractual work but the agent kept the awarding of such contracts in his hands. If owners went outside of their own town to find prime contractors, they could still place some work near home. They would have to transport the finished parts to the shipyard where final assembly took place but that was inexpensive.[37]

The sub-contracting method had two advantages. It allowed the agent better control over costs. Also it was the agent who had the financial resources to make payments to other contractors. The shipcarpenter did not. For that latter reason among others, the outfitting and victualling of the ship for the maiden voyage was handled by the agent.

The shipcarpenter as contractor was paid for his skilled labour and that of the people directly under him. He might be expected to install on board various items built or bought elsewhere by the agent. Ironwork was usually fitted by the carpenters. Spars and sails were rarely touched by shipcarpenters except perhaps on smaller jobs where in general the contractor tended to take on more tasks. The making of sails was contracted to sailmakers and spars might well be built by specialist mastmakers. The shipcarpenter turned the ship over to the agent with no rig fixed. It was the agent's job to get that work done. That meant towing the ship to a large crane where the masts and spars could be hoisted into place.

After handling the fitting of the rig and all other final details of construction, after hiring the crew and that would include the master if he was not a shareholder, and after making provision for the victualling of the ship, the owner's agent would then travel with the ship on its first voyage and any subsequent voyages so long as he had an interest in the ship. Usually the agent acted as supercargo keeping the books on board and supervising anything to do with the business of operating the ship. In the seventeenth century the supercargo disappeared as skippers gained business skills. Masters then could handle both the maritime and commercial sides of a shipping venture.[38] The agent's prominence was short-lived. This man combined most of the skills of the master except navigation and shiphandling, most of the skills of the merchant, all the talents of a bookkeeper and also had some cash to invest. He temporarily filled a vacuum created by the increasing complexity of business practice, of finance and of ship design in the sixteenth century.

V

The effect of changes in personnel on the design of ships did not conform to expectations. The owner's agent was to control all financial aspects of the venture, to maintain good communication between investors and men working on the project. So what distinguished a man for the task of agent was commercial experience and a share of the capital—not knowledge of ship design. But the buyer was expected to bring some expertise into the negotiations for the construction of the ship. More than that, if he was to handle contracting to other specialists he had to know a good deal about all phases of construction and the necessary equipment for sailing a ship. While masters acted as buyers their experience at sea made them good judges of the advantages and disadvantages of specific design features. With agents such experience was limited at best.

The loss of expertise in not having a master involved in design negotiations was not a disaster, however, for three reasons: the agent apparently had some knowledge of design, the agent did not always have the conservatism of a master which grew out of a healthy respect for the sea, and there was still at least one expert present, the contractor.

The position of the contractor in design negotiations was always a strong one, made even stronger by dealing with an agent of questionable knowledge. Shipcarpenters have long enjoyed a reputation for conservatism, of opposition to design changes. And with good cause. Along with the shipbuilding wharf, ideas were often handed down from father to son. The skill of building a successful ship was such a valuable art that knowledge was kept within the family to assure the progeny of a future income. Admittedly in Holland and Zeeland the protection of knowledge was not the same as in other parts of Europe. Builders were willing to spread skills to more men, but what they taught either to their sons or to others as apprentices was still consistent with tradition. Once a builder found a workable design he could most easily stick to it and make at most minor adjustments.

But Dutch contractors were not conservative. Sixteenth-century Dutch ships were subject to a lengthy series of major and minor design changes. Certainly the opportunity to innovate did exist because of an expanding market. But more than that there was an apparent drive to change design in ships, to implement new ideas and to use unfamiliar features. Given the way new ships were bought and sold such experimentation required the involvement of a number of individuals. Small experiments could be and indeed were carried out by builders themselves without approval or assistance from anyone. That was done despite the security offered by retention of traditional design. Major changes, it is true, had to have the approval and even the prodding of the buyers' representative. But as agents took the place of masters the contractor was given a freer hand.

The agent seems to have been as active as the shipcarpenter in experiment. The agent handled all of the sub-contracting so a number of changes were within his control. Most important was the agent's total responsibility for the rigging, and it was in rigging that the greatest, even if slow, advances of the sixteenth century came. The agent thus had to be willing to accept or even to promote changes. Moreover, he needed the support of his suppliers, sailmakers, ropemakers, blockmakers and of course in many instances the major contractor, the shipcarpenter. All in all the drive to innovate required at least the tacit consent of a sizeable number of participants within the shipbuilding industry and the vigorous support of at least one of the major participants. While the expert, the shipcarpenter, laid aside his conservatism, the businessman, the owners' agent, added the advantages of innocence.

By the end of the sixteenth century the tradition, at least in Holland and Zeeland, was one of experiment and of design change. The process became cumulative feeding on past success: hence the rapid turnover in designs of ocean-going ships in the last three decades of the sixteenth century. The businessman was more willing to take a chance, seeing the return to be got from large cargo space or faster round trips, while turning a blind eye to potential difficulties arising from a modified sail-plan or a new hull shape. Elsewhere in Europe especially in the Hanse ports development was slow just following improvements made in the Low Countries. Presumably in those towns one of the greatest blocks to experiment was loss of an expanding market for the product of the German builder and the buyer. That in turn meant a loss of flexibility and limited opportunities for innovation.[39]

Throughout the process of construction a crucial issue for the buyers of all ships was maintaining a proper degree of control over the suppliers and particularly over the builder of the hull. The problem was one of management, a problem made more acute by the high relative cost of the good being purchased. Buyers wanted guarantees of quality. At the same time both buyer and builder were interested in improving the value of the product, in bettering its productivity through technical improvement. The two aims, one leading to conservatism and the other to experimentation, were satisfied within the Dutch shipbuilding industry in the sixteenth century. Other towns, regions, countries used similar forms of management practice yet were not able to steer as successful a course between the two dangers implied by the aims. The acceptance of the agent, the businessman as the bargaining agent for owners in design discussions, a general feeling of the desirability of experiment and the cumulative effects of rising demand for shipping tonnage saved the Dutch shipbuilding industry from the trap of being tied to tradition and from the trap of using designs which were unworkable or unprofitable. The agent's specialisation, though short-lived and though it probably took its character from the economic and financial conditions of the moment, helped to bridge the gaps of wealth, status and knowledge between the men who built ships and the men with money to buy them. The unique background, skills and interests of the agent not only made possible but also in unexpected ways promoted advances in Dutch shipbuilding.

References

1 This is based in part on an unpublished doctoral dissertation, Yale University, 1971. Research was sponsored by a grant under the Fulbright-Hays related fellowship programme.

2 Richard W. Unger, 'Dutch Ship Design in the Fifteenth and Sixteenth Centuries', *Viator*, IV (1973).
See also J. van Beylen, *Schepen van de Nederlanden van de late middeleeuwen tot het einde van de 17e eeuw* (Amsterdam, 1970).

3 Jelle C. Riemersma, 'Trading and shipping associations in 16th century Holland', *Tijdschrift voor Geschiedenis*, LXV (1952), 331.

4 Of some fifty Amsterdammers who owned shares in ships in the late fifteenth century and who were identified by occupation, most were involved in trade. But there was also a baker, a miller, two carpenters, a leather worker and even a soap-boiler among others.
F. Ketner, *Handel en Scheepvaart van Amsterdam in the Vijftiende Eeuw* (Leiden, 1946), 146–7. The data come from inventories submitted to the *Weeskamer*, the public institution for the care of orphans, between 1467 and 1500. In England many owners were first landowners who invested in ships as a way to diversify.
G. V. Scammell, 'Shipowning in England *circa* 1450–1550', *Transactions of the Royal Historical Society*, fifth series, XII (1962), 119. The situation had changed by the eighteenth century, however. For a summary of management practices at that time, some foreshadowed in sixteenth-century Holland and Zeeland, see Ralph Davis, *The Rise of the English Shipping Industry in the Seventeenth and Eighteenth Centuries* (London, 1962), 81–109.
The apparent greater incidence of merchants in shipowning in sixteenth-century Holland and Zeeland relative to England was probably a result of the more rapid growth of the commercial sector in the Netherlands.

5 L. J. Ruys, 'Archief van Adrichem', *Inventarissen van Rijks—en Andere Archieven*, IV (1931), 260–74.
Askel E. Christensen, *Dutch Trade to the Baltic about 1600* (The Hague, 1941), 112–44. He relies heavily on this document collection and gives, in these pages, a business history of the firm. Various documents of the firm have been edited by H. E. van Gelder. See his 'Gegevens Betreffende de Haringvisscherij op het einde der 16de eeuw', *Bijdragen en Mededeelingen van het Historisch Genootschap*, XXXII (1911), 1–62; 'Scheepsrekeningen van Enkele der Vroegste Guinea-Vaarten', *Economisch-Historisch Jaarboek*, II (1916), 239–57, 'Zestiende-Eeuwsche Vrachtvaartbeschieden', *Economisch-Historisch Jaarboek*, III (1917), 124–290, and 'Zestiende-Eeuwsche Koopmansbrieven', *Economisch-Historisch Jaarboek*, V (1919), 136–89. Many of the papers remain unpublished, however.

6 Algemeen Rijksarchief, 's-Gravenhage, Archief van Adrichem, 12, 13. H. A. H. Kranenburg, *De Zeevisscherij van Holland in den Tijd der Republiek* (Amsterdam, 1946), 64–5, notices the same trend.

7 It was even less likely in a trade like that to the Mediterranean where the armament needed for protection might add 44 per cent or more to construction costs. Archief van Adrichem, 83. Published by H. E. van Gelder, 'Zestiende-Eeuwsche Vrachtvaart-bescheiden', 286–7. English shipbuilders and shippers seem to have been better able than their Dutch counterparts to solve the problems of trading

in the eastern Mediterranean. See Ralph Davis, 'England and the Mediterranean, 1570–1670', in *Essays in the Economic and Social History of Tudor and Stuart England*, F. J. Fisher (ed), (Cambridge, 1961), 127–32.

8 In Holland the master's share appears to have been smaller than was common at Emden and perhaps elsewhere. Jelle C. Riemersma, 'Trading and shipping associations in 16th century Holland', 331–2. For four voyages of the late sixteenth century, for which good accounts have survived, the master was simply an employee, paid a wage like any other seaman. The fact that all four voyages lost money appears to have been due to other management decisions. J. V. van Es (ed), 'Zestiende-eeuwsche vrachtvaart-beschieden. Vier scheepvaartrekeningen van 1574, 1575 en 1576', *Economisch-Historisch Jaarboek*, XX (1936), 259.

9 H. A. H. Kranenburg, *De Zeevisscherij van Holland in den Tijd der Republiek*, 59–71. He is reluctant to make the year 1600 crucial but does concede a clear trend towards increased interest on the part of herring merchants and sellers in the catching end of the operation.

10 Eli Heckscher, *Mercantilism*, Mendel Shapiro (trans), E. F. Söderlund (ed) (London, 1955), vol 11, 34–6. He chooses a rather narrow definition for support of the shipping industry and so gives only a limited view of government involvement.

11 J. H. Kernkamp, 'Scheepvaart en Handelsbetrekkingen met Italië Tijdens de Opkomst der Republiek', *Mededelingen van het Nederlandse Historisch Instituut te Rome*, reprinted in *Economisch-Historische Herdrukken* (The Hague, 1964), 206.
Jelle C. Riemersma, 'Trading and shipping associations in 16th century Holland', 332–3.

12 The average or even a common rate of return is difficult if not impossible to ascertain. The Van Adrichem family accounts show a 14 per cent return on all expenditure for a series of nineteen fishing voyages in the 1590s. H. E. van Gelder, 'Gegevens Betreffende de Haringvisscherij . . .', 5. For cargo ventures the accounts are even more sparse so statistics are even less reliable.
H. E. van Gelder, 'Zestiende-Eeuwsche Vrachtvaart-beschieden', 124–5.

13 G. V. Scammell, 'Shipowning in England *circa* 1450–1550', 117.

14 Gemeente Archief Veere, Shipcarpenters' bye-laws of 1520, paragraph 16. Repeated in the bye-laws of 1585 as paragraph 13. The ordinance suggests that the building of ships of more than 120 tons was often done by two or more master carpenters.

15 By the eighteenth century, however, shipbuilders were great merchants involved in a number of different trades. These men probably had little time for their wharves—they often owned more than one—and so turned the actual job of construction over to others.
S. C. van Kampen, *De Rotterdamse Scheepsbouw in de Tijd van de Republiek* (Assen, 1953), 178–90.

16 Gemeente Archief Edam, 237. Register van Verpachting, 1462 . . . fol 5v.

17 Gemeente Archief Veere, letter of 1605. The larger the town the more expensive land on the water and so moving was less likely. Cf Gemeente Archief Haarlem, 1075, Belastboeken, 1547–1613.

18 David Felix, 'Profit inflation and Industrial Growth: the Historic Record and Contemporary Analogies', *Quarterly Journal of Economics*, LXX (1956), 446–7. His Table I gives a comparison of wood prices with an index of all prices and all wages. Also John U. Nef, 'Prices and Industrial Capitalism in France and England, 1540–1640', *Economic History Review*, VII (1936–7), 155–85.

19 At Amsterdam from 1539 repair work was restricted to those who had proven themselves competent in building new ships. J. G. van Dillen (ed), *Bronnen tot de Geschiedenis van het Bedrijfsleven en het Gildewezen van Amsterdam*, Rijks Geschiedkundige Publicatiën, vol 69 (The Hague, 1929), vol 1, 1512–1611, no 235, paragraph 1. The regulation was made by the contractors and was probably a way of assuring them of getting what repair business came along.

20 Pieter Koster, *Hoorn in de Middeleeuwen* (Amsterdam, 1929), 44.

21 This involvement in shipping investment appears in such tax records as Gemeente Archief Medemblik, 1397. Lijsten en quohiere van de 100ste en 200ste penningen, 1569.

22 In the town of Veere in Zeeland shipcarpenters were the only legal dealers in wood and in pitch and tar, a privilege which they carefully guarded. Gemeente Archief Veere, two letters of 16 March 1590, concerning encroachment of the privilege by a sailmaker. The privilege was stated in the bye-laws of the shipcarpenters' guild, 1585, paragraph 23. And S. C. van Kampen, De Rotterdamse Particuliere Scheepsbouw . . ., 98.

23 Gemeente Archief Dordrecht, Afdeling Gilden, 247 (1383).

24 That was a privilege acknowledged in 1583. Gemeente Archief Dordrecht, Afdeling Gilden 9a (1583).

25 Gemeente Archief Dordrecht, Afdeling Gilden, 9b (1583) and 255 (1512–66), 256 (1525), 265 (1562). A similar situation prevailed at Rotterdam. Gemeente Archief Rotterdam, Resolutiën van de Vroedschap (21 March 1591).

26 Both men also held other high offices in town government. The Haarlemmer was Arend Meindertsz. Fabricius who owned and operated a shipbuilding wharf and woodyard. G. H. Kurtz, 'De stamvader van een Haarlemsche regenten-familie (Arend Meindertsz. Fabricius)', *Haarlem Jaarboek* (1938), 81–97. The Amsterdammer was Pieter Cornelisz. Boom who began as a ropemaker and was also involved in shipbuilding. In 1585 he paid a relatively high tax. J. G. van Dillen, *Amsterdam in 1585, Het Kohier der Capitale Impositie van 1585* (Amsterdam, 1941), fol 5. W. H. M. DeFremery, 'De Opkomst der Amsterdamsche Haven', *Jaarboek van het Genootschap Amstelodamum*, XXII (1925), 87 mentions another shipbuilder who rose to be mayor of the town.

27 Jan Rijxss., shipcarpenter, invested 900 guilders. He and 465 others of the 1,143 shareholders paid in less than 1,000 guilders so his contribution, though not the smallest, was very far indeed from the largest. J. G. van Dillen, *Het oudste aandeelhoudersregister van de Kamer Amsterdam der Oost-Indische Compagnie* (The Hague, 1958), 186.

28 Gemeente Archief Haarlem, Register van de Honderdste Penning, 1483–1500, alphabetical index. Gemeente Archief Hoorn, 1314 (c 1436), fol 11. Sj Devries, 'De Edamse Scheepsbouw, 1460–1625', unpublished Doctoraal Scriptie, Universiteit van Amsterdam, 17. Also at Haarlem houses in the district reserved for shipcarpenters sold in the second half of the sixteenth century for about 60 per cent on the average of the mean value of houses along the river Spaarne in the centre of town and therefore probably owned by merchants. Gemeente Archief Haarlem, 1075, Belastboeken, 1547–1613.

29 Karl-Friedrich Olechnowitz, *Der Schiffbau der Hansischen Spätzeit* (Weimar, 1960), 44–5.

30 The Van Adrichem family were connected with the construction of two ships for which travel accounts survive. Archief van Adrichem, 55, 56. In neither case did travel reach 5 per cent of the cost of construction and fitting out. Archief van

Adrichem, 12, 13, list 29 ships parts of which were owned at one time or other by the family, ports of construction being named for 19. The largest number but by no means the majority came from Rotterdam near Delft, but many came from north Holland. The ports and numbers of ships built at each were: Rotterdam 7, Enkhuizen 4, Edam 3, Elbing 2, Haarlem 2, and Medemblik 1.

31 Gemeente Archief Amsterdam, Notarial archive, 33, Jac Gijsbertsz., notary, fol 60. J. G. van Dillen (ed), *Bronnen tot de Geschiedenis van het Bedrijfsleven en het Gildewezen van Amsterdam*, no 894 (17 January 1597).

32 K.-F. Olechnowitz, *Der Schiffbau der Hansischen Spätzeit*, 99. Surviving contracts are rare even for the seventeenth and eighteenth centuries, even though more of them were undoubtedly written, Ralph Davis, 'Shipping Records', *Archives*, VII, 35 (April 1966), 137. The form of the *bestek* was similar for naval and merchant vessels. Cf Archief van Adrichem, 126 (a herring buss built at Rotterdam in 1567), 127 (a herring buss built at Rotterdam in 1598), Gemeente Archief Amsterdam, Notarial archive, 84, J. Fr. Bruyningh, notary, fols 100–2, and T. H. Milo, *Bronnen Onderzoek van de Nederlandsche Zeegeschiedenis*, Rijks Geschiedkundige Publicatiën, unpublished (naval vessels built for the Admiralties in 1600).

33 Nicolaes Witsen, *Aeloude en Hedendaegze Scheepsbouw en Bestier* (Amsterdam, 1671), 47–9. The description he offers is based on a sixteenth-century Portuguese one. Walther Vogel, 'Ein neuentdecktes Lehbuch der Navigation und des Schiffbaues aus der Mitte des 16. Jahrhunderts', *Hansische Geschichtsblätter* XVII (1911), 370–4. See also W. A. Baker, 'The Arrangement and Construction of Early Seventeenth Century Ships', *The American Neptune*, XV (1955), 259–86.

34 Gemeente Archief Amsterdam, Notarial Archive, 32, Jac Gijsbertsz., notary, fol 403 (1600). Archief van Adrichem, 56, 126, 127.

35 Archief van Adrichem, 55, 56.

36 Gemeente Archief Amsterdam, Notarial Archive, 33, Jac Gijsbertsz., notary, fol 403 (1600). Archief van Adrichem, 56, 126, 127.

37 The cost of transporting materials never came to as much as 1 per cent of the total cost of construction and fitting out in the extant full accounts of the Van Adrichems. Archief van Adrichem, 55, 56.

38 Ralph Davis, 'Shipping Records', 140. With large joint-stock companies such as the United East India Company the masters were not allowed the latitude in business dealings which became common in partnership ventures.

39 The decline in ship owning in Hanse ports in the sixteenth and seventeenth centuries is clear. Ernst Baasch, *Beiträge zur Geschichte des deutschen Seeschiffbaues und der Schiffbaupolitik* (Hamburg, 1899), 268–71.

III

FOUR DORDRECHT SHIPS OF THE SIXTEENTH CENTURY[1]

THE study of the history of ship design has in recent years turned away from the examination of specific illustrations. The publication of pictures and drawings has become less frequent and discussion of features of certain representations, even of ships from Antiquity and the Middle Ages is also rare. There are sound reasons for the change. Cataloguing and evaluating illustrations was a major task of writers in the nineteenth and early twentieth century and the job was well done, as is obvious from the early numbers of *The Mariner's Mirror*. Once the outline of development had been established from significant and authoritative illustrations it seemed that to multiply the evidence could only yield sharply diminishing returns. Moreover, in recent years nautical archaeology has supplied direct information from the ships themselves. The secondary evidence from artists as a result has lost its value of uniqueness. Despite those reasons it is an error to abandon totally the use of illustrations. The traditional methods used by maritime historians have proved highly successful and those methods still offer a great deal if used with care. Pictures as a source of knowledge about ship design have by no means been fully exhausted. And their use is valuable even where those illustrations only supplement and support present knowledge.

The four illustrations from sixteenth-century Dordrecht appear in a *keurboek* or collection of copies of laws made by Cornelius Cloott Cornelisz. This practice of compiling copies of major legislation of towns and provincial governments was a common one in the fifteenth and sixteenth centuries in the Low Countries. From Dordrecht alone sixteen such books have survived from before 1572. Most of course emphasize the laws of Dordrecht and of South Holland. This book was exceptional because of the inclusion of material dealing with trade to the Baltic, of a chronicle from Germany, 1175–1564, and of illustrations. The book itself is 220 parchment pages and is 40·5 cm by 26·5 cm.[2] Nothing is apparently known about the author. There are two illustrations of a flag carrier in full armour on folios 1v and 115v. And there are four pictures of ships, the first on folio 1r, the second on folio 156v and the other two appearing on a leaf inserted between folios 195 and 196. All the pictures are in essence line drawings done in brown ink with additional lines added in red ink and also in black pencil.

Pencil is used for example to show water and the sails. The author began the book in 1564 and on folio 195v he announced that this was the end of the book, giving the date of 14 August 1567. He then presumably added the two pictures on the following leaf. But he could not leave the book and returned to it adding another forty-eight pages of text and also adding the picture on folio 1r, that dated 1569. So all the pictures date from the second half of the 1560s.

The pictures were then done in a period of significant growth in the essential trading connections of the Low Countries. They were also done just before and at the outbreak of the revolt against the Spanish government. That led to a civil war which in time changed the trading connections and also the naval situation for the region. There can be little question that the book was done at Dordrecht. That was a relatively declining port in the sixteenth century. The loss of political favours early in the century had led to a relative decline in trade and also to a concentration on riverborne commerce. The larger ships of ocean-going trades found it increasingly difficult to reach Dordrecht. They had less reason to visit the port as well. The devotion to river traffic was combined with a failure to participate in a number of other trades such as those to Iberia, to the Baltic and also, though to a lesser extent, in the herring fishery. As a result shipbuilding at Dordrecht became more specialized in river boats and inland craft.[3] The illustrations show that despite the changes in trade and shipbuilding at Dordrecht there was still a knowledge of a number of different types of vessels used in all types of trades and apparently of the latest design.

The first ship (Fig. 1) in the book is a Portuguese caravelle. The drawing is on the title page and written above the ship is 'Cornelis Cornelisz. Kloot, 1564. Dit is het Koerbouck van Zuyt-Hollont ende der stede van Dordrecht'. It is just a statement of the author's name, the date and that this is the law book of South Holland and of the town of Dordrecht. On the side of the ship is written, 'La barque de Portugael. Dit schip sy gode bequaem de bercke Lutina is sinen naem 1 5 6 9'. He calls her the barque of Portugal, saying that this ship is in good condition and is called the barque Lutina, with the date 1569. This is perhaps the most complete of the illustrations and certainly the most unexpected from a Dordrecht source. It is an example of the largest of caravelles with four masts and with a topsail on the foremast. The rig is square rig on the foremast and single lateen sails on the remaining three masts, the distinguishing feature of the caravelle. And rather than just having the foremast and mainmast raked forward, in this case only the bonaventure mizzen does not have a marked forward rake. The foremast carries two square sails and the topmast yard is lowered halfway down the detachable topmast. This seems to have been common practice when square sails were not in use. The topmast presumably fits into the fighting top. There is a stay running to the bowsprit and there are also forestays for the foremast and mainmasts running to the bowsprit. The shrouds are obviously confused but there can be no question that there are ratlines, by this time in use in both northern and southern Europe. The bonaventure mizzen yard is sheeted to an outlicker but it lacks a brace to the higher end of the yard as is clearly shown on the other two yards carrying lateen sails. The lifts spreading out along the yard turn up in vessels, both square and lateen rigged, a century before and more.

The artist has some trouble drawing the lateen sails and especially the intricacy of the rigging. It is possible that he was not familiar with the complex nature of rigging for such large lateen sails. The yards are in one piece and not two spars fished together as was common on Mediterranean galleys. The yard of the mizzen mast is to starboard of the mast while those of the other two lateen spars are to port. But that too appears on other illustrations of caravelles of the sixteenth century. The tops of the masts carrying lateen sails are all to one side of the mast and that too was typical.[4] This was of course to make it possible to carry the yard over. There are men crawling on these yards, presumably working with the sail though it is not possible to claim that these were foot ropes. It is certainly doubtful. The device hanging from top end of the bonaventure mizzen yard appears to be a wind sock or at least a device to indicate wind direction. In addition to the two square sails on the foremast this caravelle also carried a spritsail stretched to a yard slung under the bow. At the end of the bowsprit there is a small mast which carries a cross. It is possible that it also carried a yard and sail and that this is a small spritsail-topmast. It was to be a typical feature of seventeenth century ships but whether this is just the flagstaff that that mast grew from or an early case of such a mast and sail is not clear from the drawing.

The hull form is consistent with that of a contemporary galleon. To call it a Portuguese galleon would not be incorrect. It certainly was a large enough ship for that name. The terms 'galleon', 'barque' and 'caravelle' appear to have been interchangeable by the second half of the sixteenth century. There are two wales running the length of the ship, indicated by darker colouring, for added strength. The bow is a bit confused. The bowsprit rises above a projection or *galjoen* which retains a pointed end to suggest its ancestry in a galley ram. The spritsail yard is slung under this *galjoen* and not under the bowsprit. This also is present in earlier pictures of the largest Portuguese warships. The problem of upward pressure on the bowsprit from the stays, especially acute because of the angle of the bowsprit, is solved by lines running to the *galjoen* underneath giving the effect of the much later bobstay.[5] There are two men shown working on the bowsprit and another perhaps at the location of the head. The positions of the broadside guns suggest a gun deck running the full length of the ship. The bow is only slightly built up while the stern has a sizeable addition with at least three decks judging from the location of the six stern guns which are shown as firing. The relatively short rudder indicates that the ship rode high in the water.

The second illustration (Fig. 2) is of a buss. It carries the simple and by the late sixteenth century standard rig of four square sails on three masts. It appears that the mainmast is not a pole mast but rather carries a top and that the topmast fits into it. The artist again introduces some confusion, this time in the sheeting of the sails, especially the foresail. The wind is coming over the port beam and so the yards are turned. One sheet from the mizzen sail runs to the outlicker and it is a bit surprising to see that spar on what must have been a relatively small vessel. The sheet from the other clew is obscured. The sheets on the mainsail run to points on the gunwale. For the foresail, one sheet runs to a point on the gunwale but the other goes to a bowsprit which points downward. There can be no question that that was intended by the artist. Such a downward rake to the bowsprit was not uncommon on busses. It may have been built that way to take the pressure from the foresail, especially since the foremast is raked forward. The sails in this as in the fourth vessel are strips of canvas sewed together. The mainsail has a bonnet, attached underneath and it is built up in the same way. The shape of the hull suggests that the greatest width is forward of amidships. That is not unexpected. There is no hint of a forecastle and there is at most one deck added at the stern which gives a slight taper to the stern. The hull shows no sign of reinforcement. The bows are bluff and the stempost sharply curved, typical features of a buss and necessary for handling the nets. The rack for the masts between the main and mizzen masts was used to stow those two masts while fishing. Then only the foresail was used to give a slight tension on the net. But there is a second rack in the waist, suggesting that at times the foremast too was struck. There is no question that this vessel is from Dordrecht with the arms of the town clearly shown on the stern. The flags on the tops of the masts may well have been in the form used by Dordrechtenaars in the sixteenth century. The eagles shown under the outlicker may be the insignia of the house of Hapsburg and in common use since the end of the fifteenth century when Holland and the rest of the Low Countries came to be ruled by a member of that family. There is a cross at the top of the mainmast. The Reformation would not remove such crosses from Dutch ships for another few years.

The third ship (Fig. 3) is a galley and the worst drawn of the four. It is also difficult to see because the impression of the ship drawn on the opposite side of the leaf comes through the thin parchment. The sloppily drawn galley done exclusively in brown ink makes an even greater impression on the opposite side. The date on the stern is 1564. The insignia on the pavilion at the stern and on the flag flying from the mast may again be that of the house of Hapsburg. The rig is out of proportion to the ship. And certainly the shrouds should not run from the gunwale to the lift. There is one gun mounted a bit too far forward for the stability of the ship and presumably set along the centreline. The bowsprit is also overdone. Though it is far from accurate, there is no doubt that the rig is a single mast carrying one lateen sail. That would make it a small galley by Mediterranean standards. For propulsion this galley also has ten oars to a side. The oval shape of the pavilion or cabin at the stern is consistent with contemporary or earlier Mediterranean galleys. Though much is unclear still this is certainly not as complete a vessel as galleys built in Holland forty years later.[6]

The fourth ship (Fig. 4) is a boyer from Dordrecht. The origin is established by the flag and the arms of the town on the stern. The vessel is shown at sea with water and fish indicated by pencil lines. The identification of the vessel as a boyer depends on evidence from other contemporary illustrations in engravings and maps.[7] The rigging is slightly confused but the spritsail on the mainmast is clear and more clearly shown in detail than in most contemporary drawings. The heavy ink of the galley shows through, so some of the ornamentation of the hull cannot be seen. The vessel like the Portuguese caravelle shows the increasing decoration on larger cargo vessels but it is still limited to painted armorial devices and some geometric patterns only on the aftercastle. The rig is that of a typical late sixteenth-century boyer. The bowsprit is not drawn straight but shows a slight upward curve. The forestay for the mainmast runs not to the bowsprit but rather to its base at the stempost. The stay for the top runs to the bowsprit however. Underneath the bowsprit is the spritsail. The mizzen mast is short compared to the mainmast, and the sail area of the lateen sail is relatively small. The shrouds and lifts, both rather light and simple, show the limited weight of the mizzen rig. The outlicker is needed not because of the size of the sail but because the mast is stepped so far aft. The mainmast is thick and heavy. It had to be to carry the three sails which were the main propulsion for the vessel. The topmast is detachable. The topsail is sheeted to the mainyard and hanging down from that spar is the large square sail which is half covered by the massive spritsail. The spritsail was typical for Dutch inland vessels, but it was on boyers like this one that it was used in the open sea. And the rig was the same. The tackle runs to the middle of the sprit. The artist has made a special effort to show that the sprit is not fitted to the mast but rather held close to it at the base of the sprit by a tackle. The spritsail was lowered by letting the sprit run forward and the leading edge of the sail run down along the mast.[8] The leading edge of the sail is securely fastened through its entire length to the mast. This seemingly clumsy arrangement combined with the fact that the sprit was almost as tall as the mast itself must have given rise to fear that the mast would break in rough weather. The spritsail has a bonnet which is of one piece of canvas and not of strips sewn together. Ratlines are shown only on one side of the mast and the forestay should be behind the larger square sail on the mainmast and more of that sail should be visible. But here the artist seems to have been more accurate about the rigging than in his other drawings.

The hull form shows the same bluff bows and the highly curved stempost of the buss. There are two wales running the length of the ship suggesting a rather sizeable vessel. But there was a limit set not by the shape of the hull but by the size of the sprit. The boyer is carvel-built throughout. At the stern there are raised decks, at least two and possibly three covered decks. The stern is rounded and above it there is a wing transom which is unquestionably tapered. Those features in the next century would become the distinguishing marks of the Dutch *fluit*.[9] Deadwood fills in under both the stem and the sternpost. The vessel shows no sign of armament.

The figures in this drawing as elsewhere are out of proportion to the ship. There is, as with the first vessel, an impressive figure in the waist, obviously an important man. Whether he is intended to be the owner or the skipper is pure speculation. There is a man at the stern, presumably making soundings, perhaps trying to find a school of fish. But that is the only suggestion that this might be a fishing vessel. It is more likely that it was a cargo ship and the massive flag on top of the mizzen mast is only sketched in in pencil so it is not clear what was intended.

This artist, probably Cornelis Cloott Cornelisz himself, has left a gallery of four widely different vessels. The types include one for naval action and long distance commerce on the high seas, a deep sea fishing vessel, a naval vessel for fighting and patrol along the rivers and a freighter for coastal and long distance carriage. Two of the vessels are clearly from Dordrecht and probably built there. The other two if not from Dordrecht, which was certainly not the case with the Portuguese barque, at least were known in the port and visited there. Dordrecht shipowners, merchants and shipbuilders apparently did not exploit their knowledge but rather held back from building anything beyond a boyer or a buss and building such vessels only rarely. It was seldom a lack of knowledge which prevented shipbuilders or shipowners from using a different type of vessel, but rather the combination of the technical constraints of local sailing conditions and the economic constraints of their own businesses.

These four illustrations are from a point in time where each of the types gained a certain important status. Trade with Portugal had expanded in the sixteenth century because Antwerp was the market for spices brought from the Far East. But after 1540 the visits of Portuguese caravelles to Low Countries ports fell off. The lightly built caravelles were not well suited to handling the bulky back cargoes and so Dutch ships were used more and more.[10] By the 1560s a Portuguese caravelle like that shown in Fig. 1 must have been a rare sight in any port in the Low Countries, and for that matter anywhere in northern European waters. The design of the buss had gone through a number of changes since it became the major vehicle for the herring fishery in the first years of the fifteenth century. But by the 1560s the design had settled down to a vessel of about 100 tons. This illustration shows such a buss. All that is needed to make it the typical buss of the seventeenth and eighteenth centuries is a rounded stern which was added in the years around 1600.[11]

Galleys were used for fighting in the Low Countries throughout the sixteenth century on the rivers and inland lakes. The type was imported from the Mediterranean and specifically from Spain. During the Revolt the galley gained new popularity. They were used by Spanish generals and government officials since much of the fighting was along the rivers and the estuary of the Schelde. The newly formed government of the United Netherlands responded by building galleys of their own, one being built at Dordrecht in 1600.[12] The illustration shows that the type was in use before the Revolt.

The boyer in the 1560s and 1570s was probably the most popular cargo carrier in the Netherlands. It was used extensively in trade to the Baltic as well as in the North Sea and along the coast of France. The low manning ratio, a result of the combination of fore-and-aft and square rig, gave it a low operating cost.[13] The collection of four pictures then gives additional information and confirming information about four types which hold a special position in the history of Dutch ship design in the years around 1565.

The quality of the drawings leaves something to be desired. The artist did not have a seaman's eye for the details of running rigging. But he seems to have been consistent on major features, especially of hull form and so the illustrations can be considered trustworthy. Each picture, with the exception of that of the galley, shows certain individual features which are not as obviously shown in other contemporary pictures of these types. Though surviving pictures of ships are more common for the sixteenth than for any previous century it was not until the seventeenth century that illustrations became highly accurate. So for the sixteenth century there can be no definitive illustration but rather understanding has to be built up from a wide variety of sources. More is known about Dutch vessels of the sixteenth century than for any previous century but more needs to be discovered since it is just in those years that great strides were made in improving well-known types in Holland and Zeeland.[14] These advances made possible the maritime dominance of the small Republic in the seventeenth century and these four pictures offer a supplementary and valuable addition to the understanding of just those improvements.

Notes and References

1 Research for this paper was aided by a grant from the Canada Council.

2 I am indebted to the staff of the Gemeente Archief Dordrecht for their help in finding these illustrations. The book is Gemeente Archief Dordrecht, Archief der Gemeente Dordrecht, I. De Grafelijke Tijd, 1200–1572, X, Keur- en Handvestboeken, 12. It is described in the inventory of that archive by J. L. van Dalen published at Dordrecht in 1909.

3 Richard W. Unger, 'Dutch Shipcarpenters' Guilds, *c.* 1400–*c.* 1600', *Mededelingen van de Nederlandse Vereniging voor Zeegeschiedenis*, 24 (1972), 5–11.

4 R. Morton Nance, 'Caravels', *The Mariner's Mirror*, 3 (1913), 265–271. Siegfried Fliedner, 'Kogge und Hulk,' *Die Bremer Hanse-Kigge*, Bremen, 1969, 104. Alan Moore, 'Rig in Northern Europe', *The Mariner's Mirror*, 42 (1956), 15.

5 See for example: R. and R. C. Anderson, *The Sailing-Ship*, London (n.d.), 125, 140, 179. G. S. Laird Clowes, *Sailing Ships, Their History and Development*, London, 1932, 80–81. R. C. Anderson, 'Ship-Paintings in the Alhambra', *The Mariner's Mirror*, 51 (1965), 6. The model built by Philipp Keller for the *Verkehrshaus der Schweiz* and now on exhibition there is highly consistent with the illustration.

6 John F. Guilmartin, Jr., 'The Early Provision of Artillery Armament on Mediterranean War Galleys', *The Mariner's Mirror*, 59 (1973), 260–262. B. van Breydenbach, *Bevaerden tot dat heilighe grafft*, Mainz, 1488. F. C. Lane, 'Venetian Naval Architecture about 1550', *The Mariner's Mirror*, XX (1934), 24–49.

Compare this illustration with that of a galley on a profile of Amsterdam of 1607. For a good reproduction of that galley illustration see Lionel Casson, *Illustrated History of Ships and Boats*, Garden City, New York, 1961, 121, illustration 145.

7 Alan Moore, *op. cit.*, 14. Rene van Bastelaer, *Les estampes de Peter Brvegel l'Ancien*, Brussels, 1908, number 98. Lucas Waghenaer, *Spiegel der zeevaerdt*, Leiden, 1584–1585.

8 J. van Beylen, *Schepen van de Nederlanden*, Amsterdam, 1970, 183–185.

9 R. Morton Nance, 'The Ship of the Renaissance', *The Mariner's Mirror*, 41 (1955), 295–298.

10 T. H. Milo, 'Portuguese Trade and Shipping with the Netherlands after the Discoveries', *Actas do Congresso Internacional de Historia dos Descobrimentos*, Lisbon, 1961, III, 4–7.

11 H. A. H. Kranenburg, *De Zeevisscherij van Holland in den Tijd der Republiek*, Amsterdam, 1946, 16–17. J. van Beylen, *op. cit.*, 137–139.

12 G. C. E. Crone, *Onze Schepen in de Gouden Eeuw*, Second edition, Amsterdam, 1943, 98–104.

13 G. V. Scammell, 'Manning the English Merchant Service in the Sixteenth Century', *The Mariner's Mirror*, 56 (1970), 132. Bernhard Hagedorn, *Die Entwicklung der wichtigsten Schiffstypen bis ins 19. Jahrhundert*, Berlin, 1914, 82–92.

14 Richard W. Unger, 'Dutch Ship Design in the Fifteenth and Sixteenth Centuries', *Viator*, IV (1973), 387–411.

PLATE I

Figure 1. A Portuguese Caravelle

Figure 2. A Buss

PLATE 2

Figure 3. The Galley

Figure 4. A Dordrecht Boyer

IV

Wooden Shipbuilding at Dordrecht [1)]

I. Origins of the Industry.

In the later Middle Ages Dordrecht sat at the point where inland water transport met sea-going transport. From the beginning the town served as a transshipment centre. The inland waterways gave excellent connection with the industrial and commercial towns of Flanders, with the growing towns of Brabant, with much of Germany and also with commercial centres along the IJsel and in Holland. By the fourteenth century Dordrecht was a trading town of some importance. It was the only town in Holland that was able to draw the attention of foreign traders. 2) The position of the town was so good that it drew the attention of governments as well. The counts of Holland thought Dordrecht had the potential to be another great international port, like Bruges. So from the late thirteenth century the government of Holland embarked on a policy of favouritism to Dordrecht. Trade there was promoted at the expense of all other ports in the county. In 1299 Count Jan I granted Dordrecht the so-called 'stapelrecht'. In 1355 Count William V revised and expanded the regulations to include 'Maasrecht' as well. This granting of privileges started with Count Floris V in 1276 and his successors were even more vigorous in aiding the town. The 1299 grant required that all goods which came down or up the rivers Lek or Merwede had to be sold at Dordrecht. There were no exceptions. Modifications were made in the 'stapelrecht' until in 1355 it was extended to include all goods which went upstream or downstream on the Rhine, Maas, Waal, IJsel, Lek or Merwede and with only minor exceptions. In essence the law required that all goods going in or out of Holland by river had to be exchanged at Dordrecht. The privilege was expanded and renewed again and again by different counts through the fourteenth and fifteenth centuries.
These counts did not act out of compassion. Jan I had enjoyed the loyalty of Dordrecht and so the 1299 grant was something of a reward. More than that the counts were interested in assuring their income from tolls. By forcing all trade through Dordrecht they could tax it more effectively and minimise smuggling and evasion. At the same time they were attacking the concentration of trade in Flanders and Brabant, the enemies of Holland in the fourteenth century. The grants brought almost immediate opposition from other towns in Holland. Despite their vigorous protests, sometimes even extending to violence, the counts adhered to the policy of favouring Dordrecht. The importance of toll income made the counts consistent defenders of the privileges. 3)
Dordrecht performed two functions: it was a regional distribution point and an entrepot for goods from all over Europe. Various governments, no matter their efforts, were never able to make the latter very important. English merchants preferred the towns of Zeeland while German Hanse merchants, despite a short stay at Dordrecht in 1359 and 1360, found their traditional Bruges factory better suited to their needs. 4) Goods such as Prussian grain and Scandinavian wood and English wool found their way to Dordrecht however to be traded there for the goods brought by river from the German hinterland. The 'stapelrecht' made Dordrecht the market for Rhine wine and that was the most important good in terms of volume and value for the town. Wood was unquestionably in second place. As early as the Carolingian period wood may have been shipped down the Rhine and by the fourteenth century the trade was well organized with large floats of unsawed wood being brought into Dordrecht. The wood was from Westphalia but also from as far away as Bavaria. It was largely hardwood. There were other goods traded of course including grain, herring and salt from local sources and from some distance away. 5)
In the thirteenth and fourteenth centuries Dordrecht was in an ideal position for the development of a local shipbuilding industry. Raw materials were available in large quantities. The hardwood, the most important component of any ship, had to pass through Dordrecht by law and much of it was sawed and prepared for sale in the town. Dordrecht shipbuil-

ders could get their wood then at relatively low prices. Local shippers exploited the opportunities created by the 'stapelrecht' and 'Maasrecht' and they need men at hand to build and repair their vessels. The ships from other parts of the Low Countries and from overseas also formed a market for repair work. The counts of Holland in protecting Dordrecht trade protected shipbuilding since they gave Dordrecht shipcarpenters to some degree a captive market. Dordrecht shipbuilding may date from the earliest years of the town, that is from the eleventh century. Apparently shipbuilding expanded up to the end of the fourteenth century and continued throughout the following years with the construction of sea-going vessels only finally abandoned in 1871. The industry survived for many years despite economic and geographic pressures and it survived by constant adjustment, by changing character.
The Dordrecht industry is an example of Dutch shipbuilding, an industry always important to the economy of the Netherlands, but also the example with the longest history stretching from the Middle Ages to the end of wooden construction and beyond. Dordrecht shipbuilding was also significant because it was subject to the most extensive guild regulation of any Dutch shipbuilding industry. Dordrecht was second only to Utrecht in the scope and effectiveness of guild power both in the trades and in town government. Shipbuilding included a number of industrial and commercial operations and was in turn tied to many other economic activities at Dordrecht. The complex history of Dordrecht shipbuilding and related operations requires an extensive examination but at least from certain data it is clear that the industry passed through a series of phases. The phases are all the more significant because they do not fully conform to the general pattern of development of Dutch shipbuilding and the exceptions are important. Also important is the fact that the changes at Dordrecht coincide with the changes in the policy of the town government, illustrated both by decisions of the shipcarpenters' guild and of the town government itself. The industrial development or lack of it is then inseparable from the history of legislation.

II 1300 - 1600

The trade expansion of the Middle Ages, first without and later with the aid of government protection, created work for Dordrecht shipbuilders and generated large enough numbers to make possible the formation of a guild, and by as early as 1300. In the fourteenth century Dordrecht probably experienced her greatest expansion with a population of about 10.000 in 1354. The guilds which had often been informal associations of men in a common trade became official institutions of government in 1367. Their rules and earlier grants of status were affirmed. More than 40 guilds were granted official status and the shipcarpenters' was prominent among them. Their patron saint was St. Pancras and they were committed to the maintenance of the St. Pancras altar in the 'Groote Kerk'. They like the other guilds pressed for a greater voice in town government and through the last years of the fourteenth and the first years of the fifteenth century their power to affect decisions increased.
Because of the growth of commerce it was possible to have a high degree of differentiation in trades and this was reflected in the separation of trades into a large number of different guilds. Significantly for shipbuilding by 1376 both the sawers and the wood sellers were separated from the shipcarpenters, even though the sawers had been in the shipcarpenters' guild as late as 1351. 6) The separation while convenient for administrative purposes assured protection for the specific interests of each group and effectively prevented the integration of the principal activities in shipbuilding for many years.
The fourteenth century rules of the shipcarpenters' guild reflected the expansionist nature of the industry. Shipcarpenters were not encumbered by extensive regulation. In fact there were hardly any limits on them. Like all guilds the shipcarpenters' enjoyed a monopoly and only guild members could exercise the trade. The rules for the guilds were made by the officers and guildbrothers themselves. The decision to have limited regulation then was knowingly made by the shipcarpenters themselves. Members had to attend meetings of the guild and violation of guild rules were to be judged by the officers. But beyond that the consuming interest of the shipcarpenters was their jacks. These were windlasses used to raise a ship onto a wharf or to move a ship under construction or to move large pieces of wood. The guild held a monopoly of these devices. Only those

owned by the guild could be used in Dordrecht. Guildbrothers were given equal opportunities to rent the jacks from the guild, paying a rent to the common chest and having to return them in good condition. 7) Other than conforming to the rules about use of the jacks shipcarpenters were free to go about their business as efficiently and effectively as possible.

The beginning of economic decline was marked by the St. Elizabeth's flood of 1421. That inundation shifted the waterways and made it difficult to use a number of established routes. The change favoured ports downstream, especially Rotterdam, at the expense of Dordrecht. Changes in the 'stapelrecht' and 'Maasrecht' worked against Dordrecht as well. Charles V finally conceded to the persistent demands of other Holland towns and in 1540 and 1541 he limited and modified the privileges. Among other things he freed trade in a number of goods and freed trade on the river IJsel. From that point on the privileges deteriorated. All Rhine wine still had to pass through the town but extensive compulsory shipment of a wide variety of goods through Dordrecht was at an end. The 'stapelrecht' did not finally end until 1795 but by 1601 it was not very important to Dordrecht or to the rest of Holland. At the same time Dordrecht trade was hurt by the expansion of the Antwerp market and the success of that Brabant town in attracting international trade by land and sea. The beginning of expansion at Amsterdam worked to the disadvantage of Dordrecht in the same way.

The increasing size of long distance cargo carriers made it more difficult for them to visit the port of Dordrecht. The result overall was a contraction of trade and of the size of Dordrecht itself. In 1514 the population was about 6500 and recovered to as much as 10.000 by mid-century. But that growth could not compare with Rotterdam. Dordrecht industry contracted. The number of brewers fell from as many as 30 in 1500 to only 15 in 1555. Shipping suffered too but merchants were still able to draw on their traditional connections along the upper Maas and the Rhine. They also enjoyed a certain advantage in the capital which they had generated over the years and which they could use in the wine and wood trades, both of which required sizeable investment in inventories.

In 1555 when 492 men reported their trade in a tax assessment there were only seven shipcarpenters. A tax of 1594 showed a rise in their number to 19. 8) There may then have been a revival of shipbuilding after the Revolt but overall the fifteenth and sixteenth centuries was a time of contraction. It was also a time of contraction in the size of vessels built. More work relatively was done on inland vessels and on repair to those and other ships visiting the harbour.

While business for shipcarpernters decreased the rules and regulations they had to follow markedly increased. The guilds no longer took an active interest in extending their role in town government. Rather in the case of the shipcarpenters' guild concentration was on the internal operation of the guild and on vigorous defense of any and all guild privileges. The drive throughout the fifteenth and sixteenth centuries was toward exactness and more careful regulation of all aspects of guild life and with it the activities of shipcarpenters. The meetings of the guild were regularized. And they served as a market for unemployed carpenters by bringing them together with masters at fixed times and with established hiring procedures. The meetings were subject to more strict rules. Misbehaviour, causing a disturbance, insulting a fellow member, attacking the officers for the performance of their duties were all subject to punishment by the guild. The guild was required to keep more extensive records of everything especially of any financial dealings. The guild had to maintain its altar, be sure that there were sufficient candles on it and that masses were said at it. Each member had to visit the altar at least once every Sunday. Of course after the Reformation those requirements disappeared.

There was a set of officers chosen by a formal vote and with fixed duties and with the responsibility for the efficient and orderly operation of the guild. Funerals of members, their wives and children had to be attended by all members wearing proper clothing; that meant 'tabbert en caproen', a long coat and cap used at all solemn occasions. And if they were worn while working the guildsman was punished. The day of a funeral was an enforced day off. Certain holy days were also days off. Anyone working on those days was punished by the guild. In 1467 the shipcarpenters also started the practice of making payments to guildsmen who were too sick to work. Working at night by candlelight or firelight was prohibited because the town was afraid of an outbreak of fire. Having 'tab-

bert ende caproen' was even a requirement for entry into the guild. The proceeds from all the fines plus an annual fee payable by all members and by apprentices went to the maintenance of the altar, that is up to the Reformation. 9) Though these rules on the operation of the guild and on the social activities of guildsmen were important to the lives of members and impinged on their actions there was more important legislation which dealt directly with their business operations.
Master status in the guild meant the right to contract to do shipbuiling work, either new work or repair. All other workers could not agree to take on jobs and so were always employees at best. Thus the guild could limit the potential shipbuilding operations by limiting the number of masters. The monopoly of jacks remained. Everyone in the town was prohibited from making jacks or any lifting device unless at the specific request of the guild and for the use of the guild. Entrance to the guild was made more difficult. Natives were charged lower entry-fines than people from outside of Dordrecht but in general the tendency in the fifteenth century was to charge higher entry-fines. The town had to step in and set maxima for such fees. The next step in the limitations of membership came in 1587 when the guild required a rather complex proof for admission to master status. It was the most extensive proof required by any Dutch shipcarpenters' guild. Many of them did not even bother with such a requirement. Without completing the proof a shipcarpenter could not work as a master. The proof had to be done to the satisfaction of the guild officers who were weighed down with increasing responsibilities. 10) In general the guild became more legalistic, taking more and more cases to the courts of the town or appealing directly to governments. The officers had to personally pursue these cases, petitions and negotiations. Cases covered relations with other guilds and with the members of other guilds, the development of new shipbuilding facilities, the proper payment of dues and fines by members, the enforcement of the guild monopoly and a mass of other matters. 11) The increase in legal business was only the logical outcome of the expansion of regulation.
The shipcarpenters tried throughout the sixteenth century to infringe the monopoly of the wood dealers. These efforts were vigorously and successfully resisted by the guild of wood sellers. The latter were always afraid of shipcarpenters. After building a ship the carpenter often had wood left over and he then liked to sell it at a substantial profit. Shipcarpenters were important customers for the wood dealers and they did not want those men buying on their own. The solution to this jurisdictional dispute was to allow shipcarpenters to join the guild of wood sellers. That arrangement was possible at Dordrecht but in his second guild a member had only the right to pay dues. He had no voice in decisions and could not hold any office. The wood sellers tried to get the town government to require that new members give up their first guild and with it their first trade, in this case shipcarpentry. The prohibition of shipcarpenters selling wood dated at least from the fourteenth century. And the wood sellers actively prosecuted any shipcarpenter who violated that rule. An agreement was reached by the two guilds in 1583 which favoured the wood dealers. Shipcarpenters had to sell any wood left from building a ship to a wood dealer. The only way shipcarpenters could sell it themselves was by joining the wood sellers guild and paying dues to that guild. It was in essence the same agreement which had been reached in 1437. The wood sellers appear to have been slow in admitting shipcarpenters to their ranks and so the court cases and the disputes continued into the seventeenth century. 12) The wood sellers effectively prevented shipcarpenters from extensive expansion into wood dealing. And as a result the potential profits from shipbuilding and the ability of shipcarpenters to accumulate capital were limited. The separation of trades, imposed in a period of expansion, continued despite relative decline.
Overall the great increase in legislation of the shipcarpenters' guild, generated in an atmosphere of stagnation, had the effect of confining shipbuilders. At the same time the mass of rules and the strenuous efforts of guild officers to defend privileges meant the continuing existence of those fourteenth century institutions associated with shipbuilding.

III 1600 - 1800

In the seventeenth and eighteenth centuries Dordrecht failed to participate in the most rapidly growing sectors of Dutch commerce and industry. The herring fishery required

ports closer to the sea. Trade to the Indies, both East and West, was confined by law to certain towns and Dordrecht was not among them. The expansion of whaling in the midseventeenth century caused some stir in Dordrecht but nothing like that in northern Holland. Most important of all Dordrecht was almost completely excluded from the 'moederhandel', the shipment of Baltic grain to Iberia and the Mediterranean for salt and sub-tropical goods, by the competition of towns in northern Holland. 13) The result was relative decline. Dordrecht shipping and with it shipbuilding fell further behind other towns in Holland in quantity and variety of business. The wars of the seventeenth and eighteenth centuries, the fighting in Germany, in nearby Flanders and Wallonia and the problems of trade to the southern Netherlands all certainly damaged the prospects for Dordrecht.
The town could not completely avoid the prosperity of the seventeenth century however. And there was an absolute rise in goods shipped and also probably in the number of ships built. The expansion of arable land by the reclamation of land near Dordrecht expanded the local market. Harbours were expanded and a crane was set up for canting ships to aid in repair and caulking. This 'kielpaal' was first mentioned in an ordinance of 1651 and was owned by the town. 14) It reflected the increasing importance of repair work to local builders as well as the increase in that business.
The general expansion of the seventeenth century and the stagnation of the eighteenth both allowed economic activity to be increasingly centralized in the two largest ports of the Republic, Amsterdam and Rotterdam. The growth of the two at the expense of smaller towns was obvious through the eighteenth century. The shift of the distribution centre for south Holland from Dordrecht to Rotterdam had already occurred by 1600. In shipbuilding Dordrecht shipcarpenters like all others in the Republic had to face the intense competition of the Zaanstreek industry. This highly organized and integrated industry grew rapidly in the seventeenth century and centralized the construction of sea-going ships along the banks of the Zaan. The situation changed about 1720 as Zaanstreek shipbuilding contracted rapidly and collapsed completely by the end of the century. Through the seventeenth and eighteenth centuries it was impossible for Dordrecht builders to compete in the construction of sea-going ships. Moreover, local shippers were overwhelmingly interested in inland vessels and so the Dordrecht industry specialized in building and repairing just those ships.
The decision was made even more certain by the loss by Dordrecht builders of their relative advantage in access to wood supplies. Wood from the Rhine Valley became less significant relative to wood imported from Norway and the Baltic. Even hardwood was increasingly brought in from the North. Germany became a net importer of wood. That in turn implied a growth in the importance of ports on the Zuider Zee in the wood trade. Dordrecht had long been the site of regular wood auctions. Amsterdam developed a wood market in the early seventeenth century but more important than that was the large one in the Zaanstreek beginning in the mid-seventeenth century and reaching its peak in 1716. Significant quantities of hardwood from the Rhineland found their way directly to the Zaanstreek avoiding the market in Dordrecht altogether. If that was not enough Rotterdam in the eighteenth century developed its own wood market. While as late as 1699 Rotterdam merchants and shipbuilders were buying their wood at Dordrecht that was decreasingly the case. 15) The loss of those and other customers meant a decline of Dordrecht as a centre for the wood trade and also meant that shipbuilders in other towns could get shipbuilding wood in quantities and at prices like those of Dordrecht.
The expansion in guild regulation in the fifteenth and sixteenth centuries was a reaction to the relative decline of the industry. In the following two hundred years the attitude did not change. Guild regulation was made even more stringent and comprehensive. The guild jacks continued to consume the interest of the shipcarpenters' guild. All the rules dating from 1397 were regularly repeated and expanded. The guild officers checked any new jacks to be sure they conformed precisely to the required specifications. During each year the guild officers had to check the measuring devices used by guildbrothers to be sure they were accurate. The officers were also to judge any complaints against guildsmen about the quality of their workmanship. An employee could not leave his employer without the permission of that guild master. Meetings continued to be regulated with all the usual punishments for any disruption. Moreover, they were made secret in 1627 with members penalized for disclosing anything that went on during the meeting. Workers from outside

the town, skilled tradesmen who were not guild members, were prohibited from working on ships at Dordrecht except under the most exceptional of circumstances.
The guild was also apparently reluctant to allow new members to join, especially from other guilds. Entry-fines were raised. New members had to take an oath immediately and in front of the mayor and council of the town. The regulation of the sickness fund became more extensive and clearly the guild was trying to be sure that all members paid in full so that income for the fund would al least keep up with payments out of the fund. The funerals of members still had to be attended by all guildbrothers, properly dressed. There was increasing difficulty in having brothers comply. Attendance was excused after 1684 for any shipcarpenter working on a sea-going ship. 16) Obviously this was rare and important enough for the guild to take this unique action to try to promote such work. And there were many more regulations than that. The topics covered increased as well.
The rules show that control of the guild was increasingly in the hands of a small number of men. They used that power to concentrate work in their hands as well. Only masters could do repair work. The guild monopoly was regularly reasserted and within the guild the rights of masters to be the exclusive contractors for work was strongly enforced. These masters were the men who repeatedly held guild office and enforced the rules which worked to their advantage. More of the business of the guild was handled by the officers meeting alone. General meetings attended by all members became less common. 17) The actions of the town government did nothing to stem this oligarchic trend. In fact they helped the masters to centralize the business of shipbuilding into their own hands. In Rotterdam the town chose to promote shipbuilding in the years around 1700 establishing a new harbour for shipbuilding. The area was developed at government expense and wharves were offered to shipcarpenters at reasonable prices along with low cost mortgages. The same procedure had been followed in the first years of the seventeenth century but with little success. The second time the result was a revived shipbuilding industry in Rotterdam. Dordrecht tried similar assistance. Government-owned slipways were established along the 'Spuihaven' in 1583, rented to master shipcarpenters and used for repair work. The guild operated the wharves and part of the income from rents went to the guild. Since the town and guild controlled the location of shipbuilding activity they could force builders to move their operations to these new 'hellingen'. The renter had to supply the tools, jacks and capstans. The jacks had to be rented from the guild and returned along with a new one for about every 15 rented on long term. The goal of the town was to supply efficient repair services to shippers, both domestic and foreign, to cut down their turn around time and thus make Dordrecht an attractive port. The 'kielpaal' and associated equipment had the same purpose for larger ships. There were only three 'hellingen' though and they were only useful for repairs. While the development seems to have been successful through the seventeenth century the falling rental income in the eighteenth shows that they were of decreasing value. 18)
The most extreme act of government aid to local shipcarpenters was the requirement that all native shippers had to have their ships built and repaired at Dordrecht and therefore by Dordrecht shipcarpenters. In 1657 all shippers with licenses to operate regular services from Dordrecht were made subject to the requirement. It was expanded in 1698 to all Dordrecht shippers. No one could enter the shippers' guilds unless he owned a vessel built at Dordrecht. Violation was severely punished and as with all other privileges the shipcarpenters' guild pressed for complete enforcement. To get this restriction the shipcarpenters had to agree to rapid building and repair work of the highest quality so that shippers would not suffer a competitive disadvantage as a result.
In the case of Dordrecht this prohibition led to retaliation by other towns. That in turn led to a controversy over who in fact would benefit from such a rule. A number of shipcarpenters petitioned the town government claiming that the only beneficiaries would be the owners of slipways, the masters in the guild, since they were the ones who get the work. Retaliation by other towns on the other hand would mean unemployment for the rest of the guildsmen. The logical outcome would be increased centralization of work in the hands of a few. But the officers of the guild had the support of other tradesmen such as blockmakers, ropemakers, smiths and sailmakers who wanted a protected market. The petitioners pointed to the problems of enforcement which arose in 1655 when the first efforts to confine native shippers had to be abandoned. But the town government would not be

moved. 19) The government was again faced with hard times and such restriction was the only method they saw open to them to even maintain the position of Dordrecht as a centre of trade.
The efforts of government and guild appear to have arrested the deterioration of shipbuilding. Local carpenters could build special vessels if called on to do so. A Dordrecht builder, Jacob Spaans, constructed a 60-gun warship, 'Hercules', during the Fourth English War. Such construction was rare. Of the 24 ships reported to have been built by the Schouten wharf between 1728 and 1790 only seven were of more than 100 feet in length, and almost all of the seven were built after 1771. Up to the last years of the eighteenth century the Dordrecht industry was dominated by building smaller vessels, especially vessels for use on inland waterways. The wharves themselves were small, the 'hellingen' were not large and located on a canal which did not allow access for sizeable seagoing vessels. They could not compare apparently with the new wharves built at Rotterdam in the early eighteenth century which were 180 feet long and 80 feet wide.
The number of shipbuilders operating at Dordrecht fell as well. And there was an increasing tendency for them to form partnerships, usually with family members, so that the number of functioning wharves fell even more dramatically. The guild could boast around 120 members in about 1660. But the number of men contracting for the construction of new ships was much lower. In 1662 and 1663 a total of 17 builders won contracts for a total of 54 vessels. The vessels were almost exclusively small, one half being described as 'cromsteven schuiten'. The pattern of centralization had already been established. Three of the 17 builders constructed 40% of the ships. And the pattern continued. In 1718 and 1719 contracts for only 23 vessels were drawn. The ships were built by nine groups, four of those being partnerships so the total number of builders was 12. Just two groups accounted for nearly 40% of new ships constructed. And again small vessels dominated the list. By 1786 and 1787 the number of contracts for new ships had fallen to just 13 but there were still nine groups and therefore presumably nine shipbuilding wharves still in use. The largest ship mentioned was a 'wijdschip' of only 70 feet in length. The number of wharves in 1740 may have been as high as 20 but most were small, so small that they could take on little more than repair of smaller vessels. 20)

IV The Nineteenth Century.

The revolution in 1795 and related political changes dramatically altered the situation for Dordrecht shipbuilding. The town government had protected the industry and kept it from falling on harder times. After 1798 the town could no longer exercise that type of protection. All restrictions on trade and industry were abolished as were all the guilds. So from the first decade of the nineteenth century protection as well as limitation of shipcarpenters through the guild disappeared. The guild itself was already rather weak. For example, it took the shipcarpenter, Pieter Gips, only six years from the time he became a master to rise to the highest office in the guild. And Gips was an opponent of the guilds. Obviously there was a shortage of talented new men. That was by no means the only sign of weakness. The town then had to withdraw completely from any policy on shipbuilding through the guild or through any medium. All control over shipbuilders on the islands just north of the town in villages such as Alblasserdam and Zwijndrecht were abandoned. The guild had been the agent of such control. The government of Dordrecht finally in 1831 bought the rights to the 'hellingen', paying a monthly rent to the former members of the shipcarpenters' guild and using the land for houses. 21) Shipbuilders then were left unaided by local authorities but also left to make their own decisions.
Freedom did not bring an immediate boom to Dordrecht shipbuilding. The industry expanded only under protection, this time of the national government. The importance of Dordrecht shipbuilding in the nineteenth century was more apparent than real. But still after some three or four centuries of commitment to the construction of smaller vessels and to repair work, at least two builders at Dordrecht turned their backs on that tradition and embarked on the construction of sea-going vessels of the largest type and the latest design. It was unquestionably a conscious decision. And it was made in difficult times. From 1813 to 1828 of 83 vessels averaging 400 tons built in the Netherlands only six were built at Dordrecht. 22) In those years the town was still recovering from the disastrous Napoleonic period which decimated the domestic merchant fleet and disrupted

river trade. Shipbuilding did not thrive but it did pass through a new phase, developing in an entirely different way from the past and precisely because of the deterioration of the past.
At Rotterdam shipbuilding was in the hands of a few men. They promoted that by integrating shipbuilding and wood dealing. In some cases it was hard to tell which was more important to them. Shipbuilders became involved in a number of related activities such as trade, credit dealings, owning parts of ships and so on but wood dealing was the natural first avenue of diversification. To some degree the same pattern was repeated at Dordrecht. With the disappearance of the wood sellers' guild in 1798 along with all the other guilds shipcarpenters were free to trade in wood. The town had already given up trying to prevent the combination of wharfs so builders could assemble a sizeable area for construction. The general concentration of wealth into the hands of a few, it not being unique to shipcarpenters, was obvious from the luxurious townhouses built in Dordrecht in the eighteenth century. 23)
The Napoleonic period drove out of business many of those builders who had not moved toward diversification, integration, extension. That left two builders capable of building large vessels: Schouten and Gips. There were some five other wharves devoted to the long-standing tradition of building and repairing river craft. Little is known about the firm of Schouten. Ships were built by the firm from 1709 and perhaps before. They were involved in the wood trade on an international basis. Jan Schouten took over the business including the shipbuilding wharf on from his father in 1807 at the age of 22. In the last years of the eighteenth century his father had started to build large sea-going ships and the son was even more committed to that practice. The problem was that he like almost all Dutch builders after 1815 returned to the well-known designs of prewar years and the vessels he built were exactly like those built at Dordrecht fifty years before. 24) But with government protection the firm did grow and improve its product over the following forty years.
The firm of Cornelis Gips has the advantage of having a well-documented company history. Cornelis Gips set up his own shipbuilding wharf in 1819 based on the inheritance of his wife, the daughter of a Dordrecht shipbuilder. Cornelis' father, Pieter Gips, had owned his own shipbuilding wharf since 1794 but the operation had not proven profitable and he had to take in a partner who could supply needed capital. Cornelis Gips then had no choice but to strike cut on his own, that is with the aid of his wife's money, a mortgage and his own inheritance on his father's death in 1828.
In 1825 Cornelis built his first large sea-going ship. It was a difficult job since his wharf had always been used for building inland vessels. But he was obviously convinced that innovation was needed for the business to succeed. His next contract was for a steamboat. Schouten had already built the first Dordrecht steamboat in 1825. Contracts followed quickly for Cornelis Gips and though he built no sea-going ships from 1830 to 1834 because of the revolt to the south and the blockade he returned as soon as possible to larger vessels. For more space in 1830 he bought the wharf next to his. He was doing well since he had the money to finance that purchase as well as the construction and therefore expensive ships. This did not prevent him from building a smaller ship now and again especially in hard times to keep the wharf occupied. He invested in shares of ships, at first just selling the shares when the vessel was launched but increasingly holding them to draw the income. That meant increased capital requirements, requirements which he was able to meet. In about 1840 a new firm, Cornelis Gips en Zonen, was formed incorporating his four sons into the business. In 1835 Cornelis took on the management of a wharf in Schiedam but immediately handed operation of it over to his third son. On the death of Cornelis Gips in 1843 management of the firm was taken on by the eldest son, Pieter, in cooperation with his brothers. And though temporarily subject to some difficulties it was at that point a successful integrated shipbuilding firm specializing in the construction of efficient ocean-going vessels. 25)
The growth and development of both the Gips and Schouten firms in the 1820's, 1830's and 1840's depended heavily on orders for East Indiamen. Trade to the East Indies was protected and regulated by the Nederlandsche Handel Maatschappij (NHM), a government controlled company which chartered ships to sail to the Indies paying higher than prevailing freight rates. King William I founded the company as a way to revive Dutch commerce and shipbuilding. The king first tried direct subsidies to Dutch shipbuilders, a fee of

so much per ton for all ships over 300 tons. That continued from 1824 to 1830. Import of foreign-built ships was also restricted from 1819 to 1850. More important though was the NHM chartered in 1824.
Since Dordrecht shipbuilders were supplying this special market they had to keep up to NHM standards and also to stay below 800 tons since above that point shipping rates fell. Speed was not important because of the cargoes carried and because there was a sailing only every six months. Orders for ships depended on the approval of the NHM and the first thing done by a group of owners when they decided to build a ship was to send a representative to The Hague to get a government guarantee of two return voyages to the Indies for the vessel. In a good position to trade with the Indies, more easily reached by East Indiamen than Rotterdam or Amsterdam and favoured as a smaller port by the NHM to 1848, Dordrecht got a relatively large share of shipping contracts for the East and that meant a significant share of shipbuilding contracts.
Between 1826 and 1852 some 40 East Indiamen were built at Dordrecht. That was combined with other work. The Schouten wharf for example from 1826 to 1852 built a total of 26 vessels for the merchant marine. 26) The result was overproduction and the NHM was forced at times to change its procedures. That could only increase the cyclical nature of orders for ships. In 1843 the Gips wharf employed only 34 men and the Schouten wharf 52. On the other hand by 1855 the two firms in conjunction were renting new space from the town at 'Willigenbosch', an area not used for shipbuilding since the late eighteenth century. And again by 1858 there were only 36 men working on the Gips wharf. 27) A general change in government policy around 1850 toward freer trade aggravated the situation. The change worked to the disadvantage of Dordrecht shippers who had been prospering under the protective blanket of government controlled trade. The shippers and shipbuilders of the town found themselves unable to compete in an open market. Certainly the shipbuilding firms could run down the capital built up in the 25 years of growth but they were now exposed to the extreme changes in demand for new ships which were typical of the second half of the nineteenth century.
The Gips firm had an important advantage in such a market in the long association with the shipping firm of Gebr. Blussé. Members of the Blussé family were involved in shipping combines buying ships from the Gips family from the 1820's. The firm could expect not only to build the ship but also to supply all the masts and spars and any extra wood needed. In all likelihood the firm could expect to get the contract for any repairs too. Cornelis Gips took shares in some of the ships, becoming a partner of Adolph Blussé. The ties between members of the Gips family and A. Blussé and P. Blussé van Oud-Alblas may have stemmed from shared liberal political views dating from before 1795. The Blussé combines were deeply involved in taking advantage of the lucrative trade to the Indies and that in turn meant a stream of orders for East Indiamen for the Gips wharf. And the Blussé organizations were the greatest buyers of ships in Dordrecht in the nineteenth century. 28)
It was the connection with the family Blussé which created the opportunity for the Gips firm to establish itself as a technically superior builder. The firm of Blussé wanted to have a clipper, that type of fast ship developed in America. Dutch ships were designed as they had been since the sixteenth century sacrificing speed for hold space and small crew. Certain trades demanded speed though and the Gebr. Blussé wanted to participate in those trades. By 1850 they had already been investigating the design of American clippers. In 1853 they were able to get a model of an American clipper, originally from a firm in New Jersey. Gips en Zonen was able to convert the model into a ship, the 'Kosmopoliet I', completed at Dordrecht in 1855. Dutch builders had been trying out many of the features of the new design in their own ships. And shippers had been buying clippers from overseas. Still the job for the Gips firm was not an easy one, scaling down from the model of a vessel of 1200 tons to a ship of under 800. It is likely that she was the first clipper built in the Netherlands and though a medium and not an extreme clipper it is certainly true that she impressed shippers enough to make the type popular in the Netherlands. Her speed was comparable to the best English tea clippers, decreased the sailing time to the Indies and suggested new possibilities to all Dutch shipbuilders. The Gebr. Blussé could have had their new clipper built in America but they chose to have the job done by the Gips firm which had already built some ten ships for family combines. Because of the insistence on building domestically the Blussés were able to

interest certain prominent political figures including Thorbecke who took one thirty-second part of the investment. Again liberal political ties may have contributed to business connections. Gips en Zonen themselves took two thirtyseconds. The return on that investment was far from satisfactory but they did not lose and given the novelty of the design it was not unexpected. 29) Despite their experiment with 'Kosmo', C. Gips en Zonen were still very much tied to the success or failure of Dordrecht shipping firms. Jan Schouten died in 1852 and his two sons took over the wharf. They tried to change design to be more up to date. But with the collapse in orders in the late 1850's the firm became bankrupt and was liquidated in 1860. The Gips firm did better. The crisis led to the construction of ever larger medium clippers, a compromise design where some speed was surrendered for hold space. And Dutch yards like that of Gips en Zonen could build ships of that type with significantly smaller crews. The clippers they built grew in size, the 'Kosmopoliet II' built in 1864 being 1077 tons. The 'Kosmopoliet III' finished in 1871 was 1548 tons. The firm was able to carry on because of the commitment of Mr. A. Blussé to wooden sailing ships. He claimed in a speech in 1880 that iron and steam had no future.

Despite Blussé's views the future of wooden ships was in doubt, buyers were more scarce and the future of C. Gips en Zonen was in doubt by the late 1860's. The firm had to build ships without a buyer, hoping that one would turn up before completion and even then taking on a sizeable owner's share when the vessel was sold. 30)

They did get orders for warships from Japan. It was more than an interesting sidelight for the firm since it kept the wharf on 'Willigenbosch' going in the mid-1860's. The first warship built was so large that it had to be towed to Brouwershaven where it was rigged and fitted out. These warships required more capital, more manpower, more workspace. The firm needed these orders however because it could not interest private buyers. The Japanese orders were only won by Gips en Zonen because they were made through the NHM. 31) That is not to say that the firm did not demonstrate a high degree of skill, of expertise and technical knowledge or management ability. They did over the course of the nineteenth century improve their skills, learn the new theory of shipbuilding and fulfill difficult contracts. But the contracts were won in the context of a protected market. The last big ship which they built was 'Kosmo III', that for the Gebr. Blussé.

It was obvious then by 1870 that a change had to be made. As ships grew in size they could not reach Dordrecht through the narrow and shallow waterways which connected the town with the sea. Prices of Dordrecht built ships by the 1860's were high compared to those of smaller towns in Holland and especially high compared to those of foreign builders. Restrictions on purchase of vessels from overseas were dropped completely in 1868 and the NHM changed its rate structure in the same year to the advantage of large ships. Faced with all the problems implied by those changes and with the deterioration of their capital position the firm of C. Gips en Zonen chose to convert to the construction and repair of steam-powered iron river vessels. In 45 years of building large ocean-going ships the firm on three wharves had produced more than 100 such cargo ships. The situation was once again like that of of the sixteenth century and all firms in Dordrecht had to seek refuge in work on small vessels. And it was a wise choice given the rapid expansion of shipping along the Rhine from the 1850's.

From 1852 the Gips firm had a slipway at Papendrecht for repair work. It was used by 30 to 40 ships per year and employed 100 men, more than on the building wharves except in the best of times. The decision of the town to abandon the 'kielpaal' in 1860 made the Papendrecht slipway the only place for the repair of large ships at or near Dordrecht. Contraction meant abandoning the 'Willigenbosch' wharf and conversion of the original wharf to other work. In 1910 there were still about 100 men working for C. Gips en Zonen. The firm was finally liquidated in 1916. 32) The experiments of Cornelis Gips and his successors had made possible the long survival of the business and combined with Jan Schouten and his sons, they participated in the last and the most impressive if not the most economically significant period of wooden shipbuilding at Dordrecht.

V. Competition and Government Protection.

It is simple to say that after the prosperity of the fourteenth century the shipbuilding industry of Dordrecht declined. But the decline lasted for four hundred and fifty years

The industry survived through great swings in the fortunes of the Dutch economy and through significant changes in the design of ships. The final disappearance of wooden shipbuilding and the abandoning of construction of sea-going vessels came only after the worldwide conversion to iron and to steam. It look dramatic and extensive changes in technology and geography to destroy Dordrecht wooden shipbuilding and then as late as 1871 a Dordrecht builder was constructing a sea-going clipper ship not of wood but still of the latest design and of dimensions to make it internationally competitive. The change to iron ships put Dordrecht at a disadvantage. The metal had to be imported and there were no indigenous coal supplies so any work with iron had to be relatively costly. The growth in the size of ships worked against Dordrecht as well. And with massive changes in the waterways and then the construction of the 'Nieuwe Waterweg' for Rotterdam, started in 1866, Dordrecht shippers lost any remnant of competitive strength. And a decline for native shipping concerns meant a fall off in orders for Dordrecht shipbuilding yards. Shipbuilders in the 'Noord', in the villages between Dordrecht and Rotterdam had all of the same problems. Yet they were able to continue effective competition. Importing iron, a firm of Krimpen aan de Lek was able to build river boats at a low enough cost to sell them in Germany. And the firm of Fop Smit of Kinderdijk built some of the best clipper ships produced in the Netherlands. They were of wood, of iron or composite build - it did not matter. Already in 1850 builders in the area had been experimenting with clipper design. And Smit's reputation was firmly established with the construction of the fast sailing iron vessel, 'California', in 1852. Clipper building in the 1850's and 1860's was centered in Dordrecht and in the 'Noord', and increasingly in the latter. Smit's firm continued to build clippers into the 1870's after such construction had to be abandoned at Dordrecht. More clippers were built in the 'Noord' than in any other district of the Netherlands.

The greatest problem for many builders in the 'Noord' was the size of their wharves which were too small for building the ever larger sailing vessels. 33) Lower wages relative to the cities were crucial to the competitive advantage of both Dordrecht and the 'Noord'. In the first half of the nineteenth century Dordrecht wharves could save buyers 40% or more over having their ships built at Rotterdam or Amsterdam. As the years passed the advantage of Dordrecht faded and the building of large vessels transferred to the even lower wage region of the 'Noord'. There building sea-going vessels, with a lower labour component than with river vessels and using imported iron, survived longer than at Dordrecht. For Dordrecht builders, faced with rising labour, land and material costs and the decline of their home market for big ships the only solution was to try and use their relatively more expensive resources, especially labour, in a more productive way. The change to river vessels meant effective exploitation of their more skilled labour as well as the effective exploitation of their geographic position as a major river port. It was indeed a return to the situation of the sixteenth century.

Shipbuilding at Dordrecht enjoyed a comfortable and lengthy if not always prosperous decline. In the process it kept alive associated trades such as sailmaking and anchormaking while supplying local shippers with services they needed. After all in 1846 there were still 45 ships that called Dordrecht home port. By 1871 the number had fallen to 20 and only 12 of those were of significant size. 34) The retrenchment of the shipbuilding industry was over a very long term. The guild and then the town government and later the national government by conscious acts of policy were able to keep the industry in tact. They were able to keep the shipbuilders competitive behind their wall of protection so that buyers were not heavily penalized and at the same time local shipbuilders might on occasion win contracts from outside sources, won in an open market without the advantages of protection. The protective measures of successive governments also made the industry competitive in the construction and repair of certain types of vessels. Dordrecht probably converted from being an international entrepot to being a distribution centre for south Holland and a river port as early as 1500. The specialization of the shipbuilding industry in inland vessels reflected that change. But the facts did not keep the government of Dordrecht from retaining pretensions to international status or from retaing the legal trappings of a great international port. Those legal trappings included the indulgence of a shipcarpenters' guild with many regulations and the protection of those shipcarpenters. That kept the industry from an early collapse. The shipbuilders while accepting and defending all those features of international pretension at the same

time wisely pursued the obvious economics and tended more and more to build vessels for the inland trades.
Dordrecht shipbuilders were never able to effectively exploit their excellent position. They had certain competitive advantages such as access to wood supplies, long established investment in facilities and in trained manpower, investment in harbours and contacts with suppliers and with buyers. In the Middle Ages they were close to northern Europe's busiest ports. Despite all that the industry fell behind competitors in the fifteenth century and was never able to regain that lost ground despite specialization in smaller vessels. So shipbuilding in the fifteenth and sixteenth centuries found itself tied up in the strict confines of guild rules which were designed in the first instance to protect, defend and maintain the industry. And the expansion of trade in the seventeenth century did not mean an end to the regulations of the guild but rather an increase. The industry in those years became irrevocably tied to the Dordrecht market for ships. That was the market they supplied and they were rarely, even in the nineteenth century, able to escape from dependance on the fortunes of local shipping interests. And Dordrecht merchants and businessmen rarely if ever ordered large vessels in the seventeenth century. It was not until the end of the eighteenth century that Dordrecht shipbuilders turned to the construction of large ships. The poor location relative to the fastest growing trades a fact which had been obvious since the fifteenth century, worked against the advantages enjoyed by the industry and the result was a slowing of growth and a limitation of experimentation through the seventeenth and eighteenth centuries.
In sum the rules of government from the Middle Ages to 1795 both on shipbuilding and on trade were designed for a medieval staple town. The result was just that type of trade and that type of shipbuilding. The rules served their purpose in keeping a few shipcarpenters busy with a few contracts. The industry was confined but comfortably confined. The rules kept the industry together so that at the end of the eighteenth century it was possible to succeed in the construction of sea-going ships. There were the facilities but more important there were the talented men to exploit the opportunities which appeared. The sharp departure from the past around 1800 was possible because the diversification and integration of shipbuilding and wood dealing activities and their concentration in the hands of a few men made it possible for that small group to assemble the necessary capital. The general economic deterioration of Dordrecht, the decline of the town had brought down wages and land values relative to the large towns of Holland. At the same time the Dordrecht guild had kept the training programme for shipbuilders so that skilled labour was available. The slow decline of the proceeding four centuries was turned into an advantage by the few remaining builders. And unquestionably they were handsomely rewarded from 1826 to about 1857 for their decision to break with tradition.
The combination of long term developments in the economy and of both long and short term government policies made possible the sudden emergence of a small but successful, a limited but highly skilled and nationally and internationally prominent shipbuilding industry in what by the nineteenth century was a quiet economically insignificant provincial town.

NOTES.

1) This paper was prepared in part with the aid of a research grant from the Canada Council. I am also deeply indebted to the staff of the Gemeente Archief Dordrecht for their extensive assistance.
2) Nelly J.M. Kerling, Commercial Relations of Holland and Zeeland with England from the late 13th Century to the Close of the Middle Ages, Leiden, 1954, 4.
Henk Klompmaker, "Les Villes Néerlandaises au XVIIe Siècle, Institutions Economiques et Sociales," Recueils de la Société Jean Bodin, VII, 2(1955), 579.
J.F. Niermeyer, "Dordrecht als handelsstad in de tweede helft van de veertiende eeuw," Bijdragen voor Vaderlandsche Geschiedenis en Oudheidkunde, 8e reeks, III(1942), 3-5.
Bernard van Rijswijk, Geschiedenis van het Dordtsche stapelrecht, The Hague, 1900, 8-16.
3) Van Rijswijk, op. cit., 18, 21-40. 'Maasrecht' had in fact already been granted in 1338 by William IV but the 1355 extension was complete.

Matthys Balen Janszoon, Beschryvinge der Stad Dordrecht, Dordrecht, 1677, vol. I, 75-76.
J. L. van Dalen, "Het stapel- en Maasrecht van Dordrecht," Tijdschrift voor Geschiedenis, VI(1891), 2-6.
J. F. Niermeyer, "Een vijftiende-eeuwse handelsoorlog: Dordrecht contra de bovenlandse steden, 1442-1445," Bijdragen en Mededelingen van het Historisch Genootschap, LXVI(1948), 1-59.

4) Niermeyer, "Dordrecht als handelsstad...," 188-190, 222.
Van Rijswijk, op. cit., 41-43.

5) Niermeyer, "Dordrecht als handelsstad ...," 21-36, 177-183.
W.S. Unger, "De economische en sociale structuur van Dordrecht in 1555," De Economist (1915), 969-971.

6) Niermeyer, "Dordrecht als handelsstad ...," 1, 8e reeks, IV(1943), 91, 112.
Matthys Balen Janszoon, op. cit., vol. I, 511-512.
A. van Vollenhoven, Ambachten en neringen in Dordrecht, The Hague, 1923, 4-11, 15-16.
Gemeente Archief Dordrecht (hereafter G. A. D.), Afdeling Gilden (hereafter A. G.), 244(1367), 246(1367).

7) Van Vollenhoven, op. cit., 8-9, 20-21, 26-27, Appendix 3, no. 3.
G.A.D., A.G., 248, 254.

8) G.A.D., Archief 1572-1795, 3965(1594).
W.S. Unger, op. cit., 961, 981-984.
T.S. Jansma, "De betekenis van Dordrecht en Rotterdam omstreeks het midden der zestiende eeuw," De Economist (1943), 212-213, 233, 246.
Van Dalen, op. cit., 10-11.
Van Rijswijk, op. cit., 64-98.
Niermeyer, "Dordrecht als handelsstad...," III(1942), 2.

9) Van Vollenhoven, op. cit., 21-23, 51-53.
G.A.D., A.G., 250(1467), 252(1485), 253(1488), 266(1562), 268(before 1572), 261, paragraphs 4, 9, 10, 11(1548).

10) Van Vollenhoven, op. cit., 39-40.
G.A.D., A.G., 228, pp. 1-8, paragraphs 1-9, 13(1587), 229, pp. 27-29(1594), 33-36(1551), 232, fol. VI(1437), 258(1535), 261, paragraphs 1, 2, 12, 13(1548), 263 (1554), 277(1595), 275(1589).
Richard W. Unger, "Regulations of Dutch Shipcarpenters in the Fifteenth and Sixteenth Centuries," Tijdschrift voor Geschiedenis, LXXXVI(1974).

11) G.A.D., A.G., 260(1536), 262(1549), 264(1555), 279(1597), 282(1603).

12) G.A.D., A.G., 9a(1583), 9b(1583), 247(1384), 255(1512-66), 256(1525), 265(1561), 269(1577), 272(1582), 273(1583), 285(1615), 286(1616).

13. Nina Ellinger Bang, ed., Tabeller over Skibsfart og Varetransport gennem Øresund 1497-1660, vol. I, Tabeller over Skibsfarten, Copenhagen, 1906. From 1558 to 1660 there were only 311 passages by ships calling Dordrecht home port through the Sound. By comparison all of Holland for just the ten years 1590 to 1599 had a total of 26.359. Enkhuizen for example in just one year (1590) showed a total of 594 passages.

14) J. L. van Dalen, Geschiedenis van Dordrecht, Dordrecht, 1931-1933, 396.
A. Staring, "Geschiedenis van Dordrecht," Oudheidkundig Jaarboek, V(1925), 74-76.

15) Johan Schreiner, Die Niederländer und die Norwegische Holzausfuhr im 17. Jahrhundert," Tijdschrift voor Geschiedenis, XLIX(1934), 303-306.
C.A. Schillemans, De houtveilingen van Zaandam in de jaren 1655-1811, The Hague, 1947, 84, 140-141.
S.C. van Kampen, De Rotterdamse particuliere scheepsbouw in de tijd van de Republiek, Assen, 1953, 125-137.

16. G.A.D., A.G., 229, pp. 10-20, paragraphs 1-12, 14, 22(1627), 229, pp. 20-21 (1602), 21-25(1704), 165-168(1749), 136-141(1710), 142-144(1722), 203-205(1783), 293(1642), 295(1645), 296(1646), 297(1647), 201(1652), 305(1654), 306(1655), 312 (1677), 315(1677), 319(1684), 333(1705), 337(1710), 338(1725), 339(1725), 341(1754), 344(1784).

17) G.A.D., A.G., 229, pp. 10-20, paragraph 19(1627), 317(1594-1678), 335(1706), 336(1707).

A.J. Deurloo, "Bijltjes en Klouwers. Een bijdrage tot de geschiedenis der Amsterdamse scheepsbouw, in het bijzonder in de tweede helft der achttiende eeuw," Economisch- en Sociaal-Historisch Jaarboek, XXXIV(1971), 12 and passim. The same pattern emerged in Amsterdam in the eighteenth century and the move toward oligarchy was a common one in almost all guild regulated shipbuilding industries in Holland and Zeeland.

18) G.A.D., A.G., 229, pp. 100-114(1643), 154-157(1739), 230, pp. 119-121(1768), 237(1645-85), 241(1613), 242(1595-1799), 267(1571), 349(1583).
Van Kampen, op. cit., 32-41, 115-116.

19) G.A.D., A.G., 229, pp. 46-49(1657), 84-88(1702), 160-161(1741), 180-183(1757), 323(1685), 326(1697), 327, paragraphs 1-8(1698), 330, paragraphs 9, 10, 14-17, 28-38, 78-83, 218-224(1700), 331(1700), 332(1703).

20) G.A.D., A.G., 229, pp. 100-114(1643).
Van Kampen, op. cit., 48-49.
G.A.D., Archief van de Gemeente Dordrecht 1572-1795, 3507(1740).
G.A.D., Oude Rechterlijke Archieven, Archief van het Watergerecht, 1605, 1616, 1622.
Anon., "Tentoonstelling 'Schepen en werven te Dordrecht in de 19e eeuw,' 21 dec., 1963-16 febr., 1964," Maritiem Museum "Prins Hendrik", typescript introduction, 1-4.
E.W. Petrejus, Nederlandse Zeilschepen in de 19e eeuw, Bussum, 1974, 18.

21) Van Dalen, Geschiedenis van Dordrecht, 396.
G.A.D., A.G., 262(1549), 277(1595), 304(1654).

22) T. Lekkerkerker, Hout en Schepen. Uit de geschiedenis van de houtwerkingsbedrijven Gips, Leiden, 1941, abridged edition, 4-9.
J. Oderwald, Het Nederlandsche zeilschip van 1800 tot het einde, Amsterdam, 1939, 43.

23) G.A.D., A.G., 320(1684).
Staring, op. cit., 76-77.
Van Kampen, op. cit., 183-190, 98-103.

24) Anon., "Tentoonstelling 'Schepen en werven te Dordrecht...," 2-3.
A. Blussé van Oud-Alblas, De Geschiedenis van het clipperschip in Noord-Amerika, Engeland en Nederland, Second edition, Amsterdam, 1972, 124-125.
The Schouten firm has the advantage of having a number of important events in its history recorded in accurate and surviving pictures. See Petrejus, op. cit., 16, 17.
G.J. van Oostveen, De Economische Ontwikkeling van Dordrecht 1795-1945, Dordrecht, 1946, 85, 94, 102.
G.A.D., Dordracum Illustratum, 2472, 2473 for example.

25) Lekkerkerker, op. cit., 9-37.

26) Lekkerkerker, op. cit., 28-31.
Blussé van Oud-Alblas, op. cit., 120, 130-132.
Van Dalen, Geschiedenis van Dordrecht, 367-368.
G.A.D., Archief van de Rederij van Blussé van Oud-Alblas, for example 26, Notulenboek der vergadering gehouden door de Directeuren en de Raad der Rederij van de fregatschip 'Dortenaar', 1829-1831.
J.A. de Jonge, De Industrialisatie in Nederland tussen 1850 en 1914, Amsterdam, 1968, 130.

27) Lekkerkerker, op. cit., 55-56.
Van Oostveen, op. cit., 94-95.

28) G.A.D. Archief van de Rederij van Blussé van Oud-Alblas, 26, 27, 101, 118, 138, 158, 180. Unfortunately much of the archive was lost in the flood of 1953 so secondary sources from before that date must be relied on.
Van Dalen, Geschiedenis van Dordrecht, 371-373.

29) G.A.D., Archief van de Rederij van Blussé van Oud-Alblas, 216.
Lekkerkerker, op. cit., 41-49.
Blussé van Oud-Alblas, op. cit., 126-129, 133-142.
Anno Teenstra, De Clippers. Een nieuwe geschiedenis van de snelste Nederlandsche zeilschepen uit de tweede helft van de 19e Eeuw, Amsterdam, 1945, 53-85.
Oderwald, Het Nederlandsche zeilschip van 1800..., 71-72.

Oderwald, Nederlandsche snelzeilers. De geschiedenis der snelle Nederlandsche clipperschepen, zoals zij gebouwd werden aan de grote stroomen in de negentiende eeuw, Amsterdam, 1940, 132-135.
Thorbecke's son sailed on 'Kosmo'. The father's committment may have come from a high degree of confidence in his own judgement rather than an understanding of shipbuilding. At least that was the opinion of M. H. Jansen, the naval officer who was responsible for bringing the model from America. S. P. L'Honoré Naber, ed., Het leven van een Vloothouder. Gedenkschriften van M.H. Jansen, Werken uitgegeven door het Historisch Genootschap, 3e serie, 49(1925), 265-266, 286-287, 299-300.

30) Blussé van Oud-Alblas, op. cit., 132, 147-149, 162-164.
Anon., "Tentoonstelling 'Schepen en werven te Dordrecht...," 6.
Teenstra, op. cit., 179-182, 235-266.
Van Oostveen, op. cit., 95-96, 101.

31) Lekkerkerker, op. cit., 56, 60-77.

32) Van Dalen, Geschiedenis van Dordrecht, 370-371, 395.
Lekkerkerker, op. cit., 77-79, 84.
Blussé van Oud-Alblas, op. cit., 152-155.
Van Oostveen, op. cit., 111-112.
Van Dalen, "Een gouden jubileum. D. Boest Gips, 1860-3, sept.-1910," Dordrechtse Courant, Mon., 29 Aug., 1910, Evening ed., 2.
De Jonge, op. cit., 147-148, 135-136.

33) De Jonge, op. cit., 137, 143, 148.
Blussé van Oud-Alblas, op. cit., 143, 155, 162.
Oderwald, Nederlandsche Snelzeilers ...., 132-133.
Teenstra, op. cit., 30-31, 44, 85-87.
Anon., "Tentoonstelling 'Schepen en werven te Dordrecht ...," 14.

34) Lekkerkerker, op. cit., 57-58.
Van Oostveen, op. cit., 51, 64-65. In 1811 there were still eight sailmakers in Dordrecht and in 1816 the number had dropped to four but that was a bad period for all industry.

V

Wooden Shipbuilding in Zeeland

The province of Zeeland carved out of border lands between the medieval counties of Holland and Flanders, was tied to the sea even before it had a political existence. Made up of a series of islands at the mouths of the Maas and Schelde, the residents always had to use ships. In the region the building of vessels for sailing on the rivers and the North Sea dates from perhaps before the arrival of the Romans. During the era of Roman occupation the commonly used route to Britain lay along the island of Walcheren, so shipbuilding as well as ship repair probably took place in the province. In the early Middle Ages under the influence of the short-lived Viking state established as a "Danelaw" on the north coast of the Netherlands Scandinavian shipbuilding techniques may well have been imported and copied by local carpenters. Because transport by water was so important to island-rich Zeeland, from the earliest years of the Middle Ages building ships was and remained for a long time a principal activity of what industry there was. It was principally a by-employment practiced by farmers and fishermen working on a piece of land temporarily set aside for shipbuilding. It was not professional or specialist work. The ships built were generally small but seaworthy.

From the thirteenth century growth in the population and the economy and the related general increase in the level of income per person created a widespread tendency toward the centralization of industry in the towns. The expansion of trade brought with it more trips and longer trips for merchants and shippers in the county. For example, Zierikzee by the thirteenth century had trade relations with Baltic ports. There was growing traffic to the rapidly growing towns in Flanders just to the south. Because of the demand that traffic generated for ships and new ship types it was possible for shipbuilding to become a principal occupation for some craftsmen and for the industry to move from the countryside to the towns. Shipbuilding grew in the province but it did not grow at the same rate in all towns. The pace of growth depended on the quality of the harbour, the size of the fish catch and the volume of goods that were shipped through the port. The towns of Walcheren benefitted greatly from being the foreports of Bruges in the fourteenth and fifteenth centuries and for Antwerp in the sixteenth century. The most important of those towns was Arnemuiden, where ships were built and caulked for shippers from lands along the Baltic and from Spain and Portugal. Middelburg, the largest town in the province and the political overlord of Arnemuiden, was jealous of the growing shipbuilding there and tried to impede it. The big ships from Spain carrying cargoes for Antwerp had great difficulty reaching the harbour of Middelburg because it lay at the head of a narrow waterway. Since the vessels had to lie at anchor at Arnemuiden they often used the opportunity to have any needed repairs done there as well. A canal from the Westerschelde to Middelburg, dug in the first half of the sixteenth century, undermined the advantageous position of Arnemuiden but only to a limited extent. The plan to build a large dry dock between Middelburg and Arnemuiden in the middle of the century was never implemented. In place of it a smaller dock was built, intended specially for repairs of large ships. The town council of Arnemuiden subsidized the project. The great demand for this type of work meant it paid well. Probably in an effort to compete and get some of the lucrative work, the town Middelburg dug a dock of the same type at the end of the sixteenth century.

Flushing, on the south side of Walcheren, also had a shipbuilding industry. It was very much helped by silting and shifting of sandbars in the rivers around the Zeeland islands in the middle of the fifteenth century. Those changes put the town on ships' preferred route to Antwerp. The changes, on the other hand, worked very much to the disadvantage of Veere on the north side of the island. Shipbuilding there was certainly promoted by the choice of the government to establish the Arsenal for the Admiralty of the Low Countries at Veere at the end of the fifteenth century. The decision generated work but only on a small scale since the most important function of the navy wharf was the outfitting of warships with cannon and munitions. The visits of warships also did bring some employment for Veere shipbuilders.

There was no similar support for the other towns of Zierikzee, Goes, Tholen and Brouwershaven. Those places declined in importance as more trade shifted from the northern approaches to Antwerp and went instead thorough the Westerschelde. Even more detrimental to them, the fishery, especially the herring fishery, moved to Brill and to the ports along the Maas and in northern Holland. At the end of the sixteenth century it was difficult for the town of Zierikzee to keep the harbour at its earlier depth. In Goes shipbuilding disappeared almost altogether. By the end of the sixteenth century there was only one shipbuilding wharf left in the town.

The technical capability of Zeeland shipbuilders did indeed increase in the fifteenth and sixteenth centuries. They learned how to build full-rigged ships with square as well as the triangular lateen sail. The construction of a new type of ship's hull, with a framework of ribs and timbers set up first to give strength and integrity instead of only overlapping hull planking as with the old method, had to be learned from imported

Breton shipbuilders. The first ship of that new type built in the Netherlands was launched at Zierikzee in 1460. By around 1550 Zeeland ship carpenters were building and repairing a variety of ships and fishing boats for sailors in their own province using up-to-date methods and designs. They constructed vessels for special purposes and in some harbours carpenters even built large ships, that is of over 100 tons to be used in long distance trade.

In the sixteenth and seventeenth centuries, ship carpenters guilds were established in most Zeeland towns. They typically got special rights and all those practicing shipbuilding in the towns had to belong. The goal of the guilds was the protection of the members and of shipbuilding in general in bad times. Sometimes the regulations of the guild in one town were taken over virtually word for word by the guild in another. In general the regulations were not strict. They allowed members a great deal of freedom in their dealings with others and in plying their trade. The rules were designed chiefly to assure the social welfare of the members and to lay down a system of training for the trade. Above all the guilds tried to guarantee the quality of the work done and be sure shipcarpenters got paid for the work they had done. The governments of towns, through the chartering of the guilds, sought to control and promote shipbuilding, to levy taxes on the craft and ultimately to subsidize the trade. For the members themselves the religious functions of the guilds were equally important. The organizations had their own chapels in the parish churches. They sometimes held special services and in some places took part in religious processions and plays. After the Revolt not only the charters of the guilds changed. The guilds lost their connections with religion and they changed into purely economic organizations for the oversight of shipbuilding.

In the short period around the end of the sixteenth century trade through Zeeland grew rapidly. There was a big increase in investment in shipping and shipbuilding, in part because of the migration of some wealthy merchants from Flanders and Brabant to Zeeland. The great prosperity was short-lived, however. After 1600 difficulties for Zeeland shipbuilders increased because of the silting up of their harbours. As a result they could launch ships or bring ships close to docks for repair only with difficulty. Also the development and shifting of sandbanks in the rivers and estuaries made it difficult for ships to reach a number of harbours so that supplying raw materials, especially wood which had to come from the lands around the Baltic, was made more difficult. The general improvement in the economy and the industrial growth which were characteristics of the Golden Century were not to be found in one sector, that is in Zeeland shipbuilding, because of those problems with harbours.

The transfer of the Admiralty from Veere to Middelburg soon after the Revolt and the spreading of the navy establishment over a number of harbour towns made it impossible for any one town to gain an advantage from the presence of a large wharf with a good deal of regular and predictable employment of resources. A few Holland towns, on the other hand, benefitted from having a large admiralty establishment. The Zeeland arrangement meant that shipcarpenters in war time could be moved to harbours where they were most needed. That in turn meant extra costs and uncertainty for the shipbuilding industry in most towns. The instability created in the trade was aggravated by the fact that the government often paid late for work done. All that made it hard for entrepreneurs to attract capital to the shipbuilding industry. The closing of the Schelde for shipping did have a positive effect in the short term. It did, however, demote Flushing and Arnemuiden from the function of foreports for Antwerp. Those towns had previously enjoyed caulking and repair work which employed may men on the ships with Antwerp-bound cargoes which lay in their harbours.

The greatest obstacle for Zeeland shipbuilding, however, was always the competition from Holland, where - especially in the Zaanstreek - a centre of shipbuilding was established. The Zaan shipbuilders took over their commanding position because they produced on such a large scale, had better access to raw materials and employed a number of labour saving techniques. Above all they dominated production in the world for sea going ships designed especially for carrying bulk cargo. The Zeeland shipbuilders, as a result, had to satisfy themselves with supplying the regional, the local, the nearby market. In small harbours such as Zierikzee carpenters built mostly inland vessels and fishing boats. In Middelburg and Flushing there sea going ships were still built, but they were typically lighter, smaller, and lightly armed compared to products from Holland.

During the period of rapid expansion late in the sixteenth century Middelburg had enough political and economic influence not only to get the Admiralty and its shipbuilding wharf moved to the town but also to get a chapter or office of the United East India Company. The Company set up a shipbuilding wharf at Middelburg in 1609 with its own regular salaried personnel. The wharf produced some of the largest ships built at the time, "*retourschepen*" or East Indiamen for the trips to the Far East.
Middelburg also had a chamber of the West India Company, also with its own wharf, possibly for the construction of ships for privateering, the foremost activity of the Company. Its successor, the "*Middelburgse Commerciecompagnie*," had a wharf there too. It produced fast ships built for the slave trade, the specialty of the company. The wharf also built ships for other shipowners. The size of the wharf, the already existing equipment, and the fact that the personnel were in permanent service gave the company wharves economic advantages over privately owned and operated small wharves. Private shipbuilders did not have sufficient capital to compete with the company enterprises. So

the construction of big ships in the course of the seventeenth and eighteenth centuries became more and more concentrated on the wharves of the big companies. The growth of the Admiralty wharf at Flushing during the first decades of the eighteenth century had a similar effect. This wharf represented a very large investment and included a drydock, perhaps the first true drydock ever built in the Netherlands.

The wars of the seventeenth century meant work for the shipbuilders of Zeeland. Not only did the Admiralty need men but Zeeland was the centre of highly successful Dutch privateering. Shipowners in the province pressed shipbuilders to construct small, fast ships or pressed them to rebuild or fit out ships as privateering vessels. The shipbuilding wharves on Walcheren appear to have profited most from such work. The period of peace in the eighteenth century along with the transfer of the navy wharf to Flushing meant Middelburg shipbuilding was deprived of what had become a speciality.

In the last decade of the eighteenth century only three wharves were left in Zeeland in continuous operation capable of building large sea going ships. To survive those three wharves got subsidies in one form or another from the government. The small wharves simply could not compete with the Admiralty and company wharves. In the small harbours local governments in conjunction with the guilds tried to stem the decline of shipbuilding. Wages were fixed by the government, foreign workers were prohibited and subsidies were handed out. That was the situation in Zierikzee and in Veere from early in the seventeenth century. At Middelburg the town council forbade the establishment of a shipcarpenters' guild. They took that action perhaps because most of the skilled carpenters worked for the large wharves and so there was no private industry to control or promote through a guild. During the eighteenth century the number of fixed and protectionist regulations of the trade increased, but those changes could not alleviate the economic problems and the competitive difficulties of private shipbuilders in the smaller harbours. Those men continued an established trend and went more and more over to building small boats for the fishery and for carrying goods among the islands of the province.

The disastrous consequences of the Napoleonic Wars for Netherlands trade were reflected in the further decline in the importance of Zeeland shipbuilding. The end of government support for the United East India Company seriously threatened any possibilities of building large ships. There was definitely interest in Zeeland in the problem, as shown by a lively discussion at the end of the eighteenth century in the local society of learned men, the *Zeeuwsch Genootschap der Wetenschappen*, about the optimal measurements of East Indiamen, a discussion which had a purely theoretical character.

European shipbuilding in the nineteenth century went first to the building of ever bigger wooden ships and then to the building of iron ships. The latter change brought even more disadvantages to Zeeland which already was having trouble with wooden shipbuilding. Zeeland simply did not have the raw materials to supply the energy needed for metal working. In the middle of the nineteenth century there were only two wharves left in Middelburg which could build East Indiamen and even those wharves had disappeared by the end of the century. The shipbuilding industry came to be limited to a few wharves in small harbours like Arnemuiden and Tholen, where in the 1950s wooden ships were still being built. The small capacity of the enterprises and the lack of capital and harbour facilities made it impossible to undertake any large projects in Zeeland. From the end of the eighteenth century and probably even before Zeeland shipowners preferred to buy their ships in Holland. Only through a great national effort could the condition of Zeeland shipbuilding be restored. The setting up of the *Scheldewerf* in the 1960s was the result of government action. That action was made possible, among other reasons, by the fact that the earlier Admiralty wharf at Flushing was still in existence and could be used as a basis for the enterprise. Shipbuilding like economic circumstances and the political context in Zeeland had changed dramatically over time. The scale of the project to establish the Schelde shipbuilding yard was very much different from the scale of the small wharves for wooden shipbuilding which had thrived in the towns of Zeeland during the late Middle Ages.

Sources

An earlier version and expanded version of this study appeared as the entry on "*scheepsbouw*" in the *Encyclopedie van Zeeland*, volume III, Middelburg: Koninklijk Genootschap der Wetenschappen, 1984, 25-26.

Jan van Beylen, "Zeelandiae Descriptio," *Académie de Marine de Belgique, Communications* X (1956-57): 81-114.

H. M. Kesteloo, *Geschiedenis en Plaatsbeschrijving van Arnemuiden*, Middelburg: van Benthem en Jutting, 1876.

M. F. Lantsheer and F. Nagtglas, *Zelandia Illustrata*, Middelburg: J. D. and W. Altorffer, 1879.

J. Peter Sigmond, *Nederlandse Zeehavens tussen 1500 en 1800*, Amsterdam, 1989.

Richard W. Unger, *Dutch Shipbuilding Before 1800: Ships and Guilds*, Assen: Van Gorcum, 1978.

W. S. Unger, *Geschiedenis van Middelburg im Omtrek*, Middelburg: Zeeuwsch Genootschap der Wetenschappen, 1966.

W. S. Unger and J. J. Westendorp Boerma, "De steden van Zeeland," *Archief van het Zeeuwsch Genootschap der Wetenschappen*, (1954): 1-87 (1955): 1-70 (1956): 1-33 (1957): 1-42.

J. J. Westendorp Boerma, *De Historische Schoonheid van Zierikzee*, Amsterdam: Allert de Lange, 1946.

VI

14. DUTCH DESIGN SPECIALIZATION AND BUILDING METHODS IN THE SEVENTEENTH CENTURY

The Dutch Republic in the seventeenth century was the envy of all of Europe. The economy thrived on the expansion of agriculture but more obviously on the growth in shipping. Dutch shipbuilders supplied the vehicles for the prosperity of that shipping. "The advantages gained by cheap and adaptable shipbuilding, and cheap and careful navigation, were summed up in the lowest freight rates in Europe, and the most extensive and efficient merchant marine" (Barbour 1930, 285). Many explanations have been offered for the great success of the Dutch shipbuilding industry. The Dutch, it has been said, bought their wood at lower prices than other Europeans and that despite the fact that all of it had to be imported from the Rhineland, Westphalia, Norway or the Baltic. The extensive commercial network, the lowest interest rates in Europe and low customs duties allowed Dutch builders to get their supplies for relatively low cost, even paying less for Norwegian timber than Norwegian shipbuilders (Barbour 1930, 269-274; Barbour 1950, 91). The lower prices were a result of the superior Dutch shipping industry which in turn depended on superior shipbuilding. The argument is circular. An explanation based on lower cost raw materials essentially begs the central question.

The Dutch industry had a massive centre in the Zaanstreek just to the northwest of Amsterdam. So many yards near each other and near a large market for their products and near sources of supply made for, so it was said, greater efficiency. Large quantities of timber as well as many other supplies were stored in the Zaanstreek which made it possible to build ships quickly and also to standardize some of the parts. But centralization can not be the principal explanation for Dutch success. Growth began in the late sixteenth century when the Zaanstreek was home for a few farming villages. Industry grew up there in the wake of commercial expansion. Moreover shipbuilding was carried on in places other than the Zaanstreek, in cities and towns large and small throughout Holland and Zeeland despite the presence of a centralized Zaanstreek industry.

Dutch builders were said to be better at the job because they were more economical in their use of wood. They were also said to be extremely neat, orderly, clean and industrious. They used cheaper wood which not only lowered construction costs but was also easier to work. Durability took second place to cost and the carpenters at times worked in poor pieces of wood just to save money. Those harder working and more productive Dutch shipcarpenters were also paid more than workers in other countries so Dutch labour costs were relatively high. The greater productivity was in part because of the use of mechanical devices, especially blocks and the wind-powered sawmill (Barbour 1930, 274, 277-278; Clément 1864-1865, part 1, 211, 295, part 2, 305-307; Colenbrander 1919, 2, 15). The greater investment on the dock, the labour-saving devices did serve to give Dutch builders an advantage though one which developed only slowly with the dissemination of the use of such devices and one for which they paid in higher wages.

Dutch shipbuilders committed themselves early to specialized design. That explanation for Dutch shipbuilding and shipping success seems much more clear and more firmly based on the evidence than any other. In the course of the fifteenth and sixteenth centuries as northern European ships were transformed builders gradually came to appreciate the advantages of constructing vessels to do one job or set of jobs. Specialized design made the vessel more durable, more efficient and more profitable. The greatest division was between the warship carrying guns and the sailing packet. Already by the close of the sixteenth century the Dutch were building specialized cargo ships, the most important being the fluit, which carried few or no guns and had a large open hold ideal for carrying bulk goods. The type was modified too for certain trades for even greater specialization and thus greater efficiency. Other builders, most notably English ones, were slower than their Dutch counterparts to seize the opportunity created by the evolution of ship design and thus slower to pass on significant savings to shippers. Dutch cargo ships were not highly defensible but that problem was solved by having them sail in convoys protected by specialized warships whenever there was any danger of attack (Barbour 1930, 264, 279-282; Unger 1981, 233-236, 247-252). Specialization in design coincides precisely with the growth of Dutch shipping and continued along with expansion through the early part of the seventeenth century. It seems that the residual factor, technical change in ship design and to a lesser extent on the shipbuilding wharf serves as the most acceptable explanation for the dominant position of Dutch shipbuilding.

As with the success of the Dutch economy a number of different explanations has been offered for the technical superiority in shipbuilding. The location of the Dutch Republic, on the sea and with much of its land below sea level gave it a long tradition of shipbuilding. It was at the centre of the growing northern European trading pattern. Long experience with shipbuilding, made almost a necessity by local geography, gave Dutch shipcarpenters chances to improve their designs. Dutch guilds because of their openness and their ability to train highly skilled men made them an important vehicle for the development and dissemination of better ways of building ships. The competition among towns in the coastal provinces of Holland and Zeeland was an impetus to technical change since guilds limited competition in many way but never in improving design or quality of the final product (Unger 1978, especially 63-82). Undoubtedly these and other factors played a part in the success of Dutch shipcarpenters in establishing a lead over their counterparts in other parts of Europe. There is one other explanation however that appears to be promising and worthy of investigation. Dutch carpenters may have built their ships in a different way from southern and western European shipwrights.

The greatest contribution of ship and boat archaeology over the last forty years has been to demonstrate the distinction between shell construction where the strength of the vessel came from the exterior planking and skeleton construction where strength came from the internal frames. Excavations in the Mediterranean especially have shown that in the first millenium Europeans developed a unique way of solving the problem of carrying goods by water. Skeleton building was a revolutionary method, one which made it possible to build larger, more versatile and more economical ships. Skeleton building proved to be one of the most important technical advances in making possible European domination of the world's oceans. Northern European shipbuilders adopted skeleton construction in the course of the fifteenth and sixteenth centuries. They copied the designs of southern European vessels, even using visiting ships from the South as models, or learned how to build in the new style from southern shipwrights imported by monarchs like King Henry VIII of England. With the old method, clinker-building, northern carpenters built the hull first overlapping the planks to give strength and watertightness. With the new method the frames were set up first and then the hull planking was added on, pinned to the internal timbers. The hull planks with skeleton construction were thus fitted edge-to-edge rather than overlapping as before.

Dutch shipbuilders it appears did not completely adopt the new form of skeleton building. The distinguished maritime ethnologist, Olof Hasslöf, has pointed to evidence which shows that Dutch shipcarpenters first built the shell up to the waterline and then from there on built the skeleton. In pure skeleton building the frames go up before any of the hull planking while in pure shell building the hull is all but finished before the first frame is put in. Another important figure in the study of shipbuilding, Gerhard Timmerman, follows Hasslöf in seeing the Dutch method as a transitional stage between the clinker-built shell construction of the North and the revolutionary skeleton building of the South. Dutch builders compromised, so Hasslöf and Timmerman say, between the two very different approaches. There was a number of such combinations of the two methods in the late seventeenth century, combination methods which were carried on as shipbuilding traditions well into the twentieth century in parts of Scandinavia (Hasslöf 1972, 58; Hasslöf 1963, 169; Timmerman 1979, 26-27). Hasslöf's principal contention is that the change from shell to skeleton building "does not constitute an unbridgeable abyss in practice and that a gradual evolution from the former to the latter is by no means inconceivable, but precisely what we may have to reckon with, even if it has not yet proved feasible to establish the exact sequences in their varied aspects" (Hasslöf 1963, 172). Putting up frames first was not a one-time invention but rather a gradual development. Archaeological evidence has in recent years come to the support of Hasslöf to show that even in the Mediterranean, the home of the first skeleton construction, shipwrights slowly moved from building their hull first to putting up the frames before the hull. The seventh century Yassi Ada ship was built in a combination of shell and skeleton methods. Edge-joined planking of the outer hull was carried up to the waterline and then the internal frames were inserted. After that the higher hull planks were attached to the internal frames. The same progression may have been followed in a fourth century vessel also excavated at Yassi Ada. There was then a number of hybrid forms in the evolution of the skeleton-built ship. Recent excavations in northern Europe have unearthed more examples and ethnologists have found others in south Asia. The presence of edge-joined planks thus can be no guarantee that the vessel was shell-built or for that matter skeleton-built (van Doornick 1972, 140-144; van Doornick 1976, 130-131; Greenhill 1976, 63-70). Hasslöf has found that Dutch shipbuilders practised a combination method of building so it is necessary to explore what the implications for Dutch commercial success may have been from the approach of Dutch shipwrights. At the outset the accuracy of reports on Dutch building methods must be confirmed. If in fact Dutch builders only half-heartedly adopted skeleton building then their choice deserves some explanation. Moreover if the Dutch did use the hybrid approach why did they stick to it so long while all others were fully embracing skeleton building? It seems odd that one group of shipbuilders, especially highly successful ones, should adhere to aspects of an outmoded method. Were they alone in following the compromise system? Most important though is the problem of what effect using the bastard building system had on Dutch technical superiority. Perhaps the information on construction methods can help to explain the low cost of construction in Dutch yards or the speed with which ships were built or the high quality of the products. Perhaps also the building method can help explain why Dutch ship designers so early saw the advantages of specialized design and so early committed themselves to the construction of purpose-built ships.

Some historians of Dutch shipbuilding say nothing about the unique method (Barbour 1930; Crone 1939, 32 and passim; Unger 1978). Historians certainly are fallible and can easily be guilty of oversight but the failure of some to even mention a compromise approach suggests a need to look more closely at the evidence. Information for the unique procedures of Dutch shipcarpenters come from two sources: contemporary descriptions of Dutch shipbuilding by Dutch writers and the reports of French spies sent to Holland by the finance minister of Louis XIV, Jean-Baptiste Colbert. To the first category belong the two great works of the seventeenth century by Nicolaes Witsen and Cornelis van Yk. Witsen was a prominent figure in Amsterdam politics involved in a number of other projects while working on his Aeloude en Hedendaegze Scheeps bouw en Bestier which first appeared in 1671. The 1690 edition with another title was slightly expanded. While he himself knew little of shipbuilding he reported what experts told him. Witsen knew what skeleton building was. He described the procedure exactly in a discussion of shipbuilding in 1520. He copied the information from a Portugese text, a manuscript which was at his own university at Leiden. When it came to building ships in his own day Witsen did not say that frames went up first and then the planks were added. He said just the opposite. After setting out the names of all the pieces that go to make up a ship he laid down the steps for building a warship of 134 feet. The procedure at the start was exactly what one would expect for skeleton building with the keel laid down first followed by the stem- and sternposts and the transom. At that point Witsen says that the garboard strakes are to be fitted into the rabbet of the keel. Then the bottom of the ship is to be built out to the bilges. Then come the frames followed by the ribbands. Once the frames are complete up to deck level then the rest of the hull planking is to go into place (van Beylen 1970, 15-21, Witsen 1690, 88-89, 164-165). Cornelis van Yk was a professional shipcarpenter who had been practicing his trade since the age of 12. His book is more systematic and concrete than Witsen's and deals more extensively with contemporary shipbuilding. His description is similar to that of Witsen. After keel and posts the ship got a complete midship frame. Then the bottom was laid down, gradually rising up to the bilges. After that two ribbands were put in place and two more ribs built. Then a series of ribbands was stretched from post to post over the three ribs. The remaining ribs were built up from their constituent parts inside the ribbands. After that the keelson, a heavy timber fitting over the ribs and sitting on top of the keel, was lowered into place. Only then was hull planking added (van Beylen 1970, 19-20, 38-39; van Yk 1697, 67-78).

Neither Witsen nor van Yk thought building the floor first was odd. The Swedish writer Åke Rålamb in a collection of prints with explanations published in 1691 shows a ship under construction with the floor in place but with only one rib erected. He said it was a Dutch fluit being built with klampar. The bottom planks were fitted edge-to-edge but were never attached directly to each other so they needed some support to hold them in place until all the floor timbers had been fitted. Rålamb shows some small blocks of wood, the klampar, presumably temporarily nailed to connect the planks. Hasslöf found an example of a yawl built at Stavoren in Friesland in 1955 in exactly that way with the entire exterior shell built up before the frames were inserted. Once the hull planks were firmly nailed to the frames then the small blocks were removed (Greenhill 1976, 66, 71-72; Hasslöf 1972,

52-53; Hasslöf 1963, 166; Rålamb 1691, 34-44; Timmermann 1979, 26-27). While the Frisian example is an interesting one it is not obvious that there is a direct connection to the method shown by Rålamb or the procedures outlined by Witsen and van Yk. A Dutch artist, Sieuwert van der Meulen, in the second of a series of 16 prints from the early eighteenth century shows a ship under construction. In this, the first stage of building the hull, there are no ribs: only keel, posts and bottom planks. Those planks are held in place by props on the ground, by clamps and blocks of wood most of which are small (De Groot and Vorstman 1980, 139). It is odd that neither Witsen nor van Yk explained how the floor was to be held in place. They did not see it as a problem worth consideration which perhaps suggests use of small blocks, also called klötzchen, knapper, knab, knaben, was at least not common on seventeenth century Dutch wharves. Above all though even if little blocks were used the attachment of the exterior planks of the floor to each other was purely temporary. Though part of the shell was built first it was not a case of shellbuilding.

The three spies Colbert sent to Holland in 1669, 1670 and 1671 included a well-trained shipbuilder, the son and ultimate successor of the chief naval administrator at Marseilles and Colbert's own son, presumably an experienced administrator with a promising future. Each went off with copious directions from Colbert on exactly what they were to do and find out. Each went both to Holland and to England. Each was to compare building methods in those two countries with those in France. Their first job above all was to find out proportions, the major dimensions of ships of all types but especially warships. Arnoul was the son of the Marseilles intendent des galères and the future administrator at Marseilles. He had also made a trip to Italy to examine shipbuilding practice there. Seignelay was Colbert's son. Both remarked on the quaint Dutch method of building the bottoms of their ships first. The Dutch, so they said, after laying the keel laid planks, ten or twelve of them, and then cut the ribs to fit those planks (Clément 1864-1865, 3, part 1, 211, 234-237, part 2, 296-300, 329; Colenbrander 1919, 12; Hasslöf 1972, 59-60, Hasslöf 1963, 170). Both observers contrasted the Dutch method with common practice in England and France. It is perhaps significant that at the end of the seventeenth century despite his close association with Dutch shipbuilding in part through his friend Witsen, Peter the Great still found Dutch methods backward compared to English ones. The compromise procedure as described by the two French agents may be what the Tsar of Russia had in mind. The Frenchmen reported on practice in the Amsterdam admiralty exclusively. Obviously the Dutch were very open to these emissaries and more than willing to share information with them. The Dutch may well have been trying to impress the French officials in the hope of getting another lucrative contract to build warships like the one for six such vessels won by Amsterdam and Zaandam shipcarpenters in 1666 (van Dillen 1974, nos. 1563, 1565-1568, 1570, 1571, 1574). It is doubtful that Arnoul and Seignelay were misinformed or for that matter that their observations applied only to warships. After all Witsen and van Yk talked about building the bottoms first and did not distinguish between warships and other types.

Bestekken or building contracts suggest that the bottoms of ships could have been built first. In two such agreements from the late sixteenth century for fishing vessels the descriptions of keel, posts, transom and the thickness of the bottom planks come in that order before the details about the frames. Van Yk reprints in full some typical contracts and they follow a similar progression from keel to posts to bottom to bilges to frames. On the other hand a 1640 bestek from Amsterdam for a ship with a full transom goes from keel through posts to frames in all their parts and only then is mention made of any part of the hull planking (AA, no. 126 (1567), no. 127 (1598); van Dillen 1974, no. 497; van Yk 1697, 308-315). Of course there is no guarantee that shipbuilders proceeded in the order laid down in the bestek. It was a contract to establish the dimensions of the final product and not necessarily how the builder was to go about producing that result.

Accounts of building ships do not typically say anything about the compromise method. If the method was so odd and out of step with practice elsewhere it seems logical to expect that more notice would have been taken. No matter the approach the design was based on the shape of the central frame. It was drawing a precise outline of that frame which was the designer's most important task. Rembrandt's shipbuilder, a portrait which includes the man's wife done about 1633, was absorbed exclusively with the drawing of frames. Government officials interested in intelligence about shipbuilding like Colbert in France and Pepys in England had many questions but they did not ask about the process of building. They did not see such information as important or perhaps simply assumed that there were no differences. Thus when remarks about the building method do appear they are incidental and almost off hand (Bredius 1969, 322, Br. 408; Clement 1864-1865, for example 3, part 1, 211; Colenbrander 1919, 2, 2-7). Archaeology is not likely to produce evidence on which part went up first. Both approaches produced almost exactly the same final product. With true shell construction with frames added after the hull was in place the frames had to be notched, cut to fit precisely inside the hull planks. Wrecks of clinker-built vessels in the Ijsselmeerpolders show such notching of the frames (IJ Wreck NOP, J 137 for example). With bottom planks already in place before building in the floor timbers some extra care had to be shown in shaping the parts of the ribs but there was no dramatic difference with the floors. Any difficulty was minor. The bottom was flat and, if Witsen's advice was followed, checked with a level. Moreover the frames were built up from a number of different parts so only the floor timber fit over the planks already in place. That piece was supposed to be flat as well.

It appears that Dutch shipwrights did lay down the bottom planks first before putting up the frames with the exception of the midship one. However it is not certain that they did it for all types or at all times or in all yards. Witsen did not seem to care which job was done first. He said that the progression was really not critical and suggested that builders should proceed as they saw best. Van Yk in one case shows a small vessel under construction with the frames all up but no exterior planks in place as yet. Rålamb on the same plate as the Dutch fluit shows other ships under construction with frames up first before any of the bottom planks were laid down (Rålamb 1691, plate 1; van Yk 1697, 90, Witsen 1690, 164). Rålamb's illustration suggests that Dutch practice was considered unique and not imitated in Sweden.

Dutch shipcarpenters may have had good reason to adopt a compromise method. As in all other parts of northern Europe the skeleton system had to be imported. The first instance of building in the Mediterranean style in the Netherlands dates from 1439. The Duke of Burgundy had two vessels built for him near Brussels but needed to import Portuguese shipwrights for the job. It was 1460 before any shipcarpenters in the provinces which would become the Dutch Republic tried skeleton building. Zierikzee builders tried to construct a big seagoing ship but could not handle such a large task so a Breton shipcarpenter had to be brought in to show them how to do it. It is not clear how quickly Dutch builders went over to the new techniques. The advantages of skeleton building were greater on larger ships and Dutch yards tended to turn out few big ships especially through the first half of the sixteenth century (GAZ Register op de Charters der Stad Zierikzee, no. 224; van Beylen 1970, 7-8; van Cortgene 1551, 190). Certainly Dutch shipcarpenters did gain the skill to build in the new style but at the same time they carried on with old methods. The excavated wrecks from the Ijsselmeerpolders confirm the survival of clinker-building through the sixteenth and the seventeenth centuries. There were signs of conservatism among Dutch shipbuilders, of a reluctance to plunge into something radically new. But that is certainly not surprising especially since the costs of failure were so high (IJ Wrecks NOP 028 and P33 for example; Unger 1978, 28).

Dutch carpenters had an example of building a flat floor out to the bilges before proceeding with the rest of the ship. The cog, perhaps descended from a Celtic type, was the most important cargo ship in northern Europe in the high Middle Ages. It apparently always had a flat floor with planks fitted edge-to-edge but with overlapping planking on the sides. There is no question that the cog was shell-built. Netherlands shipwrights built such vessels from the early Christian era at least down through the nineteenth century. In the course of the seventeenth century the clinkered sides were changed to having the edges of the planks abuting but the essential form remained the same. The tjalk may have been descended from the earlier cog (IJ Wrecks NOP M40 and M93 for example; van Beylen 1970, 163-165; Greenhill 1976, 69-70, 259). Dutch shipcarpenters mastered the building of a flat bottom first long before they ever tried building with the skeleton method. When they faced the problem of how to proceed with the new imported method it would have been easiest to begin in a familiar way especially for the cog, tjalk, and other related types. Also Dutch ships continued to have flat or nearly flat bottoms throughout the seventeenth century even on the largest of ships so the analogy with the cog was a strong one. All observers mentioned the square section of Dutch ships, the sharp angle of the bilges, the almost flat floor. The box-like shape of the hold had advantages in increasing carrying capacity in cargo ships. Dutch designers built warships with the same form. The explanation always offered for the flat bottoms of Dutch ships was the shallow harbours in Holland and Zeeland. There were other advantages to the design besides not running aground when approaching a Dutch port. Dutch ships rolled less, especially important when firing cannon in battle (Clément 1864-1865, 3, part 1, 336-337, part 2, 304-305; Colenbrander 1919, 2, 12-13; Crone 1939, 35, 167-169). Whatever the reasons for the flat bottoms of Dutch ships and whatever the advantages certainly if the bottom was flat it was a much simpler process to build the bottom first before the ribs.

It is not possible to say, at least at the moment, how long Dutch shipbuilders adhered to their way of combining building styles. It is difficult to find any evidence for builders abandoning the bastard system in the late seventeenth or eighteenth centuries. The Dutch shipbuilding industry declined and all but collapsed in the eighteenth century. There was a general stagnation in Dutch trade and a drop in orders for new ships at home and especially abroad. There is strong evidence that Dutch technical superiority had been eclipsed and that Dutch builders lagged seriously behind others in Europe. Under those conditions it is hard to say whether Dutch building methods changed. Efforts were made to improve the poor state of the Dutch industry through bringing in foreign, specifically English, shipwrights to improve the competence of native builders. Contemporaries wrote books about the decline of Dutch shipbuilding and suggested ways to recoup lost ground (Anderson 1947, 218-225; van Beylen 1970, 30-31, Bruijn 1972, 18-24; Unger 1978, 42-45, 60-62; van Yk 1697, 19-21). It may well be that under pressure to modernize, to change and follow practice in England and France that Dutch builders in the eighteenth century abandoned the building of the bottoms of their ships first. Just as authors largely ignored the compromise method in the seventeenth they failed to mention any change in the eighteenth. It was much less likely for authors to say how ships were built in the eighteenth century since writers turned away from being pragmatic to being rather more theoretical. The increasing number of books on shipbuilding were written to teach shipbuilders how to become designers using compasses to draw out the shape of the ship. Even in 1586 when the master shipwright of Queen Elizabeth I of England, Mathew Baker, put together his "Fragments of English Shipwrightry" he was trying to reduce the job of ship design to drawing a proper frame or frames (Baker 1585). By the eighteenth century the effort to exploit the burgeoning knowledge of mathematics and physics made books on ship-building catalogues of proportions with tables and information on doing certain types of calculations. Officials at the highest levels wanted research done on how to use mechanics and geometry to aid shipbuilding. Scholars obliged them. The resulting books would and did have a great impact in the long run. But they say little about what shipcarpenters thought in the eighteenth century and even less about what they in fact did. Witsen had included a description of how to build a ship and he implied that builders could have too many rules to follow. In the eighteenth century such scepticism about the value of the new scientific knowledge for shipbuilders was completely drowned out (Bushnell 1664; Duhamel Du Monceau 1752; Hasslöf 1963, 175; Murray 1754; Sutherland 1711; Timmerman 1963, 19-27; Witsen 1690, 223). The lack of evidence of the use of bottom-first construction then does not prove that it was not practised. Johann Röding in his marine dictionary published in 1793 mentioned the small blocks of wood used temporarily to hold planks in place but there is not the slightest suggestion that they were to be used in the construction of the hull (Röding 1793, "knapen"). He is incorrect in translating the word into English as "forelock".

It would seem very odd that only the Dutch built their ships bottom first. If the origin of the method was building cogs then all German and Scandinavian shipwrights should have followed the compromise path. Also if Dutch builders used the method they presumably took it with them when they went off to other parts of Europe. Dutch shipcarpenters by the end of the sixteenth century were in great demand because of their superior skills. They were hired away by foreign governments, most notably by the Swedish crown for the building of warships such as the Vasa, and presumably they built in the same style abroad as they did at home (Baasch 1899, 11-12, 213; van Dillen 1974, 2, no. 1368, 3, no. 946, 1312-1313, 1316, 1637). It is certain that Dutch builders did not successively introduce bottom-first construction into every country where they worked. France is the most important and obvious example. Colbert hired large numbers of Dutch shipwrights to work in naval yards (Barbour 1930, 274; Clément 1864-1865, 3, part 1, 132-133, 199-200, 211; van Dillen 1974, 3, no. 40, 1683). The agents sent from those same French naval shipyards to Holland were amazed when the saw carpenters laying down the bottom before putting up the frames so their imported Dutch carpenters obviously did not use the bastard method when they got to France.

If the Dutch were indeed alone in building in their different way then the method itself may have imparted some real basis for the superiority of Dutch shipbuilding. But the case for explaining Dutch dominance of seventeenth century shipbuilding by their building method is not a strong one. It was certainly easier for Dutch shipcarpenters to adopt the compromise method. The transition from the old shell technique to the revoluntionary skeleton one would have been made simpler, less dramatic. But the gain would have been only a short term one. Over the long period of two or three centuries Dutch builders could have comfortably taken the next step toward building completely in the Mediterranean fashion as did so many others in western Europe. For most design practices it was really unimportant in which progression the job was done. Dutch shipbuilding gained advantages from having standardized parts for their vessels. But such standardization was practical, effective and cost-saving in any building method. There was no difference in the type or quality of wood used for building in the compromise or pure skeleton method. The Dutch were free to use lower qualities of wood and to use them more economically no matter which progression they followed. It may well be that the bottom of a ship where the planks were fitted edge-to-edge first was more watertight than one where planks were tacked onto the frames after the timbers had been erected. More important though to watertightness was the skill of and execution by the shipcarpenter. No matter which method was used the bottom of the ship still had to be caulked. It was said that Dutch builders were very good at cutting their ships in half and lengthening them by inserting new ribs between the old halves (Colenbrander 1919, 2, 14). But what made that easy was the flat floor typical of Dutch vessels and not the way in which the bottom was built. The skills required of shipcarpenters were much the same with both methods. It may be that having the men fit the bottom planks first required more effort. Greater precision may have been needed to get the planks to fit exactly but a good deal of precision was needed with the other approach too. It appears that teaching carpenters was much the same no matter the method. So with training as with most other aspects laying the bottom first meant little advantage to Dutch shipbuilders.

There were some ways that made the Dutch method technically better, ways noticed by contemporaries. There was greater flexibility in the design of the ship once building was underway. If the frames went up first their shape irrevocably established the form of the hull. But in the compromise method the shape of the hull could be adjusted while the bottom was being laid or even after it was completed. The timbers were cut to fit the planking already there, the planking serving as something of a mould to dictate the shape of the lowest member of the frame. The basic design was of course determined by the length of the keel, the curve of the posts and the curve of the midship frame. That was true no matter the approach. But once that was done if the frames went up followed by the ribbands there was no room for change by the builder. Dutch designers then worked more by eye than by theory. They saw what the ship would look like once the bottom was in place and then had carpenters shape the rest of the frames to fit the bottom (Hasslöf 1972, 59-60; Hasslöf 1963, 170). Building the bottom first decreased the number of errors in design. That made it easier to train the designers of ships, the chiefs of the shipbuilding yards, since they did not have to have the same command of theory and of mathematics as designers in France or England. Presumably there was less need to make changes or adjustments once the ship was launched. Adding extra timber to improve sailing qualities was a common practice in the seventeenth century once the crankness of a ship had been determined on a shakedown cruise. If indeed the compromise method made Dutch designers surer of their products then the approach they took to design may in fact have helped them to produce ships quickly and to produce reliable ones. They may also have found themselves better able to estimate the draught of a ship before it left the ways. That was a critical skill in the Low Countries. Not until the eighteenth century could designers reliably predict how much a ship would draw using the new scientific approach so again Dutch shipbuilders may have been better off (Perrin 1918, lxx-lxxiv; Timmerman 1963, 15-16; Witsen 1690, 302-309). The Dutch drive toward specialized design began with the herring buss in the fifteenth century which had a flat bottom. It also had a high ratio of length to breadth. The compromise method could have been easily practised and probably was on the buss. The fluit design grew out of experiments with the buss and it too had a flat or nearly flat bottom. It is easy to see how the compromise approach could have been carried over from one design to the other in the process of specialization. But Dutch designers appear to have used their unique method on many different types of vessels. Warships got the same treatment as fishing boats, cargo ships and inland craft. So for the critical question of design specialization it appears that choosing to build the bottom first before the frames had little if any immediate effect.

Though it would be helpful to have more substantiating evidence still it seems likely that Hasslöf and Timmerman are right. Dutch builders in the seventeenth century probably did build part of the shell of their vessels first, just to the bilges, before they put up the frames. However they were unquestionably still practising skeleton construction. The strength of the vessel came entirely from the internal frames. The planks on the bottom were certainly fitted edge-to-edge but never edge-joined as would be the case in pure shell construction. The principal conclusion which Hasslöf draws from his examination of building techniques can not be disputed. The shift from shell to skeleton construction in northern Europe was not a single event, not one revolutionary change but rather a gradual process with many intermediate steps. In fact it may well be that there was not even an evolution at all but simply borrowing from different traditions as builders tried to find the most practical and effective way to build their boats (Greenhill 1976, 70-72, 91-95, 294). But there are still some matters that are uncertain. It is not clear how common it was for builders in the Dutch Republic or for that matter in all of northern Europe to build the bottom of their vessels first or if they used that approach only for certain types or sizes of ships. It is also unclear how long the method survived in the Netherlands.

The effect of the Dutch choice of building method on their commitment to design specialization seems to have been minimal. While there were some possible savings to be reaped from putting down the bottom first in the training of designers and avoiding mistakes in construction those savings do not appear to have been great enough to explain the massive success of the Dutch shipbuilding and in turn the shipping industry in the seventeenth century. The strengths of Dutch ships and of the industry which produced them seem to have come from other sources. On the other hand the choice by Dutch shipcarpenters to retain some of the older ways of doing things, to rely on past practice does not appear to have inhibited the industry in any way. The way Dutch designers and builders put together the most profitable ships of the seventeenth century is certainly of great importance to the history of shipbuilding. The choices which those men made indicate clearly that in general and in seventeenth century industry in particular techniques changed only gradually. Maritime ethnology has proven a valuable tool in documenting the slow and often tentative adoption of new technology. However it appears that, until further information about the value of the compromise approach or the prevalence of that approach becomes available, it will be from the well-known older explanations that the reasons for Dutch commercial and industrial success in the seventeenth century will come.

REFERENCES

A.A., Archief van Adrichen, Algemeen Rijksarchief, The Hague.

Anderson, R.C., 1947. Eighteenth Century Books on Shipbuilding, The Mariner's Mirror 33 (1947), 218-225.

Baasch, E., 1899. Beiträge zur Geschichte des deutschen Seeschiffbaues und der Schiffbaupolitik.

Baker, M., 1585. Fragments of English Shipwightry, Magdalene College, Cambridge University, Pepys Library, 2820.

Barbour, V., 1950. Capitalism in Amsterdam in the 17th Century.

Barbour, V., 1930. Dutch and English Merchant Shipping in the Seventeenth Century, Economic History Review 2, 261-290.

van Beylen, J., 1970. Schepen van de Netherlanden Van de late middeleeuwen tot het einde van de 17e eeuw.

Bredius, A., 1969. Rembrandt, The Complete Edition of the Paintings, rev. H. Gerson.

Bruijn, J.R., 1972. Engelse scheepsbouwers op de Amsterdamse Admiraliteirtswerf in de achttiende eeuw: enige aspecten, Mededelingen van de Nederlandse Vereniging voor Zeegeschiedenis 25, 18-24.

Clément, P., ed., 1864-1865. Lettres, Instructions et Mémoires de Colbert.

Colenbrander, H.T., ed., 1919. Bescheiden uit Vreemde Archieven omtrent De Groote Nederlandsche Zeeoorlogen 1652-1676.

van Cortgene, J.R., 1551. Dye Cronucke van Zeeland.

Crone, G.C.E., 1939. Onze Schepen in de Gouden Eeuw.

De Groot, I. & Vorstman, R., 1980. Sailing Ships Prints by the Dutch Masters from the sixteenth to the nineteenth century (trans. M. Hoyle).

van Dillen, J.G., ed., 1974. Bronnen tot de Geschiedensis van het Bedrijfsleven en Gildewezen van Amsterdam, vol. 3, 1633-1672.

van Doornick, F., 1972. Byzantium mistress of the sea: 330-641, in A History of Seafaring based on Underwater Archaeology (ed. G. Bass), 133-158.

van Doornick, F., 1976. The 4th century wreck at Yassi Ada. An interim report on the hull, The International Journal of Nautical Archaeology and Underwater Exploration 5, 115-131.

Duhamel Du Monceau, J.R., 1972. Elémens de l'architecture navale.

GAZ, Gemeente Archief Zierikzee, Register op de Charters der Stad Zierikzee.

Greenhill, B., 1976. Archaeology of the Boat. A new introductory study.

Hasslöf, O., 1972. Main Principles in the Technology of Ship-Building, in Ships and Shipyards Sailors and Fishermen, Introduction to Maritime Ethnology (eds. O. Hasslöf, H. Henningsen, A.E. Christensen Jr.), 27-72.

Hasslöf, O., 1963. Wrecks, Archives and Living Traditions, The Mariner's Mirror 49, 162-177.

IJ, Ijsselmeerpolder Wrecks.

Murray, M., 1754. A Treatise on Ship-Building and Navigation.

Perrin, W.G., 1918. Introduction, in The Autobiography of Phineas Pett, xv-civ.

Rålamb, Å., 1691. Skepps Byggerij eller Adelig Öfnings Tionde Tom.

Röding, J.H., 1793. Allgemeines Wörterbuch der Marine in Allen Europaeischen Seesprachen nebst Vollstaendigen Erklaerungen.

Sutherland, W., 1711. Britain's Glory or, Ship-Building Unvail'd.

Timmerman, G., 1963. Das Eindringen der Naturwissenschaft in das Schiffbauhandwerk, Deutsches Museum, Abhandlungen und Berichte, 30, 5.53.

Timmerman, G., 1979. Die Suche nach der günstigsten Schiffsform, Schriften des Deutshcen Schiffahrtsmuseums, 11.

Unger, R.W., 1981. Warships and Cargo Ships in Medieval Europe, Technology and Culture 22, 233-252.

VII

Marine Paintings and the History of Shipbuilding

Studying works of art as sources of information for the history of technology – in this case the history of shipbuilding – means evaluating artists' reliability rather than their skill. Paintings are immediately denuded of any value for style. The worth of the work, its documentation value, depends on the type of ship or boat illustrated and how precisely that job is done – nothing more. So historians of technology seek in art the lowest level of meaning. Dutch artists in the seventeenth century were highly accurate in their treatment of ships. Virtually all painters took the time to make their pictures reflect precisely what ships looked like and how they worked. It is remarkable that not only marine artists but landscape painters, and indeed all painters, typically chose to make ships look exactly like ships. Their works are the earliest in European history that are completely reliable and that can be used as a source for the study of the history of ship design and construction. This is even more remarkable since in the sixteenth century, painters working in the Low Countries produced some pictures of vessels that certainly could never have sailed. Art in the Netherlands went through a phase during which painters did not show ships accurately, in deviation from previous practice.

In the late Middle Ages the symbolic value of an object in a work of art was still most important. Artists, harkening back to early Christian thought, typically believed that there was much more to a ship than ropes, sails, and planks. It could bear a number of meanings. To carry symbolic meaning, the object had to be recognizable, to look real even if not exactly correct. The simplest thing to do was to create a reasonable facsimile of the object. Throughout the fifteenth century Low Countries' artists made their sym-

This paper was originally presented by Richard W. Unger at the 1987 conference 'Art in History/History in Art', sponsered by the Getty Center for the History of Art and the Humanities. An edited version was published as 'Marine Paintings and the History of Shipbuilding' in *Art in History/History in Art: Studies in Seventeenth-Century Dutch Culture*, edited by David Freedberg and Jan de Vries.

1. Master WA,
A Carrack, ca. 1480,
engraving.
Paris, Bibliothèque Nationale.

bolic ships look like real ships. In the seventeenth century Dutch artists, striving for realism, made their ships look precisely like the ships they saw around them. In the intervening period, struggling with a new style not yet fully understood, artists created images that give some hints of the nature of real ships but obscure some contemporary technological developments. That period fell, unfortunately, at a time of critical and even revolutionary advance in Dutch shipbuilding.

Despite variations in representational strategies, works of art are one of the, if not the most important, sources for the study of sixteenth-century shipbuilding. Paintings provide the most numerous contemporary depictions of ships. Maps such as Cornelis Anthonisz.'s *Caerte van Oostlant* (1543) usually show vessels on the seas.[1] This is especially true of the more accurate and less expensive sea charts that appeared toward the end of the century. Lucas Waghenaer's *Spiegel der zeevaerdt* offered accurate charts with a great deal of information, all at much below the usual cost. Waghenaer's book was widely available, translated into English as *The Mariner's Mirror* and sold throughout northern Europe. "Wagonner" became a generic term for such works in English. The ships shown on the charts remained the same throughout the many reprints and translations, still reflecting accurately the types in use at the time of the original publication, 1584–1585.[2] Town seals are a critical source for the study of thirteenth- and fourteenth-century shipbuilding but they had ossified by the sixteenth century.[3] Towns left seals intact for centuries; the seals even gained in authority by having obsolete ships depicted on them. Town views and topographical views of all sorts became increasingly popular in the sixteenth century, and views of ports included pictures of ships as well. Ship and boat archaeology and written records of shipbuilding (which only begin to appear in the sixteenth century) serve to confirm, at least to some extent, what artists chose to depict. A few shipbuilding contracts, mostly from the latter years of the century, have survived. Books on shipbuilding began to appear in Italy as early as the fifteenth century. Originally, they were extremely simple, giving a few dimensions and short descriptions. By the sixteenth century they had become more sophisticated, but the Netherlands would have to wait until the mid-seventeenth century for such books.[4]

Evaluating the accuracy of an artist's illustration of a ship is difficult, not only because of the medium but also because of the basis of such evaluation: the measure of accuracy must often be consistency with the work of another artist. Hence, assessment of works of art risks becoming relative only to other works of art. Fortunately, there are some highly reliable alternative sources against which to gauge paintings and drawings. The large number of wrecks uncovered on the bottom of the former Zuider Zee in the IJsselmeerpolders since the 1930s gives a new and highly accurate way of assessing the depiction of hull construction. Since the rigging is not preserved in such wrecks, masts, spars, sails, and ropes must be checked by a combination of written sources, experimental archaeology, investigation of earlier and later practices, and common sense. Once the accuracy of a work or works has been confirmed with these sources, it is possible to evaluate additional works by the same artist or other artists with some confidence. When a class of attributes or even a class of ship types is established by a combination of sources, even works that contain obvious inaccuracies, such as some sixteenth-century paintings, can serve as sources for at least certain specific information.

Reliable and informative sources are a serious matter because Dutch shipbuilders made great strides in the sixteenth century, moving from being technically backward to being the leaders in the field in Europe. In the fourteenth and fifteenth centuries Mediterranean shipbuilders had made technical advances, developing a new type of vessel, the full-rigged ship, which incorporated the advantages of southern construction methods and the triangular lateen sail with the hull form and the square sail of northern European types. These full-rigged ships proved capable of long trips in unknown waters, making possible the voyages of discovery. In a different form, they also made possible much greater commerce between the Mediterranean and northern Europe. The carrack (fig. 1) was a large ship able to carry bulk goods economically between Iberia and the Low Countries and between Iberia and India. Shipwrights built the hull in the Mediterranean style. The ribs were set up first and exterior planks were then tacked to them. This form of skeletal building was more efficient in terms of wood and more flexible than traditional northern European methods of building hulls, in which strength came from exterior planks. Shipwrights in the Low Countries learned the new method from craftsmen who were brought in to build in the new design, for example, in Brussels, where Portuguese built a ship in 1439 in the new style, and in Zierikzee, where Bretons came to show Zeeland ship carpenters how to proceed in about 1460.[5]

Dutch shipbuilders learned quickly and well. They adopted the new method but modified it to make it more consistent with what they had done in the past.[6] Using the new method, they created designs specifically suited to their own needs. By the end of the sixteenth century they developed the *fluit* for bulk trades to the Baltic. In a print from the middle of the seventeenth century, Reiner Nooms, also called "Zeeman" because of his experience as a sailor, shows two examples of this type: one designed for use in trade to the Baltic and the other for use in the Norwegian timber trade (fig. 2). But even by the middle of the sixteenth century, vessels in Dutch yards began to take on many of the features of that type. The tiered and tapering stern, shown by Pieter Bruegel the Elder in a print of 1555 (fig. 3), was a hallmark of the seventeenth-century *fluit*. The *fluit* was typically five or six times as long as it was wide, about double the ratio of the carrack. The lengthening of cargo ships may have come from experience with *busses* (boats used in the herring fishery). *Busses* had to be relatively long to control their large drag nets. Virtually defenseless, they traveled together in convoy during wartime. That practice was carried over to cargo ships. So too was design differentiation between fighting vessels and the ships they were to protect, such as the *fluit*. Dutch builders concentrated on efficiency and on specialization in design. They also made many improvements in small craft.[7] Unfortunately, the evolution of the design of types used on rivers, canals, and lakes has gone largely unrecognized, in part because it was seldom recorded by painters. But the changes in small craft proved valuable not only for the internal economy of the Netherlands but also as a source for innovations in larger seagoing vessels.

Despite the improvements going on around them and the changes, which were already having significant effects on the economy and politics, few artists appear to have noticed, or to have cared to notice, the technical advances in ship design that were taking place in the sixteenth century. None of these artists, apparently, had direct knowledge of ships or shipbuilding. While they were not purposefully unfaithful in their depictions of vessels, they typically did not place great value on technical accuracy. There were, however, significant and fortunate exceptions.

In 1520 Lucas van Leyden drew a highly accurate picture of a contemporary ship perfectly consistent with other sources (fig. 4). The depiction harkens back to the earlier tradition of incidental accuracy. The Van Leyden vessel even has four masts and is therefore one of the earliest works to show what would be a common feature of large ships by the end of the century.

VII

Marine Paintings and Shipbuilding

2. Reiner Nooms called "Zeeman," *T'Geele Fortüyn een ooster-vaerder, De Liefde een noorts-vaerder* (two specialized cargo carriers), mid-seventeenth century, engraving, 13.2 x 24.7 cm. Amsterdam, Rijksprentenkabinet, Rijksmuseum.

VII

3. Frans Huys after Pieter Bruegel the Elder, *Nef de bande ou de haut bord* (a large cargo ship with a stern typical of the *fluit* and a very efficient smaller cargo vessel, a boyer, to the right), 1564–1565, engraving, 24 x 19.4 cm. Brussels, Bibliothèque Royale Albert 1er (Cabinet des Estampes).

4. Lucas van Leyden, *La nef de Saint-Reynuyt ou la nef de la mauvaise gestation*, 1520, woodcut, 74 x 116 cm. Amsterdam, Rijksmuseum.

5. Cornelis Anthonisz.,
The Siege of Algiers, 1542,
woodcut, 37.4 x 58.3 cm.
Amsterdam, Rijksprentenkabinet,
Rijksmuseum.

6. Pieter Bruegel the Elder,
Nef de bande vue de trois quarts à droite par-derrière, armée de canons, 1564–1565,
engraving, 25 x 28.7 cm.
Brussels, Bibliothèque Royale Albert Ier
(Cabinet des Estampes).

7. **Maarten van Heemskerck,**
Panoramic Landscape with the Abduction of Helen (detail), 1536,
oil on canvas, 147 x 383.5 cm.
Baltimore, Walters Art Gallery, no. 37.656.

Many of the details are undoubtedly correct, such as the shrouds, though there might be some confusion in the way the main yard was fixed to the mainmast. In any case, there was no effort to make this vessel look like anything other than a sixteenth-century Dutch ship. In the middle of the century, Cornelis Anthonisz. did much the same, producing pictures of contemporary ships that, though lacking a great deal of technical detail, still resemble the vessels that the Dutch sailed. In a drawing of the town and port of Algiers from 1542 (fig. 5), the artist even went so far as to show ships like those used in northern Europe in Mediterranean surroundings. It is possible, however, that a ship like the one in the lower right corner of this work, which has all the characteristics of a ship from the North Sea, actually did sit in the Algerian harbor at that time.

Bruegel took the most care of any sixteenth-century artist to represent ships accurately. In 1564 and 1565 he did a series of ship prints. Since ships were the subject of the works and in most cases formed the only focus, Bruegel seems to have taken time to show the vessels as they actually were. That he could create such works and expect to find buyers for them in the Antwerp of his day suggests that already by the mid-sixteenth century a market for ship illustrations was emerging. Bruegel depicted different types of vessels – cargo ships and warships, sailing ships and oared ships. The galleys in his prints were typical of the Mediterranean, but navies used them in the early phases of the Eighty Years' War in the Low Countries, so these vessels were not strangers to local waters. He also depicted sixteenth-century northern ships in works showing mythological events such as the fall of Phaëthon and several versions of the fall of Icarus (fig. 6). He might choose a Mediterranean landscape but did nothing to make the ships look like anything other than the Low Countries' vessels of his day. Bruegel seems to have had more of a functional than a symbolic interest in ships. That was not true for many others.

In his large painting *Landscape with the Abduction of Helen*, 1536 (fig. 7) Maarten van Heemskerck showed ships as idealized and classicized, so much so that the vessels would not have been seaworthy. The painting records a classical event. The ships are also classical, or were at least given what the artist thought was a classical appearance. The vessels themselves are strange combinations with contemporary features corrupted by scrollwork and decoration that would never have been found on any ship. The rudders are much too large, the bows and sterns are decorated with carved faces, and the prows are squared off to carry dramatically curved and flowing timbers. The sails, yards, and masts often lack lines to hold them in place.

For the early church fathers, Noah's ark symbolized the Church, carrying, caring for, and protecting souls. The whole story of Noah proved a rich inspiration for medieval artists. Renaissance artists in Italy made Noah an even grander, more impressive figure, something of a creator. It was exactly this Noah that Van Heemskerck had in mind in his drawing of 1558 (fig. 8), one of a series on the story of the patriarch. The artist changed not only Noah but also the character of the ark. A drawing depicting the Flood, executed a year later, has a ship that simply could not have sailed and might not have been able to float with any stability. It, too, is classicized to the extent that it is no longer seaworthy. Lambert van Noort, working in Antwerp at about the same time as Van Heemskerck, produced a picture of the animals leaving the ark in which the vessel owes more to Roman reliefs or mosaics than to sixteenth-century shipbuilding practice. Noah's ark was beginning to be seen as a vessel that did not have to be able to sail: writers were beginning to wonder if the measurements of the ark could be known or if they mattered.[8] But artists gave a classical appearance to many different ships in many different works, not only to Noah's ark.

Italian art, of course, had a deep and extensive influence on the sixteenth-century Netherlands but not one leading to greater interest in technology, accuracy, or precision in depicting ships. Though the Italian Renaissance would, over the long term, generate a new and more precise realism, in the sixteenth century "classicizing" had the greatest influence on artists' treatment of ships. The interest in the classical past led to an effort to understand and depict vessels of ancient Greece and Rome. The ships were idealized – borrowing from classical texts and Roman artworks – and mixed with some knowledge of contemporary shipbuilding and with pure guesswork Roman ships were a feature of Baroque art, and descriptions of them were typically included in the opening chapters of shipbuilding manuals through the Enlightenment.[9] Dutch artists on trips to Italy seem to have picked up the interest in classical ships and as a result tried to make their ships look Roman. But the artists knew much more about contemporary ships and therefore borrowed features from them. The results were the strange hybrid vessels that appear in biblical and classical scenes. Artists virtually dressed the ships of their own time in classical garb. Though the treatment of ships may reveal the artists' intentions, as a source of information about six-

VII

Marine Paintings and Shipbuilding

8. Maarten van Heemskerck,
Noah Receiving Direction from God and Construction of the Ark, 1558,
brown ink and black chalk on yellow paper,
201 x 253 cm.
Copenhagen, Statens Museum for Kunst,
Den kongelige Kobberstiksamling.

teenth-century shipbuilding they are of very limited value and must be used with much greater care than the works that preceded them in the late Middle Ages.

Jan van Scorel's handling of the story of Saint Ursula from the 1540s (fig. 9) shows the change in style when it is compared to Hans Memling's handling of the same topic on a reliquary in Bruges from the 1480s (fig. 10). The "medieval" artist gives a more accurate picture of a ship. The vessels in Memling's work have highly rounded hull forms, a feature of the ships of his time. Van Scorel, on the other hand, includes scrolled bows, which shipbuilders would never have added to a contemporary ship.

The practice of depicting contemporary ships with some accuracy did not, however, entirely disappear in the sixteenth century. Pieter Bruegel the Elder, as noted, is the best, but not the only, example of a continuing concern for accuracy. Jan Sadeler in a print after Dirck Barendsz. dated 1582 (fig. 11) shows Jonah being thrown to the whale in a highly dramatic way from a ship with curves that owe more to stylistic considerations than to shipbuilding practice. The depiction of the same scene from about twenty years earlier, possibly by Pieter Pourbus, shows a much more realistic ship (fig. 12).[10]

In everyday works like charts and city views, Dutch artists tended to show ships accurately. When it came to depicting boats, Italian influence and related classicism in Holland seem to have been stronger than, for example, in Flanders. At the same time, the influence from the south was not strong enough to destroy completely the established tradition among Dutch painters and other artists of showing vessels much as they were. That older tradition may have grown out of lack of interest in ships as technical objects and was typified by inattention to detail and some resistance to novelty. Even so, it was the base from which Dutch artists started, and that tradition was eroded but never completely eradicated by the first incursions of Renaissance style into the Low Countries.

The classical garb of sixteenth-century ships in Dutch paintings seriously undermines the documentary value of such works for historians of technology. Still, there is a great deal to be learned about shipbuilding from those images. Many contemporary features were hidden beneath the artists' additions – some general and some specific – and the depiction of ships can still be of help to corroborate information from other sources. In general, archaeology confirms, as does art, that many advances and changes

VII

Marine Paintings and Shipbuilding

9. Jan van Scorel,
Ursula and the 11,000 Maidens, 1540–1550,
brown ink and wash drawing on paper,
26.9 x 36.5 cm.
Amsterdam, Rijksprentenkabinet,
Rijksmuseum, no. 11:108.

10. Hans Memling,
The Departure from Basel from *The Shrine of Saint Ursula*, before 1489,
oil on panel, 87 x 33 x 91 cm (entire shrine).
Bruges, Memling Museum, OCMW-museums.

11. Jan Sadeler after Dirck Barendsz., *Jonah Thrown to the Whale*, 1582, engraving, 24.2 x 20 cm. Amsterdam, Rijksmuseum.

12. Pieter Pourbus (?), *Jonah Cast over the Side*, early 1560s, drawing, 21.5 x 13.5 cm. Rotterdam, Museum Boymans-van Beuningen.

VII

Marine Paintings and Shipbuilding

12. Pieter Pourbus (?),
Jonah Cast over the Side, early 1560s,
drawing, 21.5 x 13.5 cm.
Rotterdam, Museum Boymans-van
Beuningen.

13. Hendrik Vroom,
The Arrival of Frederik V, Elector Palatine, and Elisabeth Stuart at Vlissingen (Flushing)
(detail of the largest warship), 1623,
oil on canvas, 203 x 409 cm.
Haarlem, Frans Halsmuseum, no. 300.

in shipbuilding practice came earlier than has been indicated in written sources. The work of artists suggests that by the mid-sixteenth century, if not earlier, Dutch shipbuilders had closed any technological gap that may have existed between them and masters in Italy, Spain, France, and Portugal. By that time they had joined the leaders in their field in Europe. The work of artists also implies that Dutch builders made many improvements in the sixteenth century, successfully exploiting the earlier great breakthrough in ship design when the best of northern and southern practices were combined in the full-rigged ship. By the mid-sixteenth century Dutch builders seemed to be technically equal to Iberian builders, a fact typically obscured by the great sixteenth-century navigational success of captains in the service of Spain and Portugal and by Dutch artists' interest in classical rather than contemporary Dutch vessels.

Holland had a close association with the sea. Certainly, by the sixteenth century her prosperity depended heavily on seafaring. The sea as a threat and an avenue for exchange and travel was already an important part of Dutch life. In the seventeenth century the association with the sea generated a school of artists who painted seascapes and pictures of ships. Hendrik Vroom has been called the father of marine painting. His ships look like the warships of Bruegel but have much greater detail and as great or greater accuracy (fig. 13). Reiner Nooms was a former sailor, and as a result, his lines and rigging are absolutely accurate. Through much of the seventeenth century the Willem van de Veldes, Elder and Younger, produced pictures of ships and sea battles so highly prized that the artists enjoyed government and royal patronage.[11] Their massive sketches of sea battles and their paintings of warships in action or sitting in harbors extolled naval power and success. Their works are valuable as sources of highly precise information about the ships themselves. But the Van de Veldes were only the most highly esteemed, widely recognized, and well paid of a large school of marine artists. In addition, landscape painters often felt obliged to include boats in their scenes, and they, too, showed the vessels as they were. The works of all of these artists attain a realism beyond that of late medieval works. Their own personal experience with ships was one reason; another was a desire to produce realistic work. The incidental realism of the later Middle Ages gives historians of technology basic information about changes in ships, but the works of art are neither extensive enough nor precise enough to rival the knowledge that can be gleaned from the seventeenth century. As the work of painters changed in the closing years of the sixteenth century, the role of art as a source for the history of technology changed. It is difficult, if not impossible, to explain exactly why artists altered their ways of thinking and painting, but more of them began concentrating on seascapes and landscapes. The development and growth of interest in the two types was closely related. In both cases artists painted from nature. They tried to make their works more consistent with life.[12] Certainly, neither the changing character of the art market nor the naval and maritime success of the Dutch Republic detracted from the value of seascapes. Such explanations, however, only begin to supply an understanding of what happened.

During the 1500s the artists' approach combined with influence from the south severely limited the value of their paintings as a source for understanding the great technological strides made by Dutch shipbuilders. The Low Countries were deeply influenced by the Mediterranean, first in the fifteenth century in shipbuilding practice and then in the sixteenth century in artistic practice. The former led to major improvements in ship design while the latter led to greater classicism. The seventeenth century witnessed greater realism but also a conscious effort to show the world ships exactly as they were and exactly as sailors handled them. It may be that part of the explanation for the high degree of reliability both in general outline and in detail of Dutch marine art in the later period is to be found in the older tradition of incidental accuracy, as in the pre-1520 practice of artists not bothering to change what they saw around them. Fortunately for the study of technology, after 1600 classicism was relegated to the introductory chapters of practical manuals on how to build ships, and artists no longer bothered to dress their vessels in the style of the Romans.

Notes

1. F. J. Dubiez, *Cornelis Anthoniszoon van Amsterdam 1507-1550* (Amsterdam: H. D. Pfann, 1969), 16–21.

2. Cornelis Koeman et al., *Lucas Jansz. Waghenaer van Enckhuysen: De maritieme cartografie in de Nederlanden in de zestiende en het begin van de zeventiende eeuw* (Enkhuizen: Vrienden van het Zuiderzeemuseum, 1984).

3. Herbert Ewe, *Schiffe auf Siegeln* (Berlin: Delius, Klasing & Co., 1972).

4. Richard W. Unger, *Shipbuilding in Holland and Zeeland Before 1800* (Assen: Van Gorcum, 1978), 42.

5. Ibid., 24–40, 189.

6. Richard W. Unger, "Dutch Design Specialization and Building Methods in the Seventeenth Century," in *Postmedieval Boat and Ship Archaeology*, ed. Carl Olof Cederlund (Oxford: B.A.R., 1985), 153–64.

7. Richard W. Unger, "Warships and Cargo Ships in Medieval Europe," *Technology and Culture* 22, no. 2 (April 1981): 233–36, 247–51.

8. Don Cameron Allen, *The Legend of Noah: Renaissance Rationalism in Art, Science and Letters* (Urbana: Univ. of Illinois Press, 1963), 60–91.

9. For example, Nicolaes Witsen, *Architectura Navalis et Regimen Nauticum ofte Aaloude en Hedendaagsche Scheeps-Bouw en Bestier* (Amsterdam: Joan Blaeu, 1690), 19–44.

10. J. Richard Judson, *Dirck Barendsz. 1534–1592* (Amsterdam: Van Gendt, 1970), 38.

11. W. Voorbeytel Cannenburg, "The Van de Veldes," *The Mariner's Mirror* 36 (1950): 185–204; David Cordingly, *The Art of the Van de Veldes: Paintings and Drawings by the Great Dutch Marine Artists and Their English Followers*, exh. cat. (London: National Maritime Museum, 1982).

12. Margarita Russell, "Seascape into Landscape," in *Dutch Landscape: The Early Years, Haarlem and Amsterdam 1590-1650*, ed. Christopher Brown (London: National Gallery Publications, 1986), 63–71; John Michael Montias, *Artists and Artisans in Delft: A Socio-Economic Study of the Seventeenth Century* (Princeton: Princeton Univ. Press, 1982), 141–48.

VIII

Shipping in the Northern Netherlands 1490-1580

Historians have not studied sixteenth century Dutch shipping as deeply or extensively as the seaborne trade of the Golden Century. In the fifteenth and sixteenth centuries the southern Netherlands were more important for international trade than the northern coastal provinces of Zeeland, Holland and Friesland. Data for their trade are scarce and almost always filled with gaps.[1] The northern Netherlands had not yet taken on the form or trappings of a state so to historians it has appeared as if the economy and the political institutions were not worthy of sustained study. Despite that the fact remains that the foundation for the trade growth of the Golden Century was already laid down around 1500. Commerce developed early in the sixteenth century in a way that would be all too common a hundred years later. Shipping was the most dynamic sector of the Netherlands economy. Above all the changes in the commerce and especially in shipping were the most important source for the striking changes in the economy and in society.

Geography gave the northern Netherlands a very favourable position for trade. The large number of rivers, lakes and inland waterways made possible rapid movement of goods among the different harbours and further inland as well. Many harbours had connections with the sea. Close by, in the south, lay Antwerp, the largest port of Europe. As a result the Netherlands found itself well placed to profit from the changes in the international structure of supply of demand. The population of Europe was, after reaching a low point at the start of the fifteenth century, rising consistently. Improvements in the climate, especially noticeable in the first half of the fifteenth century, made larger harvests and, therefore, population increase possible. The trend to urbanization made more people dependent on foodstuffs that had to be shipped to them. The already existing tension between the growing population and the supply of foods, which could not keep pace, was made even greater in the second half of the sixteenth century by a deterioration in the climate. This applied especially to southern Europe.[2] For the Dutch those changes created the possibility of shipping needed food there at a profit.

The seaborne trade between northern and southern Europe had increased markedly already back in the fourteenth century. Genoese and other Italian shippers brought wine, alum and silk to the Low Countries. Merchants from northern Europe began to orient themselves more toward the south as, for example, the German Hanse did in setting up a small colony in Lisbon in the middle of the fifteenth century. The pattern continued in the sixteenth century. The northern Netherlands, because of the advantageous geographical position, could profit not just from the growing trade with South Europe but they were also beneficiaries of the ever growing quantities of goods shipped within their old network around the North Sea. Antwerp drew the greatest profit from the increasing trade in the sixteenth century. Shipping of the northern Netherlands expanded in the shadow of the Brabant market, in which they participated to an increasing degree. At the same time the Dutch were constantly developing commerce in other ways entirely and exclusively their own.

Holland and Zeeland carriers appear to have already become specialized in the shipment of bulk goods. The costs of transport by sea were based on the amount of space the cargo took up in the ship and the distance the goods were carried. Though bulk goods were simple to handle the shipping costs were always a large part of the delivered price, and that became even more true as the distance travelled increased. The proportion of transport costs in the delivered price for the goods carried by Dutch shippers was typically 50% or more. The profit per unit was low. Still the concentration on the shipping of bulk goods appears over the long term to have proven advantageous. With the growth in European population, food and especially grain prices went through a rapid relative increase. Because of that those goods were better able to sustain the shipping costs, and therefore it was worth the trouble to transport them over ever greater distances.

The changing structure of supply and demand meant an increase in the range of shipments. Though the profit was perhaps less than with the shipment of luxury goods expansion of bulk trades in the sixteenth century led to a dramatic increase in the number of ships and with it the number of sailors to man those ships. Because of the character of the cargo and because ships went regularly, even if slowly, between the same places Dutch shipping kept its simple framework. Up to the end of the period there was still little shipping outside of Europe with only now and again a trip to the Mediterranean or the Arctic. There were some hints of the world wide trading network of the Golden Century had not yet developed but the intense network of the trade in bulk goods within Europe was by 1580 was established on a very strong foundation.

North Netherlands shippers could offer their services at relatively lower prices than their competitors. Although they were not yet generally known for their greater efficiency, they could and did keep on lowering their consistently low prices. The costs of their capital goods, the ships, were kept down because those vessels were deliberately designed for specific tasks. Capital cost formed an important part of total costs in shipping. Since trade volume expanded in a fixed pattern it was, therefore, easier over time to exploit the advantages of such specialized ship types. Since Netherlands shipbuilders had learned in the fifteenth century to build hulls with flush planking in place of the common overlapping planking, they produced an ever greater variety of vessels making good use of both old and new design features. The developments were suited to and guided by the wishes and demands of the shippers. If the ships had to be used for two purposes, such as fishing vessels which served as cargo ships out of the fishing season, the shipbuilders had the ability to accommodate that demand. They did it too. The wood for shipbuilding had to be imported and lower shipping costs in turn lowered the price of that chief raw material of shipbuilding. Improvements in the way buyers of ships were able to keep oversight of the construction of their ships also served to keep building costs under control.[3] The organizations representing works in the industry, the shipcarpenters guilds, were very loosely organized so that they did not drive up the delivered prices of the ships.

Capital costs were also kept down by the decline in losses at sea. The greater effectiveness of captains, because they were more familiar with the routes and because of improvements in navigation techniques, drove down risk. In the Baltic, shallows formed the greatest problem while in the North Sea and in the Atlantic Ocean seamen had to deal with tides and periods of bad weather. In northern Europe it was lead and line that were the typical equipment of seafarers to test the depth of water and to find out the character of the bottom. In imitation of their counterparts in the Mediterranean Dutch captains in the fifteenth and sixteenth century made ever more use of charts along with the compass,and the astrolabe. Around the middle of the fifteenth century the first rutters for the northern coasts began to appear. These became more and more reliable, accurate and were supplemented with tide tables, books about navigation and then charts. These developments reached their high point with the appearance in 1584 of *Spieghel der Zeevaerdt*, the book of charts by Lucas Jansz. Waghenaer. He united high levels of accuracy with a low price thanks to the use of a new, cheaper copper engraving technique.

Dutch shippers and merchants also learned better business methods from their southern European counterparts. In the fifteenth century Dutch merchants were far behind their Italian colleagues in ways of doing business. Improvements in such matters lead to a lowering of administrative costs, fewer losses and greater efficiency. In the fifteenth century shipping was in general still in the hands of the owners of ships which travelled from harbour to harbour looking for cargoes and loading just what there was to get. Even though this tramping never completely disappeared, more sophisticated and efficient methods of doing business evolved supplementing and supplanting the old ones.

Because ships became larger and the necessary capital to set up a shipping enterprise rose, the ships came more and more to be the property of groups of share owners organized in ship-owning partnerships. The participants were men and women from different sectors of economic life, some having no connection with shipping at all. Part ownership of ships, in order to spread the risk, already existed in classical Antiquity. In the Netherlands use of it was increasingly made so that larger amounts of capital than ever before became available for shipping. The skipper was the representative of the partnership, but also an employee. He was often a part owner as well. In order to keep the owners informed, better bookkeeping methods developed. Often a representative of the owners, a purser responsible for the business side of things, was sent along on the voyages of the ship. The ships were also chartered, that is leased by the owners to merchants for a certain voyage or a specified period of time. Because of the increasing use made of maritime insurance, the need to spread cargoes across a number of ships fell sharply. Because of the greater certainty and stability of trade relations Dutch merchants could establish overseas factors, often young men, who represented them in foreign ports. The factors oversaw the lading of cargo and kept the merchants back home informed of business conditions. That meant a decrease in turnaround time for Dutch ships in foreign ports. The available tonnage could be used more effectively which brought with it further cost reductions. To some degree trade and shipping were two clearly different economic pursuits, but it was often the same people who owned the ships, bought and sold the goods, sent pursers on board and kept factors in foreign ports.[4] This concentration made possible a decrease of risks, of shipping costs and opened the door to expansion of Dutch shipping. Labour costs were also kept relatively low. Per capita incomes in the northern Netherlands were below levels to the south and population growth increased the supply of all workers including able seamen. The expansion of the fishery, which took place at the same time, also meant that there existed an ever growing group of experienced sailors available. It is impossible to assess the precise relationship between the growth in population and employment in shipping, but it is clear that both grew dramatically in the sixteenth century.

Because many harbours and their approaches in the northern Netherlands were shallow the draught and thus also the size of Dutch ships had to be kept low. Even so large ships were to be found. For example, Enkhuizen in 1542 was the home port for eight ships between 500 and 600 tons. With the growth of shipping of bulk goods over fixed routes Dutch ships on average got bigger. In 1565 the province of Holland had more than 1800 ships and Amsterdam alone boasted more than 500 large ships. There were reports that annually 250 to 400 large vessels entered Amsterdam just with grain. The fleet of the northern Netherlands at the end of the fifteenth century consisted of some 60,000 tons and in 1570 the fleet was about 230,000 tons. The estimates are probably minima and they are consistent with data on the scale of trade.[5] While the fleet at the end of the fifteenth century was perhaps about as large as that of the German towns in the Hanseatic League, by the end of the sixteenth century the northern Netherlands had a fleet with a much greater carrying capacity than any other similar region in northern Europe. Crew size varied from 10 to 20 men per ship with the average at about 14. In the 1570s there were thus about 16,500 seamen directly employed in shipping, that is less than 2% of the total population of the coastal provinces of the northern Netherlands.

The social status and independence of those seamen declined in the sixteenth century. Because the ships became larger the size of crews increased and the work on those ships became more complex. Democracy on board, still common in the Middle Ages, disappeared completely. In the Middle Ages decisions about the handling of the ship, when and whether or not to leave harbour and other important matters were made by a vote of the crew members. Skippers, as employees of the owners, came over time to have greater responsibility and also greater powers. The changeover from democratic to autocratic authority on board ship progressed slowly but in the sixteenth century it advanced faster and faster. On fishing boats the members of the crew could still own a part of the ship or the net. On large cargo ships seamen were wage labourers, signed on for a fixed time or for a fixed voyage, which naturally lessened their social status and the certainty of their livelihood. As on board so on shore the position of the captain became more important, not because he was more commonly an owner but because he exercised command and because of his sailing experience. The changes in the organization and the scale of shipping improved the social position of the merchants and of the shipowners the most. Their houses, always in town, and the taxes that they paid, reflected their improved position. Their higher social rank in the end depended on the rising incomes they enjoyed from shipping.

The grain trade was unquestionably the most important task of the Dutch shipping in the sixteenth century. The grain was principally exported from ports along the south coast of the Baltic. When their own grain supplies proved inadequate, Netherlands merchants turned first to the valley of the Lower Rhine, to France and England. When those sources appeared inadequate in the fifteenth century the merchants went to the Baltic region. In the last decade of the fifteenth century the export of grain from the Baltic grew significantly and Dutch shippers carried the lion's share. The import of grain in terms of value was third among imports into the Netherlands after Italian silk and English cloth, but it was in terms of volume by far and away the most important.

From 5,000 tons annually around 1460 grain imports grew to 20,000 tons in 1500 to more than 110,000 tons annually by 1560.

The strong growth in grain exports from the Baltic region after about 1490 meant a change in the balance of trade for the ports of the eastern Baltic. Exports exceeded imports by a large margin, not only in value but especially in volume. While the Dutch shippers in the last years of the fifteenth century dominated the export of grain from the Baltic, in the first half of the sixteenth century merchants and traders from the Hanse towns took over an ever larger part of the growing trade. But after the middle of the century it is clear that the Dutch and especially Dutch shippers predominated.[6] An increasing portion of the grain trade was a transit trade, with the grain brought to a Dutch harbour, Amsterdam being the most common choice, off-loaded there, stored and then reexported. England, France and in large part Spain and Portugal were the principal destinations for the grain. In years of shortage grain exporting from the Netherlands was often forbidden. Such restriction elicited protests, for example, from the port town of Danzig which contended that such limitations brought great damage to their trade, now heavily dependent on the Netherlands as the grain entrepôt of northern Europe.

Amsterdam relied especially on the grain trade, specializing in it just as, to a lesser degree, did the other towns on the Zuiderzee. Ships from these towns could reach the Baltic more easily than those from South Holland or Zeeland. In the middle of the fifteenth century Amsterdam was the most important town in the northern Netherlands, in part because it had an excellent harbour. In order to compete other towns, especially in North Holland, in various way attempted to improve their harbours. Any dredging or other work had to be paid for from harbour dues, but with the growing trade it was certain that the costs for improvements would be covered.[7] The Zuiderzee harbours profited also from the fact that for them Norway, with a rapidly growing trade in the second half of the sixteenth century, was relatively easy to reach. Amsterdam dominated the developing grain trade there too. The growth in Amsterdam in the sixteenth century was a good reflection of her important position. The availability of grain was also necessary for one Dutch industry, beer brewing. Amsterdam merchants would dominate the beer trade, not only along the coast of the North Sea but also along the inland waterways of the Low Countries. As well as beer produced in their own town Amsterdam shippers carried beer brewed in a number of other towns in the northern Netherlands.

The grain ships typically followed a fixed pattern of travel leaving Amsterdam for the Baltic, arriving there in April or May, then back to Holland, after which came a trip to France or the Iberian peninsula with a load of grain, and then back to Amsterdam.

Sometimes before winter set in there was another trip, a short one, for example to England, fit into the schedule. In the sixteenth century fewer and fewer ships made the stop in the Low Countries on the way back from the Baltic. Most sailed straight to the Iberian peninsula, the so-called '*deurgaende vaert*,' and only then made their way back to the Netherlands. The ships which came from the Baltic carried other goods too among which especially was wood and, in lesser amounts, sylvan products such as pitch, ashes, tar and potash. The export of wood from Poland fell in the sixteenth century since Dutch merchants went over to importing much cheaper Norwegian wood. No matter any fluctuations, grain remained consistently by far and away the most important commodity of the Baltic trade and all other goods from the region were in a certain sense mere supplements to the grain traffic.

The available carrying capacity on any route is fixed by the good which takes up the greatest volume in ships' holds. If one cargo on the outbound trip can cover all the costs of the entire trip then the goods that are carried on the return trip can travel for nothing or, at most, for the cost of loading and unloading. If a skipper can find a return cargo so that he carries goods in both directions, he can reduce the price for his services by as much as half. The size of the saving depends on the return cargo, the harbours between which it is carried and the value per unit volume of the return cargo in comparison to the original one. Dutch shipping in the period after 1500 consisted in a certain sense of a constant search for a back cargo or cargoes for the ships that went into the Baltic for grain. The problem was made more pressing and urgent and the possibilities augmented in the sixteenth century by the growth in the volume of grain exports from Poland. In 1460 Holland ships carried only 1.8% of the total imports into Danzig but in 1498-99, after grain exports had shot up, the share had risen to almost 50%. The success of Dutch shipping depended to a significant degree on whether or not it was possible to find goods destined for ports in the Baltic. The fact that a large proportion of Dutch ships entered the Baltic in ballast indicates that the problem was never completely solved. But in their efforts to find a complement for the cargoes of the grain trade, the Dutch developed a whole range of related trades in which they were able to mount effective competition because the shipping of grain covered such a large part of the costs.

The most obvious back cargo was salt. The preparation and preservation of food created a market for salt throughout all of northern Europe. It was a bulk good and could easily be shipped in the same types of ships used for carrying grain. In the beginning of the fifteenth century Hanse ships made regular trips to the south coast of Brittany for salt from the Bay of Bourgneuf. The Dutch slowly took over the trade from the Hansards. The salt had to be brought from places along the coast of the Atlantic

Ocean which were ever further south - first the Ile de Ré and then from Brouage and later even from Setubal in Portugal and Cadiz in Spain. The Dutch continued to dominate the trade and by 1570 the tonnage of salt shipped into the Baltic was more than half the volume of grain shipped out.[8] Dutch shippers found other markets besides the Baltic for the salt that came from France and the Iberian peninsula. After being refined in the Netherlands, mostly in one of the towns on the island of Walcheren in Zeeland, the salt was shipped to the upper Rhine, the British Isles and Scandinavia and to the eastern part of the Baltic region. As a result a great part of the local production of salt in the Netherlands and for that matter in all of northern Europe was supplanted by cheaper imported salt carried north in Dutch ships.

In terms of volume herring was the second most important export commodity to the Baltic. In a certain sense this was only adding value to the salt since about 15% of Dutch salt imports were used to cure herring. In the second half of the sixteenth century Dutch herring dominated the entire Baltic market and Dutch shippers dominated the transport of the good to the Baltic region. Hoorn and Enkhuizen, important harbours for the herring fishery, were understandably also active in the salt trade. The fish as well as the salt trade contributed to the strengthening of the position of the ports along the Zuiderzee in the Baltic trades. The fishery also generated a fairly large local trade, whereby herring from the fishing villages on the coast was shipped to the most important towns and also from those big ports inland along the great rivers. In addition, salt was used extensively for the preparation of agricultural products for export, especially butter and cheese. Agriculture was still the largest sector of the Dutch economy. In all of the northern Netherlands cattle were raised and milk made into various dairy products. There existed an old, established local trade in agricultural products, especially along the inland waterways to the cities of Flanders and Brabant. This trade expanded in the sixteenth century with Dutch goods replacing locally produced commodities in the south Netherlands market. Farm products were exported to the regions around the Upper Rhine and Upper Maas as well. In this trade in agricultural goods North Holland ports supplanted towns along the Ijssel River such as Deventer. Although agricultural products were exported to the Baltic, they contributed only a little to that exchange.

Cloth was extraordinarily important for trade with the Baltic, not because of the volume but because of the value of goods supplied which helped to pay for a part of the grain. Poland had only a poorly developed textile industry which meant here was an outstanding market for western cloth in Baltic ports. The cloth was still of the heavy type, characteristic of production in the Middle Ages. Only in the second and third decades of the seventeenth century did the lighter 'new draperies' penetrate the markets of northeastern Europe. The cloth imported into the Baltic region was not produced exclusively in the Netherlands nor was it only carried in Dutch ships. Although around 1500 the share of cloth shipped to the Baltic ports by the Dutch was perhaps half of the total, in 1565 English cloth in English ships made up 70% of what was sent.

The Dutch cloth industry, with Leiden as the most important centre, produced for local and nearby German and south Netherlands markets, places which could be reached easily by river, and also for more distant regions. The industry was very much involved in finishing cloth. The raw material for that work was imported from England so Dutch shippers carried unfinished cloth and raw wool from England over the North Sea to the Low Countries. Though those two goods along with finished English cloth made up more than a quarter of imports into the Low Countries by value, they were of much less importance for shipping because of their relatively small volume and the short distance over which they were carried.

The trade with England was not limited to goods for the textile industry. Zeeland harbours because of their location had long had close trading connections with England. They also functioned as the foreports for Antwerp and not just for English goods. It was difficult to sail up the Schelde and for that reason it was sensible for skippers to offload their cargoes in Flushing or Arnemuiden and then to take on another outbound cargo there. The goods were bought and sold by merchants in Antwerp, but the goods never reached that town. The total value of the Antwerp trade doubled about three times in the period from 1480 to 1570. Port towns along the Maas also had an advantage because of their location opposite England and that helped them to take part in the growth in the shipment of goods over the North Sea. Because the towns were at the same time fishing ports, cured herring was one of their most important export products. The rise in the import of coal into the northern Netherlands brought a bulk good into the trade with England and Scotland. Most coal for north Netherlands markets, however, was shipped from the southern Netherlands downstream along the Maas from Liège. The greatest part of imports from Scotland, that is in terms of value, was taken up by wool and hides of lesser quality. The Scottish government demanded that these goods enter through a staple port.[9] The back cargoes to England and Scotland consisted of agricultural products and of goods which had been brought to the northern Netherlands from all over Europe and beyond. The trade with England and the shipment of English goods to third countries did not grow as quickly as the other trades in the second half of the sixteenth century. English shippers, supported by their government, looked after more of the carriage of their own export goods and they travelled to all parts of Europe to bring back commodities. This direction of affairs altered dramatically, at least with trade to the Baltic, when after the 1580s Dutch shippers replaced the English in the carrying of their goods and those of virtually everyone else back and forth to northeastern Europe.

When the Portuguese had opened the all sea route to the Far East, they initially imposed a monopoly on the import of spices from India and the Indonesian archipelago into Europe. Pepper and other spices were sent to Lisbon, and from there they were shipped to Antwerp for further distribution throughout northern Europe. But through the grater part of the sixteenth century spices could still travel along their traditional land route through the Middle East to arrive in Italy and then, for example, be distributed in southern Germany at the same or even lower prices than those which prevailed for spices which came around the Cape of Good Hope. The spice trade via Lisbon did continue to grow in the sixteenth century and Dutch shippers found it possible to make themselves masters of an ever increasing portion of shipping from Portugal to northern Europe. Copper from the Baltic was the most important export article to the already developing markets outside of Europe and Dutch ships brought that copper directly to Lisbon. The most important reason that the Portuguese as well as the Spanish broke the habit of trading within Europe was that it appeared much more advantageous for them to use their few ships and sailors for the more lucrative trade to the Far East and the New World. The Dutch, who sailed to Portugal for salt, quickly filled the vacuum created by the shift in effort by the Iberians. The Dutch brought an ever greater variety of goods from outside Europe to the North, distributing those goods throughout the rest of northern Europe. In the course of the sixteenth century Amsterdam took over the central place in that pattern of exchange from Antwerp.

Quantitatively the spice trade was of relatively inferior importance. That certainly did not apply to trade with Spain. The union of the Netherlands with Spain under the Hapsburg crown had two important advantages for the Dutch. It gave merchants and shippers from the Netherlands the same privileges in Spain as the Spanish themselves. The government also felt obliged to protect Dutch ships that travelled between the Iberian peninsula and northern Europe. That augmented the competitive position of Dutch shippers during the wars of the sixteenth century. Netherlanders shipped grain and cloth, wool as well as linen, to the Iberian peninsula in exchange for sub-tropical products such as wine, olive oil, fruits, and in increasing quantities, also goods from Asia and the New World.

Dutchmen also went to the New World themselves, despite the prohibitions of Iberian authorities. Their intention in excluding interlopers was to keep the profitable trade for their own merchants and shippers. Appearing in the name of Iberian merchants a substantial number of Dutchmen sailed to Brazil and the Caribbean islands. By the middle of the sixteenth century silver was the most import trade good for the Spanish. The silver was brought over from the New World and increasingly used by the Spanish to pay for imports from northern Europe. Because of the high value per unit volume it was a poor return cargo. It had little direct influence on shipping but was very much in demand in the Baltic. The silver was necessary for the maintenance of the grain trade because it covered the balance of trade payments for western Europe. Because of the distance, the long-established pattern of the trade, and the value of the goods shipped trade with Spain and Portugal was important for the continuing growth of Dutch shipping. In the course of the sixteenth century this was more and more so as the quantity of bulk goods carried to and from Iberia increased.

Wine was traded in the Netherlands throughout the Middle Ages. The trade was principally in Rhine wine, sent downstream to Dordrecht and then shipped on from there. That town fell into relative decline in the fifteenth century because the local shippers and merchants did not take part in the most rapidly growing trades, such as that to the Baltic. What is more in the sixteenth century, Dordrecht lost its staple, the legal requirement that all goods which were shipped down the Rhine had to be off-loaded in the town and offered for sale there.

The trade in Rhine wines was supplemented and expanded by a growing trade in French and Spanish wines. Ships which had gone to French harbours like Rouen, Nantes, La Rochelle and Bordeaux loaded with grain brought back the French wines. Those were of lower quality than the wines from the Rhine valley. Both were intended more for use in the Netherlands than for export, though Rhine wines of good quality were shipped to England and also used as a back cargo to the Baltic. A few Zeeland harbours, especially Middelburg, applied themselves to the import of French wine and the distribution of those wines in the Netherlands. In the middle of the sixteenth century the wine trade reached its highest point. Imports into Zeeland exceeded 10,000 tons annually, around one third of all wine imports.[10]

Shipping in the sixteenth century consisted for a large part of transit trade. After having initially shipped goods which were produced locally such as dairy products, or which were intended for local consumption such as grain, Dutch ships began more and more to take up the task of transporting goods produced outside the Low Countries and which were also intended for foreign markets. The growth of this carrying trade meant that many of the cargoes which Dutch shippers hauled were never off-loaded in the home ports of those shippers. The relatively large portion that did indeed get stored for any time in Dutch warehouses made it possible for many ports, most notably Amsterdam, to function as entrepôts. The availability of those goods also created opportunities for industries which could either refine the goods for the consumer as was the case with salt, or use the goods as raw materials for another product as was the case with grain for beer, wool for cloth, or wood for shipbuilding. The export of grain from the Baltic was not based solely on an expansion of the traffic in rye, wheat and barley but also in the long term growth of domestic industries. In 1530, and perhaps even earlier, the importance of the Baltic exchange for all of shipping and for the economy was already fully appreciated.[11]

The trade pattern made it more or less necessary for shipping to have protection and support from the government. The ever longer shipping routes fell prey more easily to disruption by enemy action. In the Middle Ages the use of force at sea was an anticipated phenomenon. It did not disappear in the sixteenth century. Losses at sea could only cause the costs of shipping to go up. Safety during the voyage was very important to the survival of the Dutch sailors making the trip but also it was critical for the competitive position of Dutch shipping. Piracy was always a threat and was even more serious a problem during wartime when governments encouraged privateering. Wars did not yet have such a destructive effect on shipping as they would in later centuries. Governments could not hold back trade with or through belligerent acts, and often it would seem they did not want to. Political divisions made it difficult to levy a total embargo in any one region. Papers could always be forged so a ship could appear to be a neutral. The Dutch had the advantage in the sixteenth century that their trade was carried on principally in the Baltic and North Sea regions and along the coast of the Atlantic Ocean. These waters, especially the former, were relatively peaceful in comparison with, for example, the Mediterranean where the Dutch began to carry on trade around 1590. In terms of the political situation and the pattern of trade the greatest dangers for Dutch shipping always lay in a war with France and with the closing of the Sound, the entry to the Baltic.

A war with France disrupted trade not only with France but also with the Iberian peninsula and with England because of the privateering by French raiders in the North Sea and along the French coast of the Atlantic Ocean. To 1559 the rivalry between the houses of Hapsburg and Valois led to wars between France and the lands of Emperor Charles V and then his son, Philip II of Spain. The danger led to greater efforts to protect shipping by the use of convoys. In the fifteenth century some towns often fitted out ships for the protection of their fishing fleets, but in the sixteenth century the organized protection of cargo ships increased. Despite resistance from a number of towns in Holland, by the middle of the century there was a government policy of protection, institutions to carry out the policy and taxes in place to pay for it.[12] With the establishment of an admiralty for the Netherlands there were signs of a general, commonly established system of protection for shipping through the device of convoys, especially for ships going between the Netherlands and Spain. Each time, though, the general effort disappeared with the end of hostilities with France. Between cargo ships and warships there was still no clear difference, especially with those ships that were fitted out for privateering. It was thus common in war time to arm ships more heavily, temporarily. That was very expensive, however, so the costs of transporting goods went up. Above all then war in the sixteenth century raised shipping costs.

The closure of the Sound by one of the Scandinavian princes or the Hanse towns was a much more serious problem for Dutch commerce because the entire trading network was to such a large degree dependent on the grain trade. This was certainly clear in the Golden Century, but the importance of keeping the Sound open for Dutch ships was already under discussion in the war Holland fought against some of the Hanse towns fought from 1438 to 1441. In that war the Dutch discovered the advantage of being ruled by a prince, in this case the Duke of Burgundy, who exercised some limited political influence in the negotiations with the Hanseatic towns at the end of the war.[13] The union with the other Hapsburg lands saw to it that the Netherlands in general had more influence in international affairs. Among other things, that meant they were in a better position to keep the Sound open through diplomatic means, except whenever such a policy came into conflict with the dynastic goals of the Hapsburgs. Such a conflict did arise in the 1530s but it was finally resolved with the Treaty of Speyer in 1544. That peace gave the Netherlanders access to the Baltic under favourable conditions. Despite the treaty, the Dutch were not free from difficulties emanating from other wars in the North, such as the Danish-Swedish War of 1563 to 1570 when the Sound was, in theory, closed by the king of Denmark, a closure which could and did threaten Dutch access to the Baltic.[14] Despite the difficulties in the 1530s it appears that support of the Hapsburgs was on balance very beneficial for the Netherlands trade with the Baltic.

Ultimately, the greatest danger for Dutch commerce lay not in war overseas but in civil war at home. The struggle after the succession of Maximilian as regent for his son, Philip, in 1482 led to the interruption of trade contacts, famine and a general disruption of trade. The Revolt starting in the 1560s and the subsequent looting by the Sea Beggars led to unemployment and food shortages in coastal towns in the northern Netherlands. This came on top of the already existing disruption caused by the uncertain situation after 1566. By 1579 the Dutch *de facto* had an independent state, engaged in a war again the southern Netherlands, Spain, and, after 1580, Portugal. The small group of northern provinces had the most powerful state and the most powerful maritime power of Europe opposing them. Once the internal problems were solved, the Sea Beggars having seized some port towns in 1572 and Spanish troops having been largely driven out by 1578, Dutch trade revived extraordinarily quickly. Commercial relations with the Baltic revitalized and the volume of goods carried in both directions returned to previous levels. Despite the war against the Hapsburg overlord there was also an increase in trade over the inland waterways with the southern Netherlands, carried on by shippers who bought licenses to do so. Even trade with the Iberian peninsula grew again after 1580, though the Spanish government now and again tried through trade prohibitions and seizures to prevent the Dutch from again capturing an over balance in the share of shipping between northern and southern Europe. In 1580 the Dutch no longer needed a strong ally for the protection of their shipping, although the government recognized that only slowly. Similarly the new state discovered it was best able to undertake the necessary political and maritime steps to safeguard shipping. It made use of the divisions among rivals and potential enemies and developed a system of protection at sea that was financed from a tax on shipping.

The rapid rise of the northern Netherlands to the preeminent position in Europe at the end of the sixteenth century came because of the economic power of the region which, in turn, was based squarely on the growth in shipping in the previous period. The government was in the hands of merchants, those who had profited most from the expansion of trade. Those men found only little opposition from the land-based aristocracy which in some cases had other ideas about the proper goal for the new government. The flexibility of the commercial system during and after the Revolt was a sign of the strong economic foundation which had been laid for trade and shipping in the sixteenth century during a period of general expansion in Europe. The Dutch appear to have been in a better position than anyone else to draw profit from the trading connections that had been put in place by others in the fourteenth and fifteenth centuries. There was, over a broad front, improvement in shipping and many industries related to shipping. The most important sector, however, was trade with the Baltic. The balanced growth of shipping was based on the domestic production of cheaper goods, the exploitation of a favourable geographical location in relation to the increasingly busy trade routes of northern and western Europe, the adjustment to new markets, and foreseeing the demand for cheap shipping of bulk goods. This all, pressed by international specialization in production as well as in consumption, resulted in dramatic growth for shipping. The same type of successful use of the available expedients appears to have been of equally great value for Dutch shippers in the Golden Century.

Notes

1. I. J. Brugmans, "Les sources de l'evolution quantitative du trafic maritime des Pays-Bas (XIIe-XVIIIe siècles)," in M. Mollat, ed., *Actes du Quatrième Collque International d'Histoire Maritime* (Paris, 1962), 417-424.

2. G. Utterström, "Climatic Fluctuations and Population Problems in Early Modern History," *Scandinavian Economic History Review*, III (1955), 7-43; E. LeRoy Ladurie, *Times of Feast, Times of*

Famine: A History of Climate since the Year 1000, B. Bray, trans. (New York, 1971), 58-79.

3. R. W., Unger, "Dutch Ship Design in the Fifteenth and Sixteenth Centuries", *Viator*, IV (1973), 387-411; *idem.*, "Selling Dutch Ships in the Sixteenth Century", *Maritime History*, III (1973), 125-146.

4. A. E. Christensen, *Dutch Trade to the Baltic about 1600 Studies in the Sound Toll Register and Dutch Shipping Records* (Copenhagen and The Hague, 1941), 113-114; J. C. Riemersma, "Trading and Shipping Association in 16th Century Holland," *Tijdschrift voor Geschiedenis*, LXV (1952), 331-336.

5. W. Brulez, "Le Commerce International des Pays-Bas au XVIe siècle; essai d'appreciation quantitative," *Belgisch Tijdschrift voor Filologie en Geschiedenis*, XLVI (1968), 1205-1221; W. Vogel, "Zur Grosse der europäischen Handelsflotten im 15., 16. und 17. Jahrhundert," *Forschungen und Versuche zur Geschichte des Mittelalters und der Neuzeit, festschrift Dietrich Schäfer*, (Jena, 1915), 268-333; R. W. Unger, "The Tonnage of Europe's Merchant Fleets 1300-1800," *The American Neptune*, 52, 4 (Fall, 1992), 247-261.

6. A. E. Christensen, *Dutch Trade to the Baltic about 1600*, 34-48; H. Samsonowicz, *Untersuchungen über das Danziger Bürgerkapital in der Zweiten Hälfte des 15. Jahrhunderts* (Weimar, 1969), 30-41; J. Schildhauer, "Der Seehandel Danzigs im 16. Jahrhundert und die Verlagerung des Warenverkehrs im Nord- und Mitteleuropäischen Raum," *Jahrbuch für Wirtschaftsgeschichte*, III (1970), 155-178; W. S. Unger, "De Sonttabellen," in *Tijdschrift voor Geschiedenis*, XLI (1926), 137-155; W. S. Unger, "De Hollandsche Graanhandel en Graanhandelspolitiek in de Middeleeuwen," *De Economist* (1916), 251-269.

7. W. S. Unger, "De Hollandsche Graanhandel...," 476-486; Gemeente Archief Enkhuizen 241 [charter 61].

8. W. Brulez, "De zoutinvoer in de Nederlanden in de 16e eeuw," *Tijdschrift voor Geschiedenis*, LXVIII (1955), 181-184; A. E. Christensen, *Dutch Trade to the Baltic about 1600*, 368-379, Spading, *Holland und die Hanse im 15. Jahrhundert. Zur Problematic des überganges vom Feudalismus zum Kapitalismus* (Weimar, 1973), 50.

9. M. P. Rooseboom, *The Scottish Staple in the Netherlands* (The Hague, 1910); T. L. M. Thurlings, *De Maashandel van Venlo en Roermond in de 16e Eeuw, 1473-1572* (Amsterdam, 1945); W. S. Unger, *De Tol van Iersekeroord. Documenten en Rekeningen 1321-1572*, (The Hague, 1939), 151-164.

10. Jan Craeybeckx, *Un grand commerce d'importation: Les vins de France aux anciens Pays-Bas (XIIIe-XVIe siècle*, (Paris, 1958), 21-43; W. S. Unger, "De Scheepvaart der Bretons naar de Walcherse Reede in de XVe en XVIe Eeuw," *Bijdragen voor Vaderlandsche Geschiedenis en Oudheidkunde*, 6th series, VIII, 225-244; H. van der Wee, "De handelsbetrekkingen tussen Antwerpen en de Noordelijke Nederlanden tijdens de 14e, 15e en 16e eeuw," *Bijdragen tot de Geschiedenis der Nederlanden*, XX (1965-1966), 267-285.

11. Rudolf Häpke, ed., *Niederlandische Akten und Urkunden zur Geschichte der Hanse und zur Deutschen Seegeschichte 1531 - 1669*, (Munich and Leipzig, 1913-1923), volume 1, paragraphs 8-10, 178 (1534), paragraphs 21, 28-30.

12. James D. Tracy, "Herring Wars: The Hapsburg Netherlands and the Struggle for Control of the North Sea, ca. 1520-1560," *The Sixteenth Century Journal*, XXIV (1993), 256-270.

13. Jacques Paviot, *La Politique Navale des Ducs de Bourgogne 1384/1482*, Lille: Presses Universitaires de Lille, 1995, 235-249.

14. A. Bijl Mz., *De Nederlandse Convooidienst (1300-1800). De maritieme bescherming van koopvaardij en zeevisserij tegen piraten en oorlogsgevaar in het verleden*, (The Hague, 1951), 6-51; E. Daenell, "Holland und die Hanse im 15. Jahrhundert," *Hansische Geschichtsblätter*, IX (1903), 20-29; P. Dollinger, *La Hanse (XII-XVIIe siècle)*, (Paris, 1964), 399-403.

Sources

The published sources illustrate the uneven character of the surviving material. There are collections of documents made by region, of which the two largest are Z. W. Sneller and W. S. Unger eds., *Bronnen tot de Geschiedenis van den Handel met Frankrijk I (853-1585)* (The Hague, 1930) with a continuation by W. S. Unger, ed., *Supplement(1447-1585)* (The Hague, 1942); and H. J. Smit, ed., *Bronnen tot de Geschiedenis van den Handel met England, Scotland, en Ireland (1485-1585)*, 2 vols. (The Hague, 1942-1950). Documents which deal especially trade with central and eastern Europe were published by Rudolf Häpke, ed., *Niederlandische Akten und Urkunden zur Geschichte der Hanse und zur Deutschen Seegeschichte 1531-1669*, 2 vols. (Munich and Leipzig, 1913-1923). Very little quantitative material has survived. For an overview of the sources see I. J. Brugmans, "Les sources de l'evolution quantitative du trafic maritime des Pays-Bas (XIIe-XVIIIe siècles)," in M. Mollat, ed., *Actes du Quatrième Collque International d'Histoire Maritime* (Paris, 1962), 417-424. Toll registers were published by W. S. Unger, *De Tol van Iersekeroord. Documenten en Rekeningen 1321-1572* (The Hague, 1939) and by P. A. Meilink, ed., "Rekening van het lastgeld in Amsterdam, Waterland en het Noorderkwartier van Holland in 1507," *Bijdragen en Mededelingen van het Historisch Genootschap*, XLIV (1923), 187-230, and N. W. Posthumus, *De Uitvoer van Amsterdam 1543-1545* (Leiden, 1971). The best source for data on trade is N. E. Bang and K. Korst, eds., *Tabeller over Skibsfart og Varetransport gennem Oresund, 1497-1660*, 3 volumes (Copenhagen, 1906-1933). The tables are not without error but an even greater problem is that records for only a few years from before 1560 have been retained in the Dutch archives and there are lacunae in those records for later years of the sixteenth century. For a summary of the data about the Netherlands see W. S. Unger, "De Sonttabellen," *Tijdschrift voor Geschiedenis*, XLI (1926), 137-155.

Some papers about shipping and commerce are published but what material still exists to be consulted is inconsistent. That means, among other things, that what there is gives little reliable information about the scale of shipping. What survives rather better serves to illustrate the trade practices of the day. Such materials are, for example, H. E. van Gelder, "Zestiende-Eeuwsche Koopmansbrieven," *Economisch-Historisch Jaarboek*, V (1919), 136-189, H. E. van Gelder, "Zestiende-Eeuwsche Vrachtbescheiden," *Economisch-Historisch Jaarboek*, III (1917), 124-290, and J. Nanninga Uitterdijk, *Een Kamper Handelshuis te Lissabon, 1572-1594. Handelscorrespondentie, Rekeningen en Bescheiden* (Zwolle, 1904). Among studies of trade and business practice and their relationship to shipping are J. C. Riemersma, "Trading and Shipping Associations in 16th Century Holland," *Tijdschrift voor Geschiedenis*, LXV (1952), 330-338, and R. W. Unger, "Selling Dutch Ships in the Sixteenth Century," *Maritime History*, III (1973), 125-146.

On climate change in the fifteenth and sixteenth centuries and the consequences of those changes for agriculture and shipping, G. Utterström, "Climatic Fluctuations and Population Problems in Early Modern History," *Scandinavian Economic History Review*, III (1955), 7-43, E. LeRoy Ladurie, *Times of Feast, Times of Famine: A History of Climate since the Year 1000*, B. Bray, trans. (New York, 1971), and H. H. Lamb, *The Changing Climate* (London, 1966).

For the ships which were used, their measurements and the size of the Dutch merchant fleet there are J. van Beylen , *Schepen van de Nederlanden van de late middeleeuwen tot het einde van de 17de eeuw* (Amsterdam, 1970), R. W., Unger, "Dutch Ship Design in the Fifteenth and Sixteenth Centuries", *Viator*, IV (1973), 387-411, W. Vogel, "Zur Grosse der europäischen Handelsflotten im 15., 16. und 17. Jahrhundert," *Forschungen und Versuche zur Geschichte des Mittelalters und der Neuzeit, festschrift Dietrich Schäfer*, (Jena, 1915), 268-333, and more recently R. W. Unger, "The Tonnage of Europe's Merchant Fleets 1300-1800," *The American Neptune*, 52 (1992), 247-261.

On the development and growth in protection of shipping, A. Bijl Mz., *De Nederlandse*

Convooidienst (1300-1800). De maritieme bescherming van koopvaardij en zeevisserij tegen piraten en oorlogsgevaar in het verleden (The Hague, 1951), and R. Degryse, "De Konvooieering van de Vlaamsche visschersvloot in de 15de en 16de eeuw," *Bijdragen tot de Geschiedenis der Nederlanden*, II (1948), 1-24. On the development of government policies of protection James D. Tracy, "Herring Wars: The Hapsburg Netherlands and the Struggle for Control of the North Sea, ca. 1520-1560," *The Sixteenth Century Journal*, XXIV (1993), 249-272, and Jacques Paviot, *La Politique Navale des Ducs de Bourgogne 1384/1482*, Lille: Presses Universitaires de Lille, 1995.

Surviving figures from the Netherlands and also from ports of the Hanseatic League has made it possible to discuss the balance of trade and the balance of payments, among others in W. Brulez, "Le Commerce International des Pays-Bas au XVIe siècle; essai d'appreciation quantitative," *Belgisch Tijdschrift voor Filologie en Geschiedenis*, XLVI (1968), 1205-1221, and by the same author in expanded form, "The Balance of Trade of the Netherlands in the middle of the 16th Century," *Acta Historiae Neerlandica*, IV (1970), 20-48. There is also H. Samsonowicz, *Untersuchungen über das Danziger Bürgerkapital in der Zweiten Hälfte des 15. Jahrhunderts* (Weimar, 1969), and J. Schildhauer, "Der Seehandel Danzigs im 16. Jahrhundert und die Verlagerung des Warenverkehrs im Nord- und Mitteleuropäischen Raum," *Jahrbuch für Wirtschaftsgeschichte*, III (1970), 155-178.

Because of the lack of reliable relevant figures for shipping general historical writing about trade and commerce is scant. The most successful studies are generally those dealing with a specific trade or with shipping from specific parts of the northern Netherlands. Regional studies are limited to H. A. H. Boelmans Kranenburg, "Zeescheepvaaart in Zuid-Holland 1400-1550," *Zuid-Hollandse Studiën*, XI (1965), 136-187, and F. C. Berkenvelder, "Frieslands Handel in de late Middeleeuwen," *Economisch-Historisch Jaarboek*, XXIX (1963), 136-187. The most extensive treatment of a specific trade is that of the grain trade with the Baltic which, typically based on the toll registers of the Sound. The complete study is A. E. Christensen, *Dutch Trade to the Baltic about 1600 Studies in the Sound Toll Register and Dutch Shipping Records* (Copenhagen and The Hague, 1941). Though his book is largely devoted to the seventeenth century, there are many observations to be found about the earlier years of development in the trade. Related, shorter works are, among others, E. C. G. Brunner, "De Ontwikkeling van het Handelsverkeer van Holland met Oost-Europa tot het Einde der 16e Eeuw," *Tijdschrift voor Geschiedenis*, XLI (1926), 353-371, A. Friis, "The Two Crises in the Netherlands in 1557," *Scandinavian Economic History Review*, I (1953), 193-217, and W. S. Unger, "De Hollandsche Graanhandel en Graanhandelspolitiek in de Middeleeuwen," *De Economist* (1916), 243-269, 337-386, 461-507. More recently, and less tied to the Sound toll records is M. van Tielhof, *De Hollandse Graanhandel, 1470-1570 Koren op de Amsterdamse molen* (The Hague, 1995).

Works about Dutch trade with England are more directed to the political and legal context than to shipping. There are N. J. M. Kerling, *Commercial Relations of Holland and Zeeland with England from the late 13th Century to the Close of the Middle Ages* (Leiden, 1954), and M. P. Rooseboom, *The Scottish Staple in the Netherlands* (The Hague, 1910). Inland shipping, a not-to-be forgotten part of Dutch traffic in trade goods, has never been considered as glorious as the trade over longer distances. A few studies about the topic are T. L. M. Thurlings, *De Maashandel van Venlo en Roermond in de 16e Eeuw, 1473-1572* (Amsterdam, 1945) and H. van der Wee, "De handelsbetrekkingen tussen Antwerpen en de Noordelijke Nederlanden tijdens de 14e, 15e en 16e eeuw," *Bijdragen tot de Geschiedenis der Nederlanden*, XX (1965-1966), 267-285. On trade with France and shipping along the Atlantic coast of Europe, Jan Craeybeckx, *Un grand commerce d'importation: Les vins de France aux anciens Pays-Bas (XIIIe-XVIe siècle*, (Paris, 1958), W. S. Unger, "De Scheepvaart der Bretons naar de Walcherse Reede in de XVe en XVIe Eeuw," *Bijdragen voor Vaderlandsche Geschiedenis en Oudheidkunde*, 6th series, VIII, 225-244, W. Brulez, "De zoutinvoer in de Nederlanden in de 16e eeuw," *Tijdschrift voor Geschiedenis*, LXVIII (1955), 181-192, and T. H. Milo, "Portuguese Trade and Shipping with the Netherlands after the Discoveries," *Actas do Congresso Internacional de Historia dos Descobrimentos*, III (Lisbon, 1961), 1-12.

About the Hanse and the connections of north German ports with the Netherlands there are older works which are still satisfactory, for example, E. Daenell, *Die Blütezeit der Deutschen Hanse*, 2 volumes (Berlin, 1905), and E. Daenell, "Holland und die Hanse im 15. Jahrhundert," *Hansische Geschichtsblätter*, IX (1903), 3-41. There are also more recent publications such as, among others, P. Dollinger, *La Hanse (XII-XVIIe siècle)*, (Paris, 1964), and K. Spading, *Holland und die Hanse im 15. Jahrhundert. Zur Problematic des überganges vom Feudalismus zum Kapitalismus* (Weimar, 1973).

IX

Warships and Cargo Ships in Medieval Europe

It is a commonplace in the history of ship design that during the 16th century Europeans developed two quite different vessel types, one for fighting and one for carrying goods. Each was so specialized that it was not advisable, and in many cases not feasible, to use them interchangeably. Once shipbuilders made the distinction, the two types were subject to separate development. As one work suggests, "One type grew definitely into a mere carrier of merchandise, while the other became a fighting-machine. . . ."[1] The lightly armed Dutch *fluit* (see fig. 1), first built late in the 16th century, is the prime example cited of a cargo ship designed solely for moving goods and ignoring problems of defense. It has been said that "the introduction of the *fluyt* marked the first clear distinction between the functions of the warship and those of the merchant packet."[2]

The implication is that cargo ships and warships were much the same during the middle ages and that ship designers and builders before 1600 were incapable of developing specialized vessels; that they were unsophisticated, blind to the potential for improvement through specialization. In fact that impression is wrong. It ignores both the presence in medieval Europe of a distinct separation between ships for fighting and ships for carrying goods and the variety of designs which existed at all times within those two large categories.

In early medieval Europe there were two distinct types of vessel built

DR. UNGER is associate professor of history at the University of British Columbia. Preparation of this article was in part aided by a grant-in-aid from the American Council of Learned Societies.

[1]Hendrik Willem van Loon, *Ships and How They Sailed the Seven Seas, 5000 B.C.–A.D. 1935* (New York, 1935), p. 113. For other similar examples, see T. K. Derry and T. I. Willaims, *A Short History of Technology* (Oxford, 1960), pp. 205, 207–8; J. H. Parry, *The Age of Reconnaissance* (New York, 1963), pp. 82–84; G. P. B. Naish, "Ships and Shipbuilding," in *A History of Technology*, ed. Charles Singer et al. (London, 1957), 3:480–81; Garrett Mattingly, *The Armada* (Boston, 1959), pp. 194–96; and R. and R. C. Anderson, *The Sailing Ship* (London, 1926), p. 135.

[2]J. H. Parry, "Transport and Trade Routes," in *The Cambridge Economic History of Europe* (Cambridge, 1967), 4:213.

Reprinted from *Technology and Culture* vol. 22, no. 2, The University of Chicago Press (April 1981).

FIG. 1.—A convoy of Dutch *fluiten* protected by a warship and a yacht (center). The broad stern and extensive armament of the warship contrasts sharply with the design of the cargo ships. From a painting by an unknown 17th-century Dutch artist now in the Museum of Fine Arts, Budapest (Reproduced from Dénes Pataky and Imre Marjai, *Ships in Art* [Budapest, 1973], no. 29.)

specifically for two quite distinct jobs. The galley, inherited from the Romans, was the warship of the Mediterranean, while the sailing ship was intended for carrying goods. Galleys could be used for moving cargo, but improvements in sailing ships further relegated them exclusively to fighting. This process of differentiation increased through the 13th century. The introduction of the great galley, for example, made light galleys into warships almost completely removed from any commercial function. In northern Europe the same kind of distinction existed between sailing ships for carrying goods and rowed vessels for fighting. Even the addition of sails to rowing barges and other design changes by Scandinavian shipwrights which made that type into a serviceable sailing ship did not eradicate the distinction between the two basic kinds of ship.

What did destroy that distinction were developments in the design of the cog in the 11th and 12th centuries. Here two types were merged into one. More precisely, the cargo ship proved to be the dominant vessel for fighting. Thus, the separation of warships from cargo ships was not a new development but was restored in the course of the 16th century, first by improvements in sailing ships and then by the addition of guns able to sink other ships.

The differentiation of the cargo ship from the warship was not a simple matter of technical change, of builders just deciding to go ahead and build two types of ships. Even if shipwrights could design and build the two forms of vessel, that by no means insured that they would be used. Throughout the middle ages technical improvements made by shipbuilders, though often incorporated into ships, could not be exploited because of external circumstances.[3] With the differentiation of cargo ship and warship that was especially so. The lightly armed cargo ship could only be used where violence was at a low level or at least where it could be contained by the use of convoys. In general, in the 16th century the channeling and organization of violence by European states made it possible to go ahead with designing and building specialized ship types. European monarchs with greater power over their subjects, approaching or attaining what would later be called absolute monarchy, could more effectively decide when and where violence would occur at sea. These monarchs could not control all of their subjects all of the time, but they still made every effort to contain and direct their violent acts. Attacks, if nothing else, became more predictable. Moreover, the monarchs were interested in increasing their cash income, and a major source of cash was taxes on peaceful trade. They were reluctant to allow indiscriminate warfare on any shippers and took a greater interest than ever before in the promotion of trade. This meant that there was an excellent chance that job-specific cargo ships would be used and that shipwrights were in a better position to build ships.

The European pattern was repeated in the 17th and 18th centuries in American trades. It was then that governments began to stabilize and to direct violence in the Caribbean and along the Atlantic coast of North America. It was precisely that change which made possible the greater use of cargo ships designed exclusively for their job.[4] The lack of order at sea, fighting, and piracy did not act as a constraint on the use of the cog and the carrack, the sailing ships of the 13th through the 15th centuries. Rather, their adoption and widespread use depended in part on the precise fact that they were so defensible. They were so capable as warships that the presence or absence of government control over violence was of only marginal importance.

The insistence on a major breakthrough in specialization of ships around 1600 is wrong, but it is excusable on two counts. First, the claim is valid for northwestern Europe and for the powers that dominated world oceans in the 17th and 18th centuries. The emphasis on the development of the *fluit* and of pinnasses and frigates, warships built

[3]F. C. Lane, "Progrès technologiques et productivité dans les transports maritimes de la fin du Moyen Age au début des Temps modernes," *Revue historique* 510 (1974): 277–302.

[4]G. M. Walton, "Obstacles to Technical Diffusion in Ocean Shipping 1675–1775," *Explorations in Economic History* 8 (Winter 1970–71): 123–40.

solely for fighting, comes from a concentration by scholars on those states and the expansion of European commercial connections over long distances and outside of Europe. Second, the claim is valid over the shorter term of the 14th, 15th, and 16th centuries. Compared with the design of ships during that period, the separation of warship from cargo ship around 1600 was a unique event. It was not unique, however, relative to the level of specialization in design in the much longer period of the entire middle ages. A summary view of just some of the ships built and used in medieval Europe and of some of the adaptations made to improve those vessels clearly demonstrates the ability of shipwrights to exploit the advantages of dividing warships from cargo ships.

The Mediterranean

In the Mediterranean in the early middle ages there was already a long-standing differentiation between ships for fighting and ships intended only for carrying cargo. Southern European vessels are usually divided into long ships, that is with length-to-beam ratios in excess of 6:1, and round ships with a ratio under 4:1 and often well under that figure. In the Roman Empire the only warships were light galleys, called Liburnians, with a high length-to-beam ratio. There were other long ships, galleys, which carried goods. Merchant galleys were of course limited to handling lightweight and valuable items since the cost of carriage was much greater than with round ships. Romans were able to build their navies exclusively of light galleys because they had cleared the Mediterranean of any potential opponents.[5] By the 1st century A.D., Roman builders produced a cargo ship, the large round ship, which was not used in fighting. They produced a warship so well suited to fighting that, in the lighter version, it was unlikely ever to be used for carrying goods. The cargo ship was not highly efficient despite its tremendous size, and it was not easy to handle. Romans used the big ships only on protected and controlled trade routes. The potential created by the distinction was not pursued because the Roman Empire, the political umbrella which made possible specialized designs, declined in strength. This signaled an end to the indirect subsidies to large sailing ships and to the peace which made the small warship effective.

With the decline and the disappearance of the western portion of the Roman Empire, the large sailing ship disappeared. Instead, small coasters typically carried cargoes. Because of challenges to Roman naval superiority in the Mediterranean, the Eastern Roman or

[5]Lionel Casson, *The Ancient Mariners* (London, 1959), pp. 213–36.

Byzantine Empire increased the force of its galleys, among other ways, by making them bigger. The Byzantine navy required and developed a complex organization. The threat of Arab fleets after the middle of the 7th century and then the need to reconquer the strategically important island of Crete in the 10th century led to the development of a large galley which could carry 300 men and more. Arab navies used the same types of ships as did Byzantine naval forces.[6] War fleets were then made up of galleys. These large oared ships which had the generic name "dromon" were also capable of competing as cargo vessels. They were, if anything, more economical to operate than Roman long ships. The smaller size of cargo vessels, though they were easier to handle than Roman round ships, did not offer the same level of competition to oared vessels.[7] Thus the distinction between warship and cargo ship was eroded through the 10th century in the Mediterranean.

The tendency was reversed in the 11th and 12th centuries by a revival in the building of round ships. Shipwrights made them much larger and more reliable than the Byzantine coasters. These cargo ships carried two masts, each with a triangular lateen sail. Using lateen sails made it possible to sail closer to the wind than with square sails. These vessels had the double advantage of being able to carry bulky goods in their holds and lighter goods above the first, but below the main deck. With heavy goods between the two decks the vessels were unstable. The presence of these sailing ships, their increasing numbers, and their capacity for handling lighter goods at a low extra cost made them a direct threat to galleys in the carriage of just that type of cargo. Indeed, the round ships had, by the 13th century come to dominate Mediterranean cargo carriage.[8] At the same time, war fleets were composed almost exclusively of galleys. There were also vessels combining the features of both types, that is, round ships with a small number of oars to give them greater maneuverability. Overwhelmingly, however, their

[6]H. Antoniadis-Bibicou, *Etudes d'histoire maritime de Byzance* (Paris, 1966), pp. 53–117; Ekkehard Eickoff, *Seekrieg und Seepolitik zwischen Islam und Abendland das Mittelmeer unter byzantishcer und arabisher Hegemonie (650–1040)* (Berlin, 1966), pp. 135–50; L. Bréhier, "La Marine de Byzance du VIIIe au XIe siècle," *Byzantion* 19 (1949): 1–16; and Lionel Casson, *Ships and Seamanship in the Ancient World* (Princeton, N.J., 1971), pp. 146–68.

[7]R. H. Dolley, "The Warships of the Later Roman Empire," *Journal of Roman Studies* 38 (1948): 47–53; A. M. Fahmy, *Muslim Sea-Power in the Eastern Mediterranean from the Seventh to the Tenth Century, A.D.* (London, 1950), pp. 116–37; and, Frederick van Doornick, "Byzantium, Mistress of the Sea: 330–641," in *A History of Seafaring Based on Underwater Archaeology*, ed. George Bass (London, 1972), pp. 133–46.

[8]Eugene H. Byrne, *Genoese Shipping in the Twelfth and Thirteenth Centuries* (Cambridge, Mass., 1930), pp. 3–11; and Hilmar C. Krueger, "The Routine of Commerce between Genoa and North-West Africa during the Late Twelfth Century," *Mariner's Mirror* 19 (1933): 430–34.

main source of propulsion was their sails. Therefore the separation of oared vessels from sailing ships was not complete.[9] After all, galleys often sailed most of the way to their destinations. But the distinction between warships and cargo ships by the 13th century was as great as it had been during the 1st and 2d centuries A.D. The crews had to be different, and not just in number. Sailors on a round ship had to have skills which were not required of the rowers, the great majority of men on a galley.

The tendency toward separating fighting ships from cargo carriers, based on Roman practice, was partially checked in the last years of the 13th century. Shipwrights at the government shipbuilding yard in Venice, the Arsenal, developed the great galley. Their goal was to get a more powerful warship by building a bigger galley, giving it more oars, and lowering the length-to-beam ratio. It was to some extent a compromise between the warship and cargo ship, but the great galley turned out to be more effective as a cargo carrier than as a warship. First in carrying pilgrims from Venice to the Holy Land, and then in trade to northern Europe through the Strait of Gibraltar, great galleys in the 14th and 15th centuries were able to partially supplant round ships, which relied exclusively on sails for propulsion. Great galleys were faster than sailing ships and in some cases were effective in moving bulk goods. On the other hand, since they had a lower ratio of crewmen to tonnage shipped than did light oared ships, they offered stiff competition to small commercial galleys. The great galley was in a sense an aberration which, though a significant development in itself, did not change the well-established long-term tendency in the Mediterranean toward the separation of warships from cargo carriers. Great galleys themselves were used for both tasks. By driving smaller rowed ships out of trade, however, they made those vessels into warships and only warships. The arrangement for rowing changed on those lighter galleys by the 13th century. To get more power, a third oarsman was added to each bench, making the warships stronger and faster.[10] It also rendered them almost useless as cargo vessels. They were too fragile, had too small a carrying capacity, and had to make even more frequent and uneconomic stops because of the larger size of the crew.

War galleys changed again with the introduction of artillery.

[9]B. Hagedorn, *Die Entwicklung der wichtigsten Schiffstypen bis ins 19. Jahrhundert* (Berlin, 1914), pp. 37–39; and Anthony Bryer, "Shipping in the Empire of Trebizond," *Mariner's Mirror* 52 (1966): 5–8.

[10]F. C. Lane, *Navires et constructeurs à Venise pendant la Renaissance* (Paris, 1965), pp. 2–24; R. C. Anderson, "Italian Naval Architecture about 1445," *Mariner's Mirror* 2 (1925): 144–45; and Björn Landström, *Sailing Ships* (Garden City, N.Y., 1969), pp. 112–15.

Though guns had already been used in naval battles by 1300, they were not in common use until well into the 15th century. By the early 16th century the result was a galley suited to amphibious operations, to cover landings, or to besiege fortresses along the coast. There was one heavy gun set on the center line of the ship to minimize instability, already a problem in these lightly built vessels. The heavy gun and perhaps one or two lighter guns flanking it were the main offensive weapons of the galley. At that point there was no question at all of using the galley for commercial service. Europeans, at least in the Mediterranean, had developed a distinct type of fighting vessel, this war galley, well before the *fluit* was first built. By the end of the 16th century, the event had lost most of its significance, however, because the galley was a warship of only limited effectiveness (see fig. 2). Except in a dead calm, most sailing ships with guns could destroy opposing galleys. Even in a calm where the oars of the galley gave it an advantage the sailing ship could still fend off the lighter and less heavily armed galley. It mattered little that the warship, the galley, was distinct from the cargo vessel, the sailing round ship. Galleys had been demoted, because of their limited potential as warships, to acting as scouting vessels, as dispatch boats,

FIG. 2.—Destruction of Spanish Mediterranean design galleys off the Flemish coast by Dutch sailing warships, October 3, 1602. From a painting by Hendrick Cornelisz. Vroom now in the Rijksmuseum, Amsterdam. (Reproduced from Dénes Pataky and Imre Marjai, *Ships in Art* [Budapest, 1973], no. 19.)

and as ships to neutralize other galleys in battle. The great galley also became a warship as it was driven out of cargo service in the 16th century. Armed with guns and called a galleass, it appeared in many naval battles. All naval powers in the Mediterranean kept a fleet of galleys if for no other reason than that other states had galleys.[11]

The history of galleys as warships is typically neglected in the discussion of the evolution of ships. Their use was limited to the Mediterranean where weather conditions made it possible to use a long ship with low freeboard and powered by oars. Only in a few other places were conditions similar; for example, in the Caribbean and the Baltic and along the shores of the North Sea and the English Channel.[12] More important in northern Europe was the design of sailing ships.

Northern Europe

After the departure of the Romans from northern Europe, and through the early middle ages, builders there produced a number of different vessel types which were specifically designed for a certain job. One type of sailing ship, probably developed during or before the period of Roman rule by Celtic shipwrights, was obviously meant to carry heavy cargoes and to cross the open sea. The cog was probably descended from this heavy Celtic sailing ship. With its flat bottom and sail, it was well suited to carrying cargo through the shoals of the southern shore of the North Sea and was ideal for landing on the sandbanks of the islands and shores there. Another type, the hulk, was also a sailing ship which may have had a Celtic origin. It was used both on the rivers of the Low Countries and across the North Sea. By the 8th century, then, there were two ship types for carrying goods, neither of which had design features to recommend it as a warship.[13] On the

[11]Lane, *Navires et constructeurs à Venise*, pp. 25–30; and J. F. Guilmartin, "The Early Provision of Artillery Armament on Mediterranean War Galleys," *Mariner's Mirror* 59 (1973): 257–80. On the technical and economic reasons for the importance of galleys in 16th-century Mediterranean warfare, see J. F. Guilmartin, Jr., *Gunpowder and Galleys* (Cambridge, 1974), esp. pp. 264–68.

[12]On a last effort to use galleys in northwestern Europe, see Randal Gray, "Spinola's Galleys in the Narrow Seas, 1599–1603," *Mariner's Mirror* 64 (1978): 71–83. Also, see R. and R. C. Anderson (n. 1 above), pp. 172–78; J. W. Sherborne, "English Barges and Balingers of the Late Fourteenth Century," *Mariner's Mirror* 63 (1977): 109–12; and J. E. G. Bennell, "English Oared Vessels of the Sixteenth Century," ibid., 60 (1974): 9–26

[13]Detlev Ellmers, "Keltischer Schiffbau," *Jahrbuch Römisch-Germanisch Zentralmuseums Mainz* 16 (1969): 73–82, and *Frühmittelalterliche Handelsschiffahrt in Mittel- und Nordeuropa* (Neumünster, 1972), pp. 59–64; Peter Marsden, "Ships of the Roman Period and After in Britain," in *A History of Seafaring Based on Underwater Archaeology*, ed. George Bass (London, 1972), pp. 119–23, and "A Boat of the Roman Period Found at Bruges, Belgium, in 1899, and Related Types," *International Journal of Nautical Archaeology* 5 (1976): 23–55.

other hand, there were rowing barges with no sails which were clearly built to move men. The crew was the cargo. The men pulled the oars, and there was little space for anything other than the crew and a few personal effects. These vessels had high length-to-beam ratios like Mediterranean galleys. Other than carrying migrants, these personnel-carrying vessels were essentially used as warships to get men to the scene of the fighting.[14] The two types with their very different forms of propulsion, manning ratios, and hull forms continued for centuries to exist side by side in northern Europe. Over time each was subjected to improvements in its design.

The Scandinavian ship, developed in the 8th and 9th centuries and used by the Vikings on their voyages of trading and raiding throughout Europe and beyond, is all too often described as a single type. The failure to appreciate differentiation in the design of Viking ships comes from a preoccupation with those vessels excavated in the late 19th and early 20th centuries which were warships. It is true that the Viking ship did develop out of the rowing barge, but even by the 9th century there was already a clear distinction between warships which would in some instances also be used for carrying goods and cargo ships which in some cases would be pressed into service as warships. Construction was similar for both types, as was the method of strengthening the hull. They both carried a single square sail and single balanced side rudder. They both had keels with ribs placed low to give support without sacrificing flexibility, which also made the bottom closer to being flat so the vessels could be easily beached. The differences were still significant, however.

The length-to-beam ratio of warships was from 4.5:1 to 7.2:1; that of cargo ships was from 2.3:1 to 4.9:1. The cargo ship had higher freeboard and a sharp angle between the bottom and the sides. On the warship there was a smoother curve. While the ribs of the cargo ships were fixed to the hull with treenails, on warships they were only lashed to protruding bits of wood left on the interior of the hull planks for that specific purpose. The difference made the warship more flexible and the cargo ship more durable. Only after 1000 did warships change to having permanently fixed ribs. Warships often relied on oars located all along both sides of the ship for propulsion. Cargo ships had few oars, six or eight, and they were placed at the ends of the ship to leave a large open space amidships for cargo. On

[14]H. Åkerlund, *Nydamskeppen en studie i tidig Skandinavisk skeppsbyggnadsknost* (Gothenburg, 1963); and R. L. S. Bruce-Mitford, *The Sutton Hoo Ship Burial* (London, 1968), pp. 40–43. For a general survey of methods of hull construction in early medieval ships, including Viking ships, see Basil Greenhill, *Archaeology of the Boat* (Middletown, Conn., 1976), pp. 174–232.

warships there was a full deck to give rowers a platform for pulling the oars. On cargo ships there were short half decks at either end so that the hold was open in the middle. The high freeboard of cargo ships meant that the oars entered the water at an inefficient angle, but it made little difference since the oars were rarely used, probably only to get in and out of harbor when there was a contrary breeze or to avoid trouble.[15]

By the 11th century differences between the two types of Scandinavian ships became, if anything, more extreme. The relatively long warship was the dominant naval vessel in northern European waters. It was used from the English Channel, in the invasion of England by William the Conqueror, to the Bosporos, in the attacks by Russians on Constantinople. Scandinavian monarchs extended the design of the personnel-carrying type, making it even longer than its predecessors. These dragon-ships, rowing barges with an auxiliary sail, reached lengths of 35 meters and more. Ships that length did lose much of the maneuverability important in naval battles, so typically the long ships were of slightly more than 25 meters. The major purpose of those warships remained the deployment of as many men as possible at the scene of a battle, be it on land or on the water.

The cargo ship also increased in size, and a capacity of 50 tons and lengths of about 25 meters were not uncommon. Such vessels made regular voyages out into the Atlantic to Iceland and Greenland, voyages which warships were not strong enough to make. The Scandinavian cargo ship was the forerunner of the keel, a variation on the basic design built in England and used all along the Atlantic coast of Europe. Keels served to carry cargoes relatively short distances but were also capable of sailing in the open sea. The names of Scandinavian ships reflected the distinction between cargo ships and warships. Both *buza* and *knarr* applied to goods carriers. *Langskip* was the generic term for fighting ships. There were other names used for both types, but in all cases they followed the difference in the design of the two basic types, just as do records of customs collectors. The design of the buss may well have been subject to even further extension, but there is a shortage of precise information on Scandinavian ships from the 12th and 13th centuries. At least it is clear that the buss served as an efficient carrier of heavy and bulk goods across the North

[15]Detlev Ellmers, *Frümittelalterliche Handelsschiffahrt*, pp. 30–45, 118–19. A similar differentiation existed among contemporary Slavic boats from the south shore of the Baltic. Also, see Ole Crumlin-Pedersen, "Kælhngen of kløften," *Handels og søfartsmuseets årbog* (1972): 63–80; A. W. Brøgger and H. Shetelig, *The Viking Ships, Their Ancestry and Evolution* (Oslo, 1971), pp. 79–111; and Olaf Olson and Ole Crumlin-Pedersen, "The Skuldelev Ships (II)," *Acta Archaeologica* 38 (1967): 73–174.

Sea well into the 13th century.[16] The character of the cargoes makes it clear that the busses were not for use as warships. Combined with the reports from sagas on the dragonships, this demonstrates conclusively that northern Europeans in the 11th and 12th centuries did not confuse warships with cargo ships.

The Deep-Sea Cog

The drive toward separation of warships from cargo ships was totally reversed by the development of the deep-sea cog. This type, which began as a coaster, was changed dramatically sometime in the 11th century by the addition of a keel. While shipbuilders did make a major change in the design of the cog, it may well be that they were imitating the use of keels on the highly effective Scandinavian vessels. Whether it was an act of invention or adaptation, the result—the cog with a keel—could survive in the open sea. It retained its single mast with one square sail and was apparently a decent sailing ship. It retained the straight posts and high freeboard of the earlier cog as well as the flat section of the bottom amidships (see fig. 3). The latter gave it an advantage over the keel or buss in having greater carrying capacity for each meter of length. Although it was always tubby with a length-to-beam ratio usually under 3:1, this greater breadth, rather than any increase in length, gave the cog better carrying capacity. The greatest advantage of the cog was that in its new form it could be built bigger and without any great loss of handling qualities. By the late 12th century, shipwrights were building cogs able to move over 300 tons. The average was below that of course, but the maximum and the average had at least quintupled over the preceding 200 years. Beam for cogs was rarely over 9.5 meters, and lengths of 25–30 meters were common, as were tonnages of around 200.

There was no question of oars. The ship was powered entirely by a large square sail, rising to 175 square meters. One or two extra pieces of canvas called bonnets could be temporarily added in good weather along the bottom once the sail had been raised to give a total sail area of as much as 335 square meters.[17] That meant that the cog could be built even bigger. With such high freeboard and rising high in the

[16] A. W. Brøgger and H. Shetelig, pp. 146–62; G. J. Marcus, "The Evolution of the Knörr," *Mariner's Mirror* 41 (1955): 115–22; Ellmers, *Frühmittelalterliche Handelsschiffahrt,* pp. 46–58; N. Bjørgo, "Skipstyper i Norrøne samtidssoger, "*Sjøfartshishtorish årbok* (1965), pp. 7–20; and K. Helle, "Trade and Shipping between Norway and England in the Reign of Hakon Hakonsson (1217–63)," *Sjøfartshistorisk årbok* (1967), pp. 7–34

[17] Paul Heinsius, *Das Schiff der Hansischen Frühzeit* (Weimar, 1956), pp. 69–93, 139; Ole Crumlin-Pedersen, "Cog-kogge-kaag, træk af en frisisk skibtypes historie," *Handels- og søfartsmuseets årbog* (1965), pp. 81–144; and Ellmers, *Frühmittelalterliche Handelsschiffahrt,* pp. 63–75, 258–59.

FIG. 3.—A cog from the second seal of Elblag, Poland, from around 1350. (Reproduced from Herbert Ewe, *Schiffe Auf Siegeln* [Bielefeld, 1972], p. 122, no. 42, drawing by Gerda Nützmann.)

water, the cog needed an extremely long side rudder just to reach the water when heeled over. The answer to this problem was to hang the rudder on the straight sternpost of the cog. The rudder did not have to extend below the keel to stay in the water at all times. The adoption of the sternpost rudder by the mid-13th century made cogs easier to handle. This modified and improved cog was easily as efficient as the contemporary buss in carrying goods. It eventually became the dominant cargo type in northern Europe, especially in the long-distance carriage of bulk goods in the North and Baltic Seas. A number of other types remained in use on shorter hauls.

Incidentally, German shipwrights also built in the cog a warship superior to any contemporary type in northern Europe. Because the cog rode so high in the water, its deck was above that of any other vessel. Fighting at sea consisted of hurling missiles at the enemy followed by boarding. Being above the enemy was almost a guarantee of victory. If nothing else, the cog was virtually impregnable to attack except by another cog. To try to overcome the advantage of cogs, Scandinavian longships were fitted with castles, small fighting platforms at the bow and stern which took them up to the height of the cog. Builders simply added castles to cogs to eradicate the gain. Scandinavian monarchs had used a muster system, a naval militia in each district, to raise their fleets of warships. The introduction of cogs forced the abandonment of that system in the course of the 12th and 13th centuries. Longships were no longer worth having, and the kings had to change to a method for supplying and equipping cogs.[18] The cog was not only higher than any other warship, it was also bigger. As cogs grew in size and as the castles became permanent parts built into the ships, those vessels could carry to battle many more soldiers than any potentially competing type. The heavy sail, the short length-to-beam ratio, and the lack of oars made the cog much harder to handle than a Scandinavian warship, although the sternpost rudder was some help. But even though the cog could not be maneuvered easily, its greater size and height made that disadvantage unimportant.

From the 12th, and certainly from the 13th, century, for the first time in northern Europe the cargo ship and the warship were merged into one type. As one scholar has said, "There was in the thirteenth century, and for many hundred years afterwards, no essential difference between the trading ship and the warship."[19] The design of the cog proved to be the best for the long haul of bulk goods and best for warships. The design was not, as it turned out, job specific. Conversion from one task to the other took a matter of minutes. In some cases additional reinforcing might be added to the castles or more cabins created for the soldiers and officers on board to make the cog a better warship. The changes were not necessary, however. Once the standard cargo ship took on troops it became a warship, and the change back to commercial use was just as simple. The crew could and did remain the same since no special skills were needed. Navies turned into fleets of cogs, whose major function was to move manpower in large groups to battles. Governments did not have to maintain war fleets since they could simply rent cogs from merchants and shippers in wartime. The monarchs of northern Europe in some cases did own cogs to be used as warships. They paid for those ships in part by leasing them to commercial shippers, that is, by using their naval vessels as trading ones during peacetime. Battles at sea became land

[18]G. O. Harbitz, "Leidangan og det gamle Norske sainfunn," in G. P. Harbitz, S. Oppegård, and R. Schech, *Den Norske leidangen* (Oslo, 1951), pp. 11–160; and Ole Crumlin-Pedersen, "The Vikings and the Hanseatic Merchants: 900–1450," in *A History of Seafaring Based on Underwater Archaeology*, ed. George Bass (London, 1972), pp. 190–92.

[19]F. W. Brooks *The English Naval Forces, 1199–1272* (Manchester, 1932), p. 27.

battles fought across the decks of the large cogs grappled together. In the 100 Years War between England and France in the 14th century, for example, the navies of both sides were made up largely of cogs, often impressed from merchants.[20]

In the Mediterranean the eradication of the distinction between warships and cargo ships did not occur. The cog had been introduced to southern Europeans, and by the early 14th century shipwrights in the south were building cogs, improving them, and accommodating them to local conditions. In the Mediterranean, however, the cog remained largely a ship for transport, especially of goods. The light breezes of the Mediterranean and the absence of the heavy seas of the Atlantic and northern Europe took away some of the advantages of the cog as a warship. Galley crews were typically larger than those of cogs, and galleys with their oars could always outmaneuver one of the large, clumsy cogs. In light airs the cog could not move, while the galley could sail or row away. That is not to say that cogs were not used in the south as warships. Mediterranean navies began to add them to their fleets soon after they were introduced there, but the numbers were always small, and they were transports rather than fighting ships. The only cases where cogs might have been used for fighting as they were in northern Europe was, for example, in wars between Barcelona and Genoa. In the 14th century, for commercial reasons, both ports had sizable fleets of cogs.[21] In the Mediterranean the cog could not and did not dominate the war galley. The latter remained the typical warship and certainly distinct from cargo carriers.

On the other hand, rowing barges as warships did not completely disappear from northern Europe. Though their use was limited to coastal patrol work and fighting other galleys, monarchs in the north still felt they needed such ships to have a complete naval force. Whether modeled on contemporary Mediterranean galleys or on Viking longships, their number remained small. They could form something of a strike force or be used to defend inland rivers where their maneuverability was a great advantage over large sailing ships in narrow passages. In general, the experiments with oared galley fleets

[20]For example, H. J. Hewitt, *The Black Prince's Expedition of 1355–1357* (Manchester, 1958), pp. 33–39, and *The Organization of War under Edward III, 1338–62* (Manchester, 1966), pp. 77–92.

[21]Lane, *Navires et constructeurs à Venise*, pp. 34–37. Cogs had the added advantage of offering more protection behind their gunwales for men firing a new weapon, the crossbow. Also, see F. C. Lane, "The Crossbow in the Nautical Revolution of the Middle Ages," in *Economy, Society and Government in Medieval Italy*, ed. D. Herlihy, R. S. Lopez, and Vsevolod Slessarev (Kent, 1969), pp. 161–65; and Hagedorn (n. 9 above), pp. 39–42.

failed.[22] Naval superiority in the north depended not on rowed ships but on the large cogs, and the sailing ship was the dominant vessel for all uses in northern Europe by 1400. It was precisely the differentiation of that sailing ship into two types which was the highly regarded technical change of the late 16th century.

The Technology of Differentiation

The first step toward building two distinct types came in improvements in the sailing ship itself. Those improvements formed the greatest single technical advance ever made for wind-powered ships. In the early 15th century builders changed rig to a composite one with both square and lateen sails. Typically there were three masts instead of the previous one or two. The fore- and mainmasts carried one or two square sails each, while the mizzen carried a triangular lateen sail. A relatively small square sail was often slung under the bowsprit to act as a headsail for better control. The rigging as well as construction of the hull required further modification to exploit this invention fully. Shipbuilders made those changes over the following 200 years.[23] The new full-rigged ship was easier to handle than its predecessors. The combination rig gave masters the ability to exploit the divisibility of square sails and to get more power from having a larger sail area. It also gave them the capacity to sail closer to the wind, a feature of lateen sails. Hulls of full-rigged ships were built in the same way as those of Mediterranean ships. Planks were fitted edge to edge, the strength coming from an internal structure of ribs and frames. Repairs were easier and the ship was lighter for each meter of length. Construction was simpler than in the north, where overlapping planking and extensive internal reinforcement was still common.

Overall, the full-rigged ship was more efficient if for no other reason than that it was more reliable, less likely to get into trouble, and better able to get out of trouble. With the lateen sail it was not confined in harbor waiting for just the right winds. Turnaround time was probably decreased. All that was a gain for the cargo ship, but the greater maneuverability had a destabilizing effect on naval use of the sailing ship. Captains were in a better position to run away, to avoid a

[22]M. Oppenheim, *A History of the Administration of the Royal Navy* . . . (London, 1896), pp. 3–8; Antonio de Capmany y de Montpalau, *Memorias historicas sobre la marina y artes de la antiqua ciudad de Barcelona* (Madrid, 1779–92), 3:81–90; Charles de la Roncière, *Histoire de la marine Française* (Paris, 1909–20), 1:399–407; and J. T. Tinniswood, "English Galleys, 1272–1377," *Mariner's Mirror* 35 (1949): 276–315.

[23]R. Morton Nance, "The Ship of the Renaissance," *Mariner's Mirror* 41 (1955): 180–92, 281–98, gives the best summary of the development and modification of the full-rigged ship.

fight. Boarding was more difficult. Naval battles between sailing ships were still fought in the traditional style; however, there was no longer the same guarantee of a battle between opposing naval forces. The sailing skill of the ships' masters became a consideration in naval warfare. Still the full-rigged sailing ship, like its predecessor, the cog, was better than any potentially competitive design as a warship. In fact, the new type even began to erode the advantage enjoyed by the galley as a warship because it was both highly defensible and maneuverable. The multiplication of sails made it possible to build bigger ships, in fact the biggest wooden vessels ever constructed. The largest of full-rigged ships were called carracks. The Portuguese used carracks of up to 2,000 tons to carry goods to India. They were such giants that they were almost impervious to attack. Though they may have been hard to handle, Portuguese carracks were strong enough to dominate any group of smaller vessels.[24] Used in the Mediterranean, the large carracks, though they could not perform all of the functions of galleys, still were of sufficient quality to decrease the value and importance of rowed vessels.

The second and more important step in generating two different types of sailing ships came with the introduction of heavy guns on board. That this was unquestionably the major factor in creating the distinction has long been widely recognized. Heavy guns changed naval warfare from hand-to-hand combat to fighting at a distance, but the change was by no means a rapid one. The guns used on sailing ships in the 14th and 15th centuries were light antipersonnel weapons, no more than missile launchers used like slings or bows from the upperworks of a ship and of no value in a stormy sea. The essential character of fighting remained the same. Large siege guns, on the other hand, had the potential to sink an enemy vessel. The exact date of the introduction of those big pieces is uncertain, but it must have occurred sometime around 1500. The big guns were put in the waist of the ship since the higher decks could not support them. Ports which opened and closed like those already in use for cargo ships made it possible to put the heavy guns down on the lower decks with the ports piercing the hull of the ship. Protected there, the gunpowder and matches were not exposed to the elements and firing was more reliable. In 1513 for the first time in Europe a ship was sunk by the use of gunfire. It was also in the early 16th century that the first battles fought and won entirely with guns took place in the Indian Ocean between Arabs and Portuguese. The firepower of guns gave Portuguese fleets an indomitable advantage.

[24]C. R. Boxer, *The Portuguese Seaborne Empire* (London, 1969), pp. 207–9; and Landström (n. 10 above), pp. 90–95.

It took some time for both shipwrights and ship captains to learn what to do with these heavy guns, partially because the guns themselves were subject to improvement over time.[25] The bronze guns of the 16th century proved surprisingly accurate, while the iron guns used in the 17th century were less so. Navies used the less accurate iron guns because they were cheaper and many more could be deployed. The broadside became the major offensive weapon for the warship, and all navies had to carry heavy guns.[26] Since the pieces, especially the bronze ones, were so heavy that decks had to be reinforced, the design of the ships changed to accommodate the guns. For guns to be carried on the main deck, the width of the ship was made markedly less at deck level than at the waterline. This tumble-home of the sides gave more support to the deck. Unintentionally, it also made boarding more difficult since, even grappled together, two warships with extensive tumble-home would still have their main decks and upperworks far apart. There was no longer the possibility of fighting a running land battle across the decks of two ships.

The warship of the late 16th century with modifications needed for artillery was not likely to be a good cargo ship. It certainly was not an efficient one. The large number of guns took up much of the space below decks, and those guns also needed many men to handle them. Changes made in the hull were matched by changes in the rig to increase maneuverability. These design modifications combined to give such vessels a high ratio of manpower to potential carrying capacity. There was a simultaneous improvement in the design of cargo ships, however, and shipwrights found ways to lower the construction and operation costs of those vessels. The change was most noticeable in the Netherlands, but the process went on in other parts of Europe as well. The rising volume of trade in the 16th century made for an increase in the number of ships and allowed for potentially greater returns from their operation. These profits gave builders more opportunities to experiment with improvements. The greater returns to shipping and the more efficient design of cargo ships made it easier for shippers to sustain charges for fitting out and maintaining convoy ships. Those armed vessels offered the necessary protection which was no longer incorporated into the design of cargo ships. In fact, the

[25]R. and R. C. Anderson (n. 1 above), pp. 127–30; and Carlo M. Cipolla, *European Culture and Overseas Expansion* (Harmondsworth, 1970), pp. 65–70. The latter is in part a reprint of his *Guns and Sails in the Early Phases of European Expansion, 1400–1700* (New York, 1965).

[26]Guilmartin, Jr. (n. 11 above), pp. 268–73; and Peter Padfield, *Guns at Sea: A History of Naval Gunnery* (London, 1973), pp. 19–56. Effective iron artillery was developed in northern Europe at the end of the 16th century.

greater frequency of convoys further released shipwrights from the constraint of building defensibility into their products. At the same time, the market for warships created by the demand for vessels for convoy duty accelerated the process of designing ships purely for fighting.[27] The final result, so often noted, was the distinction of warships and cargo ships in the 17th century.

In the late 15th and in the 16th centuries a number of European governments began the practice of granting subsidies for the construction of more defensible ships. The subsidies were sometimes indirect, but they usually went directly to those buyers of ships who complied with government wishes. The subsidies were either in cash or a rebate on the customs duties on the first voyage, which was almost as good as cash. Governments wanted heavy ships which could carry guns and could be changed into warships quickly. Rather than lay down specific requirements which would be hard to administer, they usually just gave subsidies for the building of comparatively large ships. The interest in supporting privateering had something of the same goal: to keep a stock of functioning powerful ships for the navy with the government not having to pay to build or maintain them. The English government in the reign of Elizabeth I was especially effective in changing the character of the merchant marine through subsidies and the support of certain trades and privateering. The average size of ships increased, and the number of very large ships rose even more rapidly.[28]

The granting of subsidies demonstrated that the difference between warships and cargo ships was not yet complete, even in 1600. It was still possible for shipwrights to build vessels which incorporated sizable carrying capacity along with maneuverability and the ability to carry heavy armament. It was that technical potential which governments sought to exploit. The granting of subsidies also shows that there was already by the late 15th century a distinction between warships and cargo ships and, moreover, that it was great enough for governments to try to slow the process of differentiation. Builders were interested in pursuing the savings to be obtained from designing for a specific task. Their interest was undoubtedly whetted by the demands placed on them by buyers of ships. The shipbuilders of the

[27]On the development of convoys in the Low Countries, see A. Bijl Mz., *De Nederlandse convooidienst (1300–1800)* (The Hague, 1951), pp. 1–48; and R. Degryse, "De konvooieering van de Vlaamsche visschersvloot in de 15de en 16de eeuw," *Bijdragen voor de geschiedenis der Nederlanden* 2 (1948): 1–24.

[28]Lane, *Navire et constructeurs à Venise*, pp. 99–105; Oppenheim, pp. 167–75, 472; de la Roncière, 1:245; and Ralph Davis, *The Rise of the English Shipping Industry in the Seventeenth and Eighteenth Centuries* (London, 1962), pp. 5–8.

16th century were in exactly the position as shipbuilders working in the years before the development of the cog in northern Europe. Both groups of men found in differentiation potential savings in shipping costs. Governments in the 16th century stepped in to get defensible ships and to stop, or at least deter, the process which led to differentiation. Not incidentally, where governments and their admiralties, which often administered or at least promoted the granting of these subsidies, were weakest, the design of low-cost cargo carriers moved ahead most quickly. For political and historical reasons, the authorities in the Netherlands never had the kind of control over shipbuilders which existed in France, England, Iberia, or even in Scandinavia. It is not surprising that Dutch shipbuilders pursued the possibilities of designing cargo ships earlier and more completely than builders anywhere else in Europe in the 16th century.

Conclusion

The concentration by writers on the history of ship design on northern Europe, on the 16th and 17th centuries, and on the development of the *fluit* unfortunately tends to suggest that the distinction of warships from cargo ships was the result of some long-term development. The concentration tends to suggest that technologists up to that point were not capable, for reasons of incompetence or traditionalism, of building different types of ships for different jobs. In fact the *fluit* only marked a separation of warships from cargo ships for vessels which relied exclusively on sails for power. Its significance is only relevant for the development of ship design during the 15th and 16th centuries and specifically for ships in northern Europe. Certainly the development of the *fluit* by Dutch builders did prove in time to be an important source of increased productivity in European shipping, but that fact does not mean that shipwrights in the middle ages were incapable of making significant contributions to productivity. Medieval shipbuilders were far from being primitives doing just as their fathers had done before them. Quite to the contrary, they had separated the design of warships from that of cargo ships long before 1600 and had continued over the centuries to expand and explore the possibilities created by that distinction. In their drive to find the optimal design for a cargo vessel in the 11th, 12th, and 13th centuries, northern European shipwrights built a cargo vessel which, incidentally, was an excellent warship. They just overshot the mark. This phenomenon is quite common in the history of technical change. No one has ever been able to predict with anything approaching accuracy what the results of a specific improvement might be or how the new device or method will be used. Medieval shipbuilders were in an even worse position. They had no theory and so could not anticipate with any precision what the effects of even a small change in design on the handling qualities of the ship would be. Technical change has moved at varying paces, pursuing one avenue and then another, attempting to deal with the specific needs of the moment. Technologists often fall short of those goals and sometimes overreach them.

Past emphasis on the crucial changes in ship design around 1600 has the unfortunate side effect of obscuring how specific transport needs were handled in a wide variety of ways by ship designers throughout the middle ages. It also overemphasizes a single event which in itself was a combination of many adjustments and improvements. Techniques used in the middle ages, as in all other periods, depended on more than just contemporary knowledge. Use depended on the potential advantages to be gained by the user. It is now clear that the long-term presence and use of different technologies was a distinguishing feature of medieval Europe. The survival of many traditional methods may have been the result of the incremental nature of technical change with advances being small enough that no single type of hardware or idea was overwhelmingly and obviously better than all others. In shipbuilding, technical change in the middle ages was only rarely a history of major advances. Rather, it was typically a history of the rise to relative dominance of one technique and then another, of adaptation and adjustment of various designs, with no one ever fully disappearing. That dominance often depended not on the quality of the design so much as on economic and political circumstances external to shipbuilding. Through that process medieval shipwrights could and did, in many cases, generate the potential for improvements which were significant and long lasting.

X

Admiralties and Warships of Europe and the Mediterranean, 1000-1500

While a great deal has been and is still being learned about European warships of the Middle Ages from archaeology, iconography and government documents, the history of the admiralty as a government office in medieval Europe is still something of a mystery. No author has ever tried a general survey of the evolution of the institution. The word admiral is merely a corruption of the Arabic term 'emir'. When the Normans conquered Sicily in the late eleventh century they simply retained the office for the town of Palermo. Over time the *amiratus Palermi* became an important figure in the administration of government income and also expenditure, particularly for the navy. There were a number of admirals in twelfth century Sicily, but one became preeminent and usually he combined executive duties in fiscal administration with military command.

By the late twelfth century the admiral was a naval commander who had nothing to do with fiscal administration. That man was Margaritus of Brindisi, a Greek who led the Sicilian fleet to victory in Spain, Africa, Greece, and Syria while also defeating the Pisans and Genoese all in the short space of the ten years before 1194. Margaritus' fame led Italian maritime republics to imitation, Venice having an admiral as early as 1208, Pisa in 1210 and Genoa in 1226. The office spread north from Italy to France, in 1246, and to England at the end of the thirteenth century.[1] In those cases the admiral was a military commander, expected to lead the fleet in battle. The office was temporary and could disappear in peace time, at least in northern Europe in the thirteenth century. In Sicily the admiral retained administrative duties and by 1239 was also responsible for jurisdiction over many criminal attacks carried out at sea. In the fourteenth century jurisdiction was added to the tasks of admirals in France and England.

That gave the office permanence and, more important, income from the fines and fees paid in litigation. Kings favoured an increase in the competence of admirals because it meant in fact an expansion of their own claims of sovereignty, including the sea as well as traders and ports and coastal districts. The office of admiral was thus made into another royal tool to undermine the power of their mighty subjects.[2] At least in fourteenth century northern Europe it became one of the great offices of state and was held by men from the highest social stratum, men not necessarily recognized for their fighting ability.

By about 1500 most admiralties had three functions: naval command, jurisdiction over maritime affairs and provision of a fleet.[3] The last, the administrative function was attached to the office of admiral as part of his job as naval commander. The administration of military forces at sea, the size, extent and complexity of government offices responsible for that task, their functioning and funding, while usually treated as part of naval history, is rather normally taken as just the pre-history of the administrative structures which emerged in the sixteenth century. It is certainly true that from then on the administration of the navy was typically associated with the office of admiral and naval administration did generate highly complex organizations in the following two centuries. But construction and care of the fleet was never fully integrated with the other two functions of the admiral's office since these required different skills and different organizational forms. To some degree then even though the task of administering the fleet was placed in the hands of the naval commander, the admiral, that function still had a distinct history after 1500 just as it did in the Middle Ages.[4]

It was not just the states of early modern Europe which developed offices for administering their military forces at sea. The Byzantine Empire in the course of the seventh century established a series of 'themes' with both military and administrative positions. This system was extended to include the navy by the end of the ninth century, with commanders in charge of leading ships in battle, building and maintaining ships and collecting taxes for their support.[5] The growth in naval administration, the establishment of permanent government departments to deal with fighting ships in sixteenth century Europe can just be attributed to the introduction of artillery on board ship. The guns made the job of preparing a fleet more difficult hence the expansion of bureaucracy responsible for the task. A similar claim could be made for Byzantine organization since that fleet relied heavily on artillery for offensive action, and not just Greek fire used both at a distance and at close range, but also stone-throwing machines, relying on the principle of balanced weights and on tension.[6] Certainly the use of artillery is a partial explanation of the scale of naval administration. Other systems of naval administration existed—extensive, complex and continuing—with navies where artillery was not used. For example, Scandinavian monarchies in the tenth and eleventh centuries had a system of muster ships. Districts were responsible for supplying on short notice a specified number of ships and crews when called on by the king. These *leidangen,* the muster fleets, also had to be stored in yards and kept ready for sea.[7] Italian maritime republics, Iberian and French monarchs in the thirteenth and fourteenth centuries built and expanded government-owned shipbuilding facilities which were the exterior representation of a fixed permanent organization for support of the navy. For those vessels as with the earlier Norse naval units, though to a lesser degree, artillery was of minor importance both for offense and defense. Artillery on board then was only one feature of changes in ship design which explain the evolution of naval administrations.

In the second half of the Middle Ages, from 1000 to 1500 permanent departments of government to deal with naval administration appeared and disappeared. There was a decline in the establishments of about 1000 and then a gradual re-emergence of offices for supplying and supporting fighting ships, but there were also variations and fluctuations within that long term trend. The changes can in large part be explained by technological developments in ship design.

Simply the contention here is that the existence of a permanent naval administration within governments was a function of the extent to which navies were made up of ships which could be used as cargo vessels in peace time. The composition of naval forces depended on what ship type available at any point in time was most effective for doing violence at sea. That in turn dictated what governments did to their navies. There are a number of other reasons for the level of naval administration, that is, for the size and scope of the 'admiralty.' The naval ambitions of the government, the perceived and the real threat to the kingdom or state from hostile naval forces, the degree to which fiscal or police powers were combined with naval administration, the general level of administrative competence of the government of the day, the ability of the government to afford a navy, and simply the stability of government all contributed to decisions on the number of men and quantity of resources which could be devoted to administering military forces at sea. But all those considerations were political and above all they were short term, often very short term. The list contains explanations which are known and often mentioned by naval and administrative historians.[8] Over the long term it was rather the advances in ship design which was the principal force in dictating the pattern of development in naval administration. It is an explanation generally ignored when dealing with one state or one period of time. Here it is only possible to sketch briefly the general tendencies and suggest some of the more obvious connections between changes in ships and naval organization.

The Byzantine navy was made up of both regional, territorial fleets and a central fleet based at Constantinople. For all the fleets there was a body of permanent officers responsible for having the ships manned and ready when needed. The fleets contained a variety of different ships. The major fighting vessel was the dromon, a large oared ship equipped with a ram and carrying up to 300 men. In the larger forms these descendants of Roman war galleys were not fast compared to smaller oared vessels. Byzantine fleets even included transport vessels (*chelandia*) of basically the same design. Certainly Byzantine dromons had more commercial potential simply because of their size than did the smaller galleys. Those were little more than rowing shells having extremely shallow draught.[9] With the relatively large complement of rowers, about 70, those ships could never be effective cargo ships, that is except for moving luxuries, goods light in weight relative to value. Instead relatively small coasters probably carried most cargoes. The difference was obvious from the relative length of the two types, the oared vessels having length-to-breadth ratios of 5:1 and usually higher while the ratio for cargo ships was typically under 4:1.

This distinction between warships and cargo ships was well established in the Mediterranean in Roman times and before. Historians have always acknowledged the two ship designs, but they have failed to recognize a similar contemporary variation in northern Europe, especially in Scandinavia. The warships are much better known thanks to excavations of ninth century Viking long ships. Even by that date however there were two separate designs within the Scandinavian tradition: the cargo carrier, more heavily built, a small number of oars, say six, at the bow leaving a deep, open hold amidships and a length-to-breadth ratio of from 2:1 to 5:1, and the warship, a personnel carrier with 15 to 20 oars to a side, shallow draught and low freeboard, a full deck and a length-to-breadth ratio ranging from 4.5:1 up to and over 7:1.[10] The warships, like contemporary Mediterranean oared ships, were of little value for carrying freight, that is, with the exceptions of luxuries and people. In the eleventh through the fourteenth centuries the cargo ship developed into an even larger vessel, highly seaworthy and able to maintain regular trading between Scandinavia and Iceland and Greenland.[11] Evolution in the design of the warship further differentiated it from that cargo vessel. Draught remained relatively shallow but the lines became even sharper. Oared ships were built longer, reaching more than 50 oars to a side. But warships of about 25 oars per side proved the most consistently effective. The muster system was generated to guarantee Scandinavian kings a supply of vessels of precisely that size and design. The kings went even further to assure themselves of a fleet of warships which could be deployed immediately. In the eleventh century Danish monarchs built at least four sizeable naval bases in their kingdom complete with earthworks, landing area for the ships and buildings to house crews.[12]

By the fourteenth century the naval bases were gone and the muster system had been abandoned. The indirect cause for those changes in naval organization was the development in the design of the cog. Used as a coaster in the Low Countries in the early Middle Ages, the cog was significantly changed by the addition of a keel, that some time in the eleventh century. The new cog then was capable of sailing in the open sea. A sailing ship carrying a single square sail with straight post, high freeboard, a low length-to-breadth ratio rarely exceeding 3:1, the cog could carry more goods for each unit of length than any of its predecessors. Moreover the cog could be built bigger than its predecessors with no significant loss in handling qualities. By about 1200 the cog was the dominant cargo vessel in northern Europe as well as the most sought after warship. Because the cog rode so high in the water it served as an excellent platform for hurling missiles down on an enemy as well as a good base from which to launch an attack on another ship lying alongside. The addition of castles at bow and stern increased that height above an opponent. Cog hulls were thick and there was no contemporary vessel which could be used to sink it. Fighting at sea was done between cogs at close range with victories won by boarding enemy ships. That in turn meant that as many men as possible had to be put on board. The job of naval forces became armed transport, moving troops on these large cogs. It took only a few minor changes to make a trading cog into a warship. The biggest change was the increase in the complement of fighting men.

The effectiveness of the cog in northern fleets led to significant changes in government and naval policies. In Denmark in 1304 the *leidang* system was dropped. Instead of supplying long ships the districts had to send the king a much smaller number of cogs. In some cases this was just converted to a cash payment to the crown which in turn arranged for the supply of cogs.[13] In England in the twelfth and early thirteenth centuries the king relied on the Cinque Ports for naval forces. Those ports along the southeast coast received privileges from the crown in exchange for supplying a fixed number of ships for a definite number of days per year, when called on to do so. The vessels were sailing ships, related in design to Scandinavian cargo ships. By the mid-thirteenth century the king used other sources for fighting ships. The contract with the Cinque Ports had the two disadvantages of creating a semi-independent political force within the kingdom which might prove uncontrollable—it did on more than one occasion—and of not giving the king the most reliable fighting ships of the period: cogs. By the reign of Henry III (1216-1272) the king owned cogs himself. They served as armed transports carrying troops to France. In peacetime the crown leased them to merchants. The English navy was made up largely of impressed ships since all vessels were required to serve in something like a naval militia. At times orders for the general arrest of ships set a minimum tonnage for impressed vessels. Obviously the crown was interested in getting just bigger ships, especially cogs. By the late thirteenth century the English king was able to administer his fleet with a warden to oversee the Cinque Ports, a bookkeeper to keep track of who had his own ships and to see that correct payments were made to owners of impressed ships and also a small number of specially appointed temporary commissioners responsible for impressing ships at the start of a war.[14] The king of France relied on a similar levy of all ships, especially from Norman ports. The French crown had a simple system of naval administration very much like the English one. Even naval commanders were impermanent, the crown importing them from Italy, especially Genoa, for the duration of wars with England.

Both England and France expanded their naval administrations on occasion in the thirteenth and fourteenth centuries. Such growth was not incidentally associated with the building and maintaining of galleys. Such vessels were not suited to the heavy weather and the roll in the Atlantic and the North Sea, but they were useful for raiding in the Channel. The possession of a galley fleet by one king obliged his royal competitor to do the same. In England, King John (1199-1216) established shelters for his galleys at different ports and appointed permanent officers to see to their maintenance. The practice continued in the reign of Henry III. Apparently the plan was to use the galleys for police work in peace time as well. At the end of the thirteenth century King Edward I (1272-1307) built a fleet of 30 galleys, ordering certain ports to construct fixed numbers of vessels. The development of the galley fleet to fight the French came at the same time as the appointment of England's first admiral and at the time when the royal office of Clerk of the King's Ships became permanent.[15] Edward I was reacting to the French king's opening in 1294 of a permanent shipbuilding yard in Rouen, the *Clos des Galèes*. These smaller yards were also built at Boulogne and Abbéville. They were to construct and maintain galleys and sailing barges. Taxes for their support led to some resentment including in one case a tax revolt in Rouen (1382). The royal yards in Normandy had permanent staffs including salaried shipbuilders, some of them brought in from Italy.[16] To those royal officers were added the administrators of tax money used to support the establishments. In both France and England, then, the presence of warships with no commercial use, that is, galleys, meant a drain on royal treasuries and led to comparable expansion in naval administrations.

The *Clos des Galèes* at Rouen was based on a Castilian model which in turn had been copied from the Venetian arsenal. The word is Arabic meaning 'house of work' but in Romance languages it came to mean a naval shipyard. Those buildings and slipways were an obvious representation of a continuing naval administration. The most famous, the Venetian arsenal, was just a warehouse for some naval stores until about 1300 when it got the job of building and maintaining a galley fleet. That change coincided with the development of similar arsenals elsewhere in southern Europe, for example at Seville and Barcelona. The arsenals and the concomitant growth in naval administration were the result of a change in naval warfare in the Mediterranean. It was no longer possible to rely on merchant ships for fighting and states like Venice had to support two fleets, one for peace and one for war.[17] The latter was made up of light galleys, vessels of no value as cargo carriers. Galleys lost their limited role as cargo ships in the thirteenth century because of improvements in sailing ships. Those vessels could carry heavy goods in the hold, and lighter cargoes, those traditionally reserved for galleys, between decks. Sailing ships became more defensible because of the increasing use of the crossbow, a weapon which could be used more effectively from the higher sailing ship which also had high gunwales to offer protection for archers. Valuable cargoes, luxury goods could then be carried on sailing ships with greater security. Improved navigation techniques lengthened the season for sailing ships and gave them an advantage over galleys. The introduction of the cog to the Mediterranean in the thirteenth century provided southern European merchants with a big sailing ship, not vulnerable to galleys and capable of carrying bulky goods.[18] That made it possible to design Mediterranean cargo ships more for speed and that in turn formed a threat to the traditional shipping markets for galleys. The general displacement of galleys from trade led to the development of the great galley, first built at the Venetian arsenal in the last years of the thirteenth century. The great galley was a large Mediterranean sailing ship equipped with oars for occasional use. It was valuable in certain special trades, such as carrying pilgrims to the Holy Land, which in turn helped displace smaller galleys from trade.[19]

The construction and expansion of arsenals was a result of the same developments that led to the building of the great galley. By the fourteenth century small galleys were almost exclusively fighting ships. Shipwrights made them even lighter and with shallower draught. Governments had to build those vessels themselves. No private builder would. The arsenals and all the personnel that went with them had to operate continuously if states were to have galley fleets when war came. By the fifteenth century, in Venice for example, the government arsenal had a monopoly of the construction of galleys. Private builders, the few that were left, worked exclusively on sailing ships.

The full-rigged ship, developed in the fifteenth century, combined the advantages of square and lateen sails. That and other features made for a major improvement in sailing ships. The full-rigged ship was more maneuverable than its predecessors and so captains were better able to get away from attackers. It was hard enough to catch an opponent using cogs and so there were few battles in the open sea, at least in northern Europe. The full-rigged ship made engaging an enemy ship even more difficult and that in turn made reprisals, privateering and piracy less useful as part of naval policy. Governments commonly in the Middle Ages relied on seamen's traditions of greed and violence to allow them to inflict losses on an enemy. With full-rigged ships more of the task fell to governments themselves. The larger a full-rigged ship the greater were the advantages gleaned from the new design. Moreover size was still at a premium because it made ships more defensible. As before fighting in the fifteenth century was done at close quarters and ships usually ended up grappled together as crews fought across the decks. Governments began a policy of granting subsidies for the construction of larger merchant vessels. Though probably started in Italy, by the mid-fifteenth century subsidies were being granted in Spain, France and England.[20] It could be that governments were also trying to counteract a general tendency toward smaller ships which resulted from a fall in the volume of trade in the second half of the fourteenth century. Subsidies had the added advantage of being easily paid and requiring no permanent staff. The larger cargo vessels were impressed into service in wartime just as cogs were in the thirteenth century.

The introduction of heavy guns on board ship made it difficult to convert merchant ships to warships. The use of guns on board pre-dated the development of the full-rigged ship, but they were originally only anti-personnel weapons mounted on the castles or the masts. In the early sixteenth century heavy siege guns were placed on the lower decks and that gave sailing ships the ability to sink other sailing ships and at a distance. The combination of full-rig and heavy guns meant that sailing ships used tactics in battle, something reserved for galleys to that time. Big guns gave big ships even more of an advantage and so governments enlarged the subsidy programmes. They also expanded their own shipbuilding yards, both in number and in size. Ships with heavy guns could be used for trade but less easily over time as the guns took up more space and the vessels had to be specially designed to carry the extra weight. By the second quarter of the sixteenth century northern European monarchs had to have ships of their own, built by them and for them, used only for fighting. They also had galley fleets with the same limitations.

The result of changes in ship design in the fifteenth and sixteenth centuries was a marked growth in admiralty organizations. Permanent departments of state took on the task of administering the fleets, in peacetime as well as wartime. The best-known example is the establishment of the Navy Board in England in the 1540's.[21] Admiralties extended their competence as well. Those government departments were certainly the products of earlier institutions. Organizations and officers for the administration of naval forces came and went through the Middle Ages, and the sixteenth century departments were based to some degree on the older patterns. More important, those permanent bodies for administering European navies were a product of changes in ship design which by that time had made it extremely difficult if not impossible to use warships economically as cargo carriers.

Of course it is only possible to offer a quick survey of changes between 1000 and 1500, changes in ship design, in the economics of shipping and especially superficial of changes in naval administration. The final conclusion of the survey may be almost banal and obvious on reflection, that governments did not need an extensive organization to administer their navies if the ships could be used to carry cargo in peace time. Fortunately the obvious is in this case given strong support by the evidence of changes in ship design. The development of admiralties in Europe in the second half of the Middle Ages did not follow a straight line. Certainly they did emerge with the general and long term increase in the scope of government activity in the high and late Middle Ages. But the growth of government offers neither a complete nor a specific explanation for the pattern of change in naval administration in the second half of the Middle Ages. Certainly there can be no single or unique reason for the ways governments chose to implement naval policies. On the other hand it seems clear that changes in the technology of shipbuilding, changes in the design of ships made a difference to the way wars were fought at sea. War was one of the principal functions of government if not the most important one, so it is not surprising to find a connection between the technology of warfare, in this case war at sea, and the level, character and scope of at least one set of public institutions: the admiralties.

FOOTNOTES

[1] Evelyn Jamison, *Admiral Eugenius of Sicily* (London, 1957), 33-55; Fredric L. Cheyette, "The Sovereign and the Pirates, 1332," *Speculum*, 45 (1970), 50.

[2] Fredric L. Cheyette, "The Sovereign and the Pirates, 1332," 40-68; Armand le Hénaff, *Étude sur l'Organisation Administrative de la Marine sous l'Ancien Regime at la Revolution* (Paris, 1913), 27-36.

[3] Michael Lewis, *The Navy of Britain* (London, 1948), 337-341, on the functions of the admiral and the office of admiral in the Middle Ages.

[4] See for example, J. R. Bruijn, *De Admiraliteit van Amsterdam in Rustige Jaren, 1713-1751* (Amsterdam/Haarlem, 1970); Daniel A. Baugh, *British Naval Administration in the Age of Walpole* (Princeton, New Jersey, 1965).

[5] Helène Ahrweiler, *Byzance et la Mer* (Paris, 1966), 21-117; Hélène Antoniadis-Bibicou, *Études d'histoire maritime de Byzance* (Paris, 1966), 53-58, 99-116.

[6] See for example, F. W. Brooks, *The English Naval Forces, 1199-1272* (London, 1962²), 54-63.

[7] Edv. Bull, *Leding Militær-og Finansforfatning i Norge i Ældre Tid* (Christiana, 1920); G. P. Harbitz, "Leidangen og det Gamle Norske Samfunn," in G. P. Harbitz, S. Oppegard and R. Scheen, *Den Norske Leidangen* (Oslo, 1951), 11-160; A. Schück, "Ledung och Konunghamm," *Sjöhistorisk Arsbok* (1950), 97-128.

[8] The most obvious example is the expansion of the English navy after King John lost Normandy to the King of France. F. W. Brooks, *The English Naval Forces, 1199-1272*, 87-89; Michael Lewis, *The History of the British Navy* (Harmondsworth, 1957), 16-17.

[9] R. H. Dolley, "The Warships of the Later Roman Empire," "*The Journal of Roman Studies*, 38 (1948), 47-53; Louis Bréhier, "La Marine de Byzance du VIIIe au XIe Siècle," *Byzantion*, 19 (1949), 10-14; Lionel Casson, *Ships and Seamanship in the Ancient World* (Princeton, New Jersey, 1971), 146-168.

[10] Detlev Ellmers, Frühmittelalterliche Handelsschiffahrt in Mittel-und Nordeuropa (Neumünster, 1972), 30-45, 118-119; Olaf Olsen and Ole Crumlin-Pedersen, "The Skuldelev Ships (II)," *Acta Archaeologica*, 38 (1967), 106-118; A. W. Brøgger and Haakon Shetelig, *The Viking Ships* (Oslo, 1971²), 76-90, 106-112.

[11] G. J. Marcus, "The Evolution of the *Knörr*," *The Mariner's Mirror*, 41 (1955), 115-122.

[12] For a summary of information on the military camps see, Johannes Brønstad, *The Vikings* (Harmondsworth, 1965), 173-185.

[13] Detlev Ellmers, *Frühmittelalterliche Handelsschiffahrt*.63-75, 258-259. Paul Heinsius, *Das Schiff der Hansischen Frühzeit* (Weimar, 1956), 69-93 for discussions of the development of the cog design in the high Middle Ages; Ole Crumlin-Pedersen, "The Vikings and the Hanseatic merchants: 900-1450," in *A History of Seafaring Based on Underwater Archaeology*, G. Bass, ed. (London, 1972), 190-191.

[14] F. W. Brooks, *The English Naval Forces, 1199-1272*, 64-68, 79-140, 168-195; F. W. Brooks, "William de Wrotham and the Office of Keeper of the King's Ports and Galleys," *English Historical Review*, 40 (1925), 570-579; H. J. Hewitt, *The Black Prince's Expedition of 1355-1357* (Manchester, 1958), 33-42. Ships smaller than cogs were used in war, barges and balingers for example. But they like cogs could be and were used as cargo ships in peacetime, the greatest distinction between those used for trading and those used for fighting being crew size. J. W. Sherborne, "English Barges and Balingers of the Late Fourteenth Century," *The Mariner's Mirror*, 63 (1977), 109-114.

[15] R. C. Anderson, "English Galleys in 1295," *The Mariner's Mirror*, 14 (1928), 220-241; J. T. Tinniswood, "English Galleys, 1272-1377," *The Mariner's Mirror*, 35 (1949), 276-292; Michael Prestwich, *War, Politics and Finance under Edward I* (London, 1972), 137-140; F. W. Brooks, *The English Naval Forces, 1199-1272*, 147-158.

[16] Charles de la Roncière, *Histoire de la Marine Francaise* (Paris, 1909-1920²), volume I, 399-407, volume II, 70-72; H. A. Miskimin, "The Last Act of Charles V: The Background of the Revolts of 1382," *Speculum*, 38 (1963), 437.

[17] Louis Nicolas, *Histoire de la Marine Francaise* (Paris, 1961²), 6; F. C. Lane, *Navires et Constructeurs à Venise pendant la Renaissance* (Paris, 1965), 126-135; Florentino Perez-Embid, "Navigation et commerce dans le porte de Séville au bas Moyen âge," *Le Moyen Age*, 1 (1969), 269. The Seville arsenal dates from 1252 while the Barcelona Atarazanas date from the first half of the thirteenth century.

[18] Hilmar C. Krueger, "The Routine of Commerce between Genoa and North-West Africa during the late Twelfth Century," *The Mariner's Mirror*, 19 (1933), 430-431; F. C. Lane, "The Crossbow in the Nautical Revolution of the Middle Ages," in *Economy, Society and Government in Medieval Italy* D. Herlihy *et al.*, eds. (Kent, 1969), 161-166; F. C. Lane, "Progrès technologiques et productivité dans les transports maritimes de la fin du Moyen Age au début des Temps modernes," *Reveue Historique*, 251 (1974), 287-297.

[19] F. C. Lane, *Navires et Constructeurs à Venise* 2-24; R. C. Anderson, "Italian Naval Architecture about 1445," *The Mariner's Mirror*, 11 (1925), 144-145.

[20] *Inter alia*, Charles de la Roncière, *Histoire de la Marine Francaise*, volume II, 152-153, 363-365, 462-465; Michael Oppenheim, *A History of the Administration of the Royal Navy* (London, 1896), 25-26, 37-38; F. C. Lane, *Navires et Constructeurs à Venise* 97-100.

[21] C. S. L. Davies, "The Administration of the Royal Navy under Henry VIII: the Origins of the Navy Board," *English Historical Review*, 80 (1965), 268-288; Michael Lewis, *The History of the British Navy*, 39-44.

XI

Regulation and Organization of Seamen in the Netherlands and Germany before the Industrial Revolution

From the later middle ages through the eighteenth century seamen in northern Europe experienced a fall in their status, in their social, legal and economic position. Their treatment by authorities on board ship as well as on land worsened from the high middle ages on. It is true that this was part of a general pattern of degradation of individuals and the erosion of the religious and philosophical basis for respecting all human beings. More important to seamen though were internal changes in shipping during the period. Technical developments in ships and in shipping in general were the most important source for their personal degradation. While the general outline of that long term fall in the status of seamen is usually acknowledged the goal here will be to add to the discussion by outlining the technical basis for it and, more important, to suggest a possible new source for better documenting what happened to the legal position of men who worked on northern Europe's ships.

In the thirteenth century sailors were still participants in shipping ventures. They took an active part in financing the construction and operation of ships. Often they owned a part of the ship, typically a small share. Invariably they had a voice in the management of the vessel, having a vote in deciding such critical issues as when the ship would sail. By the eighteenth century sailors were wage labourers, signed on for the duration of the voyage and paid a fixed amount while the ship was at sea. Often in the middle ages seamen, in lieu of wages, were allowed cargo space. They were shippers and traders as well as workers and thus had a direct interest in voyages. Their position as traders was protected in exactly the same way as that of other men who freighted goods on ships.[1] In the middle ages there was a distinct ". . . element of mutual dependence and cooperation in times of crisis. . . . All had a stake in the success of the shipping venture."[2] The change in the status of seamen did not come all at once nor did it proceed at the same pace everywhere. It seems to have come first in larger ships and in the fishery before finally reaching small coastal craft. Moreover the process appears to have started later in the Netherlands and Germany than, for example, in the Mediterranean.[3]

Technical change and especially in ship design lay behind the legal and economic change for sailors. In general technical change means that the same quantity of goods and services can be produced with less effort, that is, with fewer inputs. Typically technical change means that all participants in the productive process become more productive though each specific technical change tends to favour one factor of production more then another, saving more of one and thus making it relatively more productive. For seamen the critical question was what effects technical change had on the productivity of labour and the quantity of labour, the number of sailors, to be used in shipping.
In the fifteenth and early sixteenth century while all factors were made more productive by the introduction of the full-rigged ship to the North and Baltic Seas crew sizes changed little. The typical pattern was some small increase in the number of tons for each man and an absolute and also small relative

increase in the productivity of the ship, the capital invested in shipping. More important though was an overall increase in the total production of shipping services. After about 1550 the pattern of technical change took a new path. It tended to be much more labour saving. In the northern Netherlands a series of more efficient cargo ships was produced. The fluit enjoyed a justified international reputation for offering low cost transportation and those low costs were achieved by reducing crew size. The boyer, a predecessor of the fluit and in widespread use in coastal shipping from the second half of the sixteenth century, achieved slightly more impressive labour productivity than even the fluit. At their best boyers could carry twenty tons for each crew member, about the highest ratio achieved before the nineteenth century.[4]

The more efficient sailing ships of northern Europe made possible lower cost shipping services. That led to an overall rise in total shipping since goods could be moved for less. Moreover there was a rise in the amount of capital relative to the amount of labour used in shipping. The investment of capital was principally in the ships themselves but extended also to new docks, cranes and other port facilities. A greater total quantity of shipping services and a relative rise in the amount of capital used meant that labour became much more productive. In an ideal and perfect world that would lead to a rise in the wages of men working on board ship. It is not clear that wages did rise by much. At least wage payments became more reliable, an important improvement for seamen. Payment was guaranteed for a voyage. The advantages to seamen from technical change were mitigated by the character of technical change which tended to decrease the skills needed to be a seaman. With little training needed going to sea became a possibility for a larger segment of the population. During the sixteenth and early seventeenth century population was rising and the greater supply of potential sailors kept down the wages of men on board. Improvements in navigation which came at the same time as the improvements in ship design did save lives and made it safer to go to sea.[5] For shipowners they meant a decrease in risk and thus for the shipping industry as a whole a fall in capital costs. There were productivity gains but ones which in the end did not work in favour of common sailors.

If sailors' wages rose in the sixteenth and seventeenth century--and they may well have--they fell relatively. The owners of ships, the owners of the capital were able to appropriate much of the rise in income because of the rising population. Even greater beneficiaries than shipowners were the suppliers of materials, especially wood. So many of the advantages created by improvements in ships were passed on to the owners of trees and other resources in the form of higher prices for their goods. In all there was a rise in the amount of capital used relative to the amount of labour. Especially important to sailors was that the change in the capital-labour ratio made the management function much more important. The man responsible for this capital, the ship's master, was given not only a higher income but more authority so that he could effectively manage his ship. The improvements in navigation meant that the master had to be more skilled, able to do simple mathematics, read charts as well as have an extensive knowledge of sailing practice in the waters where he sailed. The more skilled masters were worth much more to any shipping venture. Their income and social status and legal status on board all reflected their enhanced position. It was the captain of the ship who gained the most from technical change in the sixteenth and seventeenth centuries. The degradation of the crew was largely a degradation relative to the master's status. His productivity rose much more than that of the members of the crew.

The standard source which confirms the change in the status of seamen is the codes of sea-law used in the high middle ages. They form a contrast with what is known to have been common practice in the eighteenth century. For the latter there are many different sources. The codes of sea-law were collections of decisions taken by specific courts. While these codes were used in many different places and though they bear different names there were very few of them, each translated and altered slightly for local use. The earliest codes used in the North and Baltic Seas date from the thirteenth century but some were not codified until the fifteenth. In all cases the Code of Oléron formed a basis if not a part of the law. A typical feature of that code was precisely the cooperative nature of the shipping venture, a cooperation in which the sailor was fully expected to take part.

The provisions of the laws of Oléron are quite rightly contrasted with the typical position of wage-earning seamen in the seventeenth and especially in the eighteenth century. The problem however is that it is not possible to see any progression, any series of steps or pattern through which the law changed over the intervening four or five hundred years. At least evidence from sea-law, the Code of Oléron and the other codes related to it or descended from it, is consistent with the changes which are expected on a theoretical basis. But since the codes were copied one from the other and since they in turn were based on earlier court decisions it is extremely difficult even to date when the code was made up let alone to try to decide when a certain practice became commonplace and commonplace enough to be acknowledged by a law court.[6] In fact what would be most fruitful would be to follow the shift in the law, to see the pattern of degradation of seamen decade by decade and then compare that development over time with technical changes which in turn led to the changing relative productivity of workers in shipping. Especially valuable of course would be to see how the authority of the master expanded under the law through the years from the fifteenth through the eighteenth century. It is not possible to make an effective case for technical change as the principal source for the changing status of seamen until some accurate chronologies of developments in ship design and in the law can be compared. There is an alternate potential source which may make such a comparison possible.

The regulations of guilds of skippers, of masters are the most likely source for expanding our knowledge of the legal status of all men in shipping during the sixteenth and seventeenth centuries. Medieval sea-law was built up from case law. The obvious source for similar decisions in a later period is the decisions of admiralty courts. Serious problems exist with admiralty law, however, problems which make the records of guilds a more likely source. First the mass of admiralty cases come too late, that is from the seventeenth century on. It is possible to have great success with the decisions of admiralty courts in building up an understanding of how the shipping industry worked.[7] But for finding out about the status of sailors it is necessary to examine a large number of cases, ones often having little or nothing to do directly with the status of seamen. While the volume of information increases in the seventeenth century so too does the complexity of the material and difficulty with using it. Moreover there are serious problems in the Netherlands and in Germany with the use of records from admiralty courts. In sum there is simply not the extensive and highly centralized body of documentation which exists, for example, for the English admiralty. At the same time cases to do with the sea often wandered into other courts. That is always a problem but with the rather less clear position of the monarch and his rights relative to the law of the sea the chances in the Netherlands and Germany for cases to go elsewhere was much greater.[8] That would only serve to complicate the job of using case law to trace the progression of the declining status of sailors.

Guild regulations were laid down by those local institutions set up by town governments. They were done by each town for each town. They are continuous and accessible in a body in each of the towns. There are serious limitations with any set of guild laws. Despite those problems they can be extremely helpful in the study of development in an industry.[9] The records of skippers' guilds offer a body of evidence which is extensive covering both a wide range of topics as well as a long period of time. The rules date from precisely the right time, that is from the late fifteenth century through the seventeenth, after medieval law codes lose their value and before all other sources become abundant. The guilds periodically codified their rules so it is not strictly necessary to go through many different separate decisions. Most important the record is continuing. Thus it just might be possible to see a progression, a pattern of change in legislation on masters and seamen.

Skippers' guilds were founded in the fourteenth but principally in the fifteenth and sixteenth centuries. Many towns along the shores of the North and Baltic Seas set up such guilds.[10] The guilds were by 1600 confederations of ships' masters. The principal guildsmen were then the individuals responsible for both the sailing of the ship and the operation of the business of freighting goods. Those men could be owners of the ships they sailed or more likely part-owners taking one of the shares. Their position as owners might vary but what was important was their position as masters at sea. It was exactly those men who were most likely to gain, at least theoretically, from the technical changes of the period. While the guild records may give little help in revealing what masters' relations were like with owners, that is with their superiors, they may well tell a good deal about relations of masters with the men who worked on board.

The guilds had sets of rules, by-laws which were periodically renewed. When towns agreed to a reaffirmation of the guild regulations they often insisted on some revision. The guildsmen themselves could also be a source for such changes in the rules. The topics of the regulations vary of course from port to port and over time yet it is surprising the degree to which the same topics recur. The guild regulations always include provisions for the internal organization of the institution, for collecting monies and making payments and for relations with the superior political authority, that is the town. It was after all the towns which sponsored the guilds and it was the towns which sanctioned the guild legislation and ultimately gave it force. In a sense then the rules of the guilds were codes of town legislation just like the sets of laws which came from towns in the middle ages, from Amsterdam, Lübeck and Hamburg for example.[11] In going through the body of legislation for skippers' guilds, at least for the northern Netherlands there is the unmistakable impression that there was a rise in legislation, more frequent issuing of laws and revisions of codes, during the late fifteenth century and then a decline in the amount of new law from the late sixteenth century on. The change may be a result of the increasing importance of other authorities in dealing with the legal problems of skippers. In any case there is a large body of material and especially from the sixteenth century.

A sample from the records of two towns should serve to suggest some of the possibilities which do exist for the use of the records of skippers' guilds. The sample is also taken from material that is easily available which even more dramatically illustrates one of the great advantages of such guild records. The information comes from Amsterdam and Middelburg, the former the largest town in the northern Netherlands and the principal port of Holland, the latter the largest town in the other coastal province, Zeeland, and one with thriving commercial ties with France and England as well as to Antwerp up the river Scheldt. There are other reasons why this pair is a good starting point. For example Amsterdam did not have just one skippers' guild but three. The names are confusing and changed over time. Essentially though there was a guild for skippers of the smallest vessels, boats used to carry goods to nearby towns and villages and for lightering in the harbour. The next guild was for skippers in inland trade and carrying on regular services between Amsterdam and other Netherlands towns along the rivers and canals. That guild had about 200 members in 1570. The next was the guild of skippers' who went outside the Netherlands. Obviously those were the men commanding the biggest ships. There was a related guild of masters who sailed to Norway, the Bergenvaarders. There is one bad feature about Amsterdam however. Most of the skippers involved in long distance commerce, sailing to the Baltic in the Mother Trade for instance, came not from Amsterdam but from the villages just north of the river IJ.[12] Even if the ship were owned in Amsterdam often the captain was not an Amsterdammer. So while it was the centre of trade for the northern Netherlands Amsterdam was not home to as many skippers as might be expected. Further, surviving material from the guild of skippers who traded within the Netherlands is of little help since it offers little information about the status of sailors. In fact with the guild of skippers' of the smallest vessels there is no mention of sailors at all. That is probably because such boats had no crew or members of the skippers' families made up what crew there was. Middelburg does not suffer from any of those difficulties. There was one guild for all skippers. Unfortunately records at Middelburg were destroyed in 1940 so nothing survives about the skippers' guild except published material from before 1572. That leaves out the end of the story. For the earlier years however there is a valuable body of information.

From this little sample it is possible to see that there are definite possibilities for the use of guild legislation. The small size of the sample means of course meagre results but results which are still impressive given the problems with and limited scope of these sources. Most of the guild legislation deals with doing business, of relations between skippers and their customers. That is not surprising since after all skippers were small businessmen. Medieval law codes were the same dominated by rules on similar issues. Such legislation is of course interesting in itself but of no help in finding out about the status of sailors. Sailors were included in the guild. They were forced to join the guild at Middelburg.[13] Seamen at joining had to take an oath before the guild officers or they were not allowed to sail. The wages of sailors were fixed in some cases. If they were not fixed the guild took steps to control the income of skippers through fixing freight rates. No matter the original goal this had a tendency to limit or control the wages of seamen.[14] The increasing frequency of such measures to regulate income for skippers and, in conjunction with that, to regulate seamen's wages indicates the drift of sailors to being wage labourers. All was not negative about the guilds. In the seventeenth century through separate funds the guilds made payments to members who were sick or too old to work. Seamen as guild members took part in religious activities such as processions on certain holy days.[15] One striking feature of the rules though is the lack of legislated protection for seamen. It is notoriously difficult to argue from negative evidence but medieval law codes protected sailors for example from physical abuse by the master.[16] There was no such regulation in even the earliest skippers' guild rules from the fifteenth century. It is impossible to avoid the suspicion that sailors were not treated well in guilds simply because the guilds were run by skippers. It was the boss on board ship who made the rules and there is no reason to believe that those skippers would have been especially kind to their employees. Moreover there is no evidence for it. Towns relied on the skippers to give them advice on legislation and not on anyone else. In fact their participation in the making of the rules is evidence in itself of the improved position of masters relative to seamen.[17]

These few results are admittedly not dramatic. They do not form a new interpretation of the position of seamen before the industrial revolution. The sample is sparse however and the space here is extremely limited. Skippers' guilds did exist and throughout the Netherlands and Germany. Each of those guilds had its own rules. Guilds were influenced by other guilds as well as by changes in shipping. Each guild dealt with complaints, with violations of their own rules and with petitions to the authorities above them. The disciplinary actions taken by the guilds form something like a body of case law which in turn can be added to the legislation which is all that is mentioned here. A survey much more extensive than this one is clearly necessary to explore all the promise of this body of material. That survey should be spread over more than just two towns and over the entire period from the fifteenth through the eighteenth centuries. Such a deeper and fuller examination of guild archives could yield positive results. Moreover such a survey will yield information not only about the status--economic, social, legal--of seamen but also more generally about the life of skippers and of the men they employed. The men who worked ships, whether they were on board or on shore, were in guilds and those guilds--economically, socially and religiously--were very much a part of the lives of those northern European sailors before the industrial revolution.

1. Karl-Friedrich Krieger, Ursprung und Wurzeln der Rôles D'Oléron (Köln, 1970), pp. 18-22.
   Timothy J. Runyan, "The Rolls of Oléron and the Admiralty Court in Fourteenth Century England," The American Journal of Legal History, XIX (1975), pp. 99-103.
   Travers Twiss, ed., The Black Book of the Admiralty (London, 1871-1876), vol. II, pp. 254-401.

2. Runyan, "The Rolls . . .," p. 101.

3. E. H. Freshfield, trans., A Manual of Later Roman Law . . . including the Rhodian Maritime Law (Cambridge, 1927), p. 59.
   See also the paper of Professor Lopez.

4. On improvements in Dutch ships see R. W. Unger, Dutch Shipbuilding before 1800 (Assen, 1978), pp. 24-62.
   Changes in armament used by seamen could and did affect their status. See for example F. C. Lane, "The Crossbow in the Nautical Revolution of the Middle Ages," in Economy, Society and Government in Italy, Essays in Memory of Robert L. Reynolds, D. Herlihy et al., eds. (Kent, 1969), pp. 161-172.

5. On improvements in the techniques of navigation, D. W. Waters, The Art of Navigation in England in Elizabethan and Early Stuart Times (London, 1958). Also his "Science and the Techniques of Navigation in the Renaissance," in C. S. Singleton, ed., Art, Science, and History in the Renaissance (Baltimore, 1967), pp. 187-237.

6. Krieger, Ursprung . . .
   R. F. Wright, "The High Seas and the Church in the Middle Ages," The Mariner's Mirror, LIII (1967), pp. 3-5.

7. The best example is Ralph Davis, The Rise of the English Shipping Industry in the Seventeenth and Eighteenth Centuries (London, 1962).

8. On French and English kings using the admiralty and admiralty law as a vehicle for sovereignty claims see F. L. Cheyette, "The Sovereign and the Pirates, 1332," Speculum, XLV (1970), pp. 40-68.
   For example some cases to do with privateering and other matters to do with shipping went to the Netherlands high court in Mechelen in the years 1465-1531. While admittedly there was no admiralty court in the period still that suggests that many courts heard cases to do with the sea. J. Th. De Smidt and Eg. I. Strubbe, eds., Chronolgische Lijsten van de Geëxtendeerde Sententiën en Procesbundels (dossiers) berustende in het archief van de Grote Raad van Mechelen (Amsterdam, 1966, 1971).

9. R. W. Unger, Dutch Shipbuilding . . ., 63-108.

10. A complete list of all skippers' guilds may be impossible to compile but certainly such guilds were commonplace in Dutch ports and also in Hanse towns.

11. Twiss, Black Book . . ., vol. III, pp. 1-33, vol. IV, pp. xxvi-cxi, 55-132, 335-355.
Phillipe Dollinger, La Hanse (Paris, 1964), pp. 184-185.

12. A. E. Christensen, Dutch Trade to the Baltic About 1600 (The Hague, 1941).
J. G. van Dillen, ed., Bedrijfsleven en Gildewezen van Amsterdam, (The Hague, 1929, 1933, 1974), vol. I, #578 (1570).

13. W. S. Unger, ed., Bronnen tot de Geschiedenis van Middelburg in den Landsheerlijken Tijd (The Hague, 1923, 1926, 1931), vol. III, #376 (before 1496), paragraph 15, #377 (1497), paragraph 22.

14. van Dillen, ed., vol. I, #819 (1593), paragraphs 8, 9.
W. S. Unger, ed., vol. III, #376 (before 1496), paragraph 10, #823 (1565).

15. van Dillen, ed., vol. I, #1127 (1608), vol. II, #649 (1620).
W. S. Unger, ed., vol. III, #377 (1497).

16. Runyan, "The Rolls . . .," p. 100.
Wright, "The High Seas . . .," pp. 26-30.
van Dillen, ed., vol. III, #278 (1637).

17. As early as 1418 skippers got the Hanse towns to agree to branding seaman who deserted and also to legislate against seamen combining to try to raise their wages. G. Asaert, Westeuropese scheepvaart in de middeleeuwen (Bussum, 1974), pp. 123-125.

XII

PORTUGUESE SHIPBUILDING AND THE EARLY VOYAGES TO THE GUINEA COAST

Early in his life Avelino Teixeira da Mota spent four years in what was then Portuguese Guinea. In that time he gained a lasting interest in the region and the role of Portugal in the history of that part of Africa (Teixeira da Mota, 1969: 4). It was only natural given his major contribution — work on Portuguese cartography in the age of discovery — that the centre of his study of west Africa should be the early voyages of Portuguese explorers and traders along that coast.

The history of those early years of Portuguese expansion is only a part of the complex and varied history of the growth of Europe which began in the High Middle Ages and continued to this century. Portugal's role was important and so has been subjected to extensive scrutiny over the last one hundred years. Portuguese historians have done much of the work and men such as Teixeira da Mota have brought the results of those labours to the attention of scholars outside Portugal. It might be helpful to review what can be learned about one aspect of Portugal's part in early expansion from available material in English, in a place far from the centre of studies on that history. Readers limited to libraries and repositories with a small range of works on Portuguese history and limited largely to works in English can and do find it difficult to discover the facts about the ships, the design and building methods used that made possible those early voyages to the Guinea coast. It is even more difficult to uncover the explanations for the behaviour of Portuguese shipowners and builders during those years. This review then can not be definitive nor can it be complete. Perhaps though it can show the effects of Portuguese scholarship and point to those problems where historians of shipping and shipbuilding outside Portugal need some help in their understanding of the achievements of early exploration.

The study of Portuguese shipbuilding and exploration invariably and immediately becomes involved in the general discussion of early Portuguese expansion, why it was done, what effects it had. The discussion of ship types has become deeply imbeded in the controversey over the role of Prince

Henry (1394-1460), called the Navigator, in carrying out the voyages of discovery. If he was the driving force behind exploration then it is possible, so it has been argued, that it was he who had a new type of vessel, the caravelle, designed and built specifically for the voyages. The importance of the caravelle and the association with the role of Prince Henry explains the attention given to the vessel type. The works of the specialists Quirino da Fonseca and Gago Coutinho offer the most exhaustive studies of the characteristics of caravelles (Teixeira da Mota, 1960a: 179). They followed Henrique Lopes de Mendonça who, writing in 1892, claimed the caravelle was a Portuguese invention created especially to explore the African coast (J. Cortesão, 1950: I, 327, 348, 498; Leite, 1958: 182). Writers in English always discuss the caravelle and even attribute development of the type to Prince Henry. Obviously they have been deeply influenced by the work of Mendonça especially as presented by Fonseca. Fortunately now work is underway on a new edition of his standard book and the process of bringing the results of some forty years of Portuguese scholarship to the attention of scholars outside of Portugal is underway. Unfortunately little of the Portuguese literature is known outside of Portugal (Boxer, 1969: 27; Bell, 1974: 77; Penrose, 1952: 35, 269-270; Barata, 1975: 404).

The most important date in the study of fifteenth century Portuguese shipbuilding seems to be from the eighteenth century. The Lisbon earthquake of 1755 destroyed masses of documents and especially documents related to the government. Since the royal family took an active part in shipping the loss of that material leaves the study of shipbuilding with a weak base. Two chronicles form the most commonly used source of information. For some years and voyages Gomes Eannes de Azurara (1404?--1474), usually just called Zurara, and Alvise de Cadamosto (1434-1488) give the only material available. The former was apparently biased while the latter covered only a short period and was interested in a wide variety of subjects, seafaring being only one among many (Beazley and Prestage, 1896 and 1897; Crone, 1937; Diffie and Winius, 1977: 48, n 4, 96). An early sixteenth century chronicler, Resende, is at times cited as well. A few documents from outside of Portugal, for exemple papal bulls, give a rare passing reference to Portuguese ships. The normal principal source of design detail for medieval ships, illustrations, is sorely lacking. Most pictures come from manuscripts and few Portuguese manuscripts have survived. It is that gap which makes it so difficult to describe with any accuracy the development of ships used in early voyages to the Guinea coast. Maps did often have pictures of ships and the early sixteenth century Portuguese cartographic masterpieces offer illustrations of many different types of vessels (for example A. Cortesão and Teixeira da Mota, 1960: 37-38, 55-61). But for the fifteenth century and especially for the early voyages maps do not yield information about the ships.

While chroniclers say little about the ships they do report the step-by--step progress of exploration down the coast. Voyages to the Canary Islands were already made in the fourteenth century. Starting in the early fifteenth sailors set off from Portugal to push their way down the coast of Morocco. The area was known to fishermen and so there was no dramatic change in knowledge until Gil Eannes doubled Cape Bojador in 1434. Nuno Tristão rounded Cape Branco and explored the region of Arguim in 1443. Dinis Dias doubled Cape Verde in 1444. Pedro de Sintra visited Sierra Leone in 1460. Portuguese sailors were travelling along the coast of Ghana by 1471. They crossed the equator by about 1475 and in 1483 Diogo Cão reached the mouth of the Congo River. In the 1430's they were visiting the Azores. About 1460 they visited the Cape Verde Islands for the first time and explored the islands of the Gulf of Guinea between 1471 and 1475 (Teixeira da Mota, 1969: 3; Morison, 1940: 11-14). It was an impressive record of accomplishment.

The chroniclers say the sailors used four different types of vessels for their explorations: barque, barinel, fusta and caravelle. Of these the caravelle was the most important. For the other three there is only scanty information. Barques were commonly mentioned in the fourteenth and fifteenth centuries and not just in Portugal. Because the word was used so frequently it is very difficult to say exactly what a barque was. It was in a *barcha* that Eannes doubled Cape Bojador (Beazley and Prestage, 1896: 33). The term, used at times interchangeably with barge, applied to many different types of vessel from ship's boats to seagoing commercial and military craft. Barques could range in size from 25 to 240 tons. It may be what distinguished barques was that they carried both oars and sails. In the fourteenth century there seems to have been one mast with a single sail but the number of masts appears to have risen to three by the end of the fifteenth century (Burwash, 1947: 108-117). Barques were known in the Mediterranean as well. At Ragusa for example the word was used for a small boat. In Sicily the *barcha* was a small ship used on rivers, for fishing and for carrying goods. The size of such vessels varied a great deal (Gatti, 1975a: 37; Gatti, 1975b: 73; Bresc, 1975: 12-14). At Genoa they were small boats which were no more than coasters. But at Barcelona where the word *barca* was in use by 1415, it was a vessel of reasonable size and capable of making voyages throughout the Mediterranean (Heers, 1958: 107-108; Capmany, 1779-1792: III, 90; Dufourcq, 1966: 38).

The barque of Portuguese exploration may well have been a descendant of the Scandinavian sailing ship developed in the eighth century. Vessels like those earlier Viking ships were built in western Europe throughout the Middle Ages. Called keels, they even survived in one form or another into the twentieth century. It is doubtful that the *barca* was descended from Celtic boats or that it was flatbottomed as has been suggested. But it certainly was true that by the thirteenth century *barcas* carried heavy cargoes such as wine and coal along the Atlantic shores of western Europe (Lewis 1975: 23-24). Bordeaux barques of the early sixteenth century had a keel about 10 metres long and were rated at 35 tons. Such vessels had a ful deck with a cabin at the stern and a forecastle. But such barques already had three masts (Bernard, 1968: 356-358). Barques were known along the Biscayan coast of Iberia and Portuguese vessels for exploration were probably most like those. The barque was a common visitor to Basque ports, usually carrying about, 40 tons but ranging up to 120 and even 150 tons (Childs, 1977: 158). The seals of the towns of Santander (first half of the thirteenth century) and from San Sebastian (thirteenth century) show vessels very similar to the keels on contemporary seals of southern English ports (Ewe, 1972: 187). The dimensions of the Basque vessels may have been about 15 metres long, 4 metres wide and $2^1/_2$ metres deep (Soto, 1975).

The *barca* of the earlier explorers was probably a boat of about 25 tons but as large as 50 tons. They had partial decks and perhaps a crew of 15. There was certainly a single mast with a square sail and perhaps at times even a second small mast at the stern also carrying a square sail. Length would have been between 21 and 27 metres, the width about 8 metres giving it a length to breadth ratio of around 3.5, similar to that for Scandinavian cargo ships of the High Middle Ages (Penrose, 1952: 269; Fonseca, 1934: 619; David, 1966: 153). The vessel used was a well-known and common type, not particularly different in any way from fishing boats which worked along the coast of Morocco or made voyages to the Canary Islands.

The second type of ship mentioned by Zurara was the barinel. Barinels are presumably related to what northern European sailors called balingers. They were originally propelled with both oars and sails but by the fifteenth century as the type got bigger the oars were abandoned. Balingers could get very large, even up to 600 tons with a crew of 109. The barinel probably originated in the Bay of Biscay since it was most common there. It was usually a vessel of moderate size used for both fishing and carrying cargo, sometimes over long distances as far as from England to Spain (Burwash, 1947: 106-108). The balinger was also known in the Mediterranean where it seems to have been more closely related to the galley, a relatively long oared vessel which was descended from a classical ship type. Four *balenieri* showed up in Sicilian registers of 1460 (Bresc, 1975: 15; Dufourcq, 1966: 40). The barinel of the voyages of the 1430's probably still had oars. It was typically larger than the *barca* that travelled with it but had the same or similar rig. There were two masts with two square sails (Penrose, 1952: 269; David, 1966: 154). The barinel is very problematical since mention of it is extremely rare, even by the chroniclers. Explorers used the barinel for over a century but it was never the most popular vessel for voyages to Africa.

Zurara also mentions the fusta or pinnance and three vessels of unspecified type. The fusta is even more of a mystery. It too is mentioned as a small ship in the Mediterranean (Heers, 1958: 107). It turned up in a number of fleets for exploration but seeminly only as an auxiliary, as one ship among many.

The caravelle is most closely associated with Portuguese voyages of exploration. It is first mentioned as part of a voyage of 1440 past Cape

Bojador to Rio d'Oro. After that caravelles dominated the fleets sent out from Portugal, to Africa. From 1448 to 1460 no other type is mentioned (Diffie and Winius, 1977: 118). Certainly Prince Henry did not invent the caravelle. A careful critical examination of the evidence shows that there is no substantial base for the grandiose claims for Henry's contribution to advances in seafaring (Diffie and Winius, 1977: 113-119). It seems he financed no more than one-fourth of all the Portuguese voyages to the African coast before his death (Scammell, 1981: 228). His famous centre at Sagres may not even have existed. As for the caravelle, the type was known long before Henry was born.

The caravelle appears in mid-thirteenth century documents as a coastal fishing boat (Fonseca, 1934: 597; Landström, 1961: 100; Godinho, 1962: 33). By the late fifteenth or early sixteenth century the term was so widely used for smaller vessels it may even have become a generic term, not describing any specific type. It is possible that chroniclers were guilty of identifying any ship sent out on a voyage of exploration as a caravelle (Haring, 1918: 263; Beazley and Prestage, 1899: 150-153; Gille, 1966: 175). The confusion about the essential features of the caravelle is increased by the use in northern Europe of the term *carvel* or *karveel* or other variations to describe a ship of framefirst construction and with the exterior hull planks fitted edge-to-edge. That form of building had been developed in the Early Middle Ages in the Mediterranean. Shipbuilders in northern Europe began to use the method in the fifteenth century. Breton shipcarpenters seem to have been the first in the North to perfect using the new approach so many ships along the Biscay coast got called carvels. In a port like Bordeaux term could refer to the type of construction or to the design of the ship. Size did not matter (Hagedorn, 1914: 56-59; Anderson, 1932: 189; Bernard, 1968: 359-365). *Karveel* continued in use down into the seventeenth century to describe inland vessels of frame-first construction (van Beylen, 1970: 166).

The Portuguese caravelle apparently always had frame first construction The design came originally from the Mediterranean, related to the Arab *carib* or *caravo* which carried two or three lateen or triangular sails on two or three masts. The lateen sail had been known in Antiquity. Romans used it on smaller vessels (Casson, 1959: 219). It gave the sailor the ability to sail closer to the wind. Lateen-rigged Arab ships could get to within 60-65° of the wind and Portuguese caravelles probably approached that (Parry, 1975: 13-14, 24-26). What seems to have distinguished the Portuguese caravelle was the design of the hull. It was long, narrow, light and had little freeboard and little draft. In those features it revealed its connection with the whaling, fishing and river boats of the Biscay coast of Iberia The ratio of length to breadth was at least 3:1 and often higher, even reaching 4:1. At most there was a single deck. Smaller caravelles had only a half deck or were completely open. There was no forecastle which gave it a rakish look. The lack of upperworks also decreased resistance in a cross wind. The deck at the stern was very short to make handling the yards easier. There was a tall stempost and a straight keel. The stern like the bow was rounded, the latter looking in some cases almost bluff (J. Cortesão, 1961: between 80 and 81; Diffie and Winius, 1977: 118-119; Bernard, 1966: 203; David, 1966: 154-155). Caravelles on exploration voyages were relatively small, at most 60-70 tons and about 20-35 metres long with a breadth amidships of 6-9 metres. Such ships would have crews of more than 20 men. The lateen sails were more difficult and dangerous to handle than square ones. They were better on long runs with favourable winds than on short stretches where the sail often had to be changed from one side to the other. Portuguese builders improved the handling of the lateen by making the yards shorter, fitting them closer to the mast and setting them more upright. The sail combined with the hull form made the type highly maneouvrable, especially along coasts with changeable winds and unknown hazards. In fact caravelles had an excellent record, very few being lost (Parry, 1961: 23-24; Marques, 1976: I, 134).

Though the caravelle evolved over time the essential features always remained. In the fifteenth century it got larger with more decking and perhaps even a third mast with a third lateen sail. The rig limited the size of the ship since the yards could carry only one sail and the yards could not get too long. Otherwise they would have become too difficult to handle. The type was called a *latine de Portogau* or *caravella latina* as well as simply a caravelle (Bernard, 1968: 367). It proved to be a very fast ship making speeds on long reaches out in the Atlantic matched only in the nineteenth century by clipper ships (Penrose, 1952: 270; Chaunu, 1969: 287). Pope Nicolas V praised the caravelle in a bull of 1454 and Cadamosto in his chronicle wrote that the caravelle was the best ship on the seas (David, 1966: 155-156; Leite, 1958: 182). While Cadamosto may have exaggerated, for his voyages going down the coast of Africa and to the Atlantic islands it was far better than any contemporary ocean going ship.

There was a number of reasons for the extensive development and effective evolution of the Portuguese caravelle during the fourteenth and fifteenth centuries. Certainly one reason for the success of Portuguese shipbuilding was the wide variety of influences that met along the Atlantic coast of Iberia (Marques, 1976: I, 134; Chaunu, 1969: 284; Parry, 1975: 24). During the Middle Ages European shipbuilding was divided into two distinct and separate technical traditions. Mediterraean shipwrights built their vessels differently from workers in western and northern Europe. Portugal, because of its location and because of changes in ship design in the thirteenth and fourteenth centuries, became the place in Europe where the two traditions met and merged.

Sailing ships from northern Europe visited the Iberian coast in the Roman period and before. In the Early Middle Ages Vikings in their longships raided the coast and even travelled into the Mediterranean. During the Crusades northern Europeans travelled to the Holy Land via Portugal in in their new deep sea cogs. There were vessels from the Low Countries in the Holy Land in the First Crusade in 1097, vessels which must have made their way south along the Atlantic coast of Europe. A fleet of more than 160 ships on their way to the Second Crusade in 1147 from England and Germany stopped in Portugal long enough to help the king take Lisbon from the Muslims. The pattern was similar in the Third Crusade in 1189 and again in a new crusade in 1217. Those trips were the precursors of regular commercial contact between Portugal and the shores of the North Sea and the English Channel. For example there were Portuguese merchants at Bruges by 1200. In the fourteenth century trading relations and contacts between Portugal and northwest Europe improved markedly. Regular trading connections had been firmly established with vessels from the Low Countries going to Lisbon and then travelling directly to the Baltic before returning home (Vogel, 1915: I, 125-129, 295; Engelbrecht, 1940: 1-9; Diffie, 1960: 55-56, 75-76). Large hulks from towns in the Baltic made their way to Lisbon in the fifteenth century, that is if English pirates did not stop them. These were hulks or *urcas* of up to 300 tons, closely related to cogs and among the largest of northern European ships. They were certainly much larger than contemporary Portuguese ships (Marques, 1959: 479-480; Marques, 1960: 329-332; Diffie, 1960: 12, 17-18, 26; Lewis, 1976: 144, 148-158).

In the High Middle Ages Portugal was also exposed to direct influence from the Mediterranean. The crown imported Italian, especially Genoese, sailors to improve naval forces for use against the Muslims. Above all those Mediterranean experts brought knowledge of how to build and use galleys. By the late of the twelfth century Mediterranean galleys could make their way out through the Straits of Gibraltar against the prevailing winds and current as well as back. By the late thirteenth century galleys were making regular trips to England and the Low Countries passing by and even stopping along the coast of Portugal on the way. Certainly galleys were an integral part of Portuguese naval forces in the fourteenth and fifteenth centuries (David, 1966: 157; Diffie, 1960: 48-55, 64; Prestage, 1933: 10; Lewis, 1976; 139-140). The regular commercial contacts between Genoa and Portugal led to the involvement of Italians in trading voyages and voyages of discovery to the Canaries and even down the coast of Africa (Verlinden, 1980: 5; Lewis, 1976: 158-160). Some features of Mediterranean shipbuilding, only known in specific places, still turn up in traditional Portuguese boatbuilding (Johnstone and Tilley, 1976: 20). But the Italian technicians who came to Portugal came to serve in the galley fleets with a view to coastal and naval operations and so they did not intervene in the development of the Portuguese sailing fleet (Teixeira da Mota, 1957: 128).

Taking advantage of knowledge of both northern and Mediterranean practices Portuguese shipbuilders produced vessels with a great deal of versatility. In fact it may have been Portuguese or at least Iberian shipwrights who first hit upon the combination of square sails and lateen sails on a large cargo carrier to create the full-rigged ship. The earliest illustration of the new highly practical ocean sailing ship comes from a Spanish-Moorish plate dated to about 1400. The ship depicted is Portuguese (Landström, 1961: 96; David, 1966: 167-168). The new type was a product of modifications made in the cog which had been brought from northern Europe. The resulting carrack or *nau* as it was called in Portugal had the advantages of manouevrability from the lateen sail and easy expansion or contraction of the driving power from square sails. The full-rigged ship dominated ocean sailing until the advent of steam. In building *naus* Portuguese shipwrights proved themselves capable of creating a whole new type of ship which could perform tasks more effectively than its predecessors. The development of the new type was based on past experience but more important on very different traditional patterns of technology. Obviously Portuguese shipbuilding was vibrant, active and in a condition to receive knowledge from elsewhere and turn it to advantage.

The choice to develop the caravelle and the *nau* however was the result of the perceived needs of shipowners and buyers. In that Prince Henry may have had some role to play. In the fourteenth century the rising income of artisans, the general growth in trade and navigation in the face of falling noble incomes made investment in commerce more attractive. For Henry, a relatively poor member of the royal family, trade and the fishery looked rather attractive (Marques, 1976: 108-109, 144; Godinho, 1969: 40-41). Henry may simply represent a general trend among the nobility in the years around 1400 toward investment in commerce, colonization and expansion.

In the fourteenth and fifteenth centuries the Portuguese economy was based on the export of wine, fish and especially salt. Cork, olive oil and subtropical fruits were of lesser importance. Imports included precious metals, especially gold, other metals, cloth, wood and spices like pepper from the Mediterranean. But increasingly and especially after 1400 in terms of volume the major import was grain. The increasing bulk of Portuguese exports led to a growth in the average size of ships built (J. Cortesao, 1950: 192-193; Godinho, 1969: 37-38; Marques, 1976: I, 92-94; Marques, 1959: 482-484). The two industries of primary importance to exports were salt--making and fishing. Both typically were carried on by small units in the many little towns and villages along the coast. Production was traded through the major ports. Imports came through those same ports. The dominance of Oporto and Lisbon may have been because of the size of the ships involved in long distance trade. The development of the *nau* may have been a response to the need for a bulk carrier.

The one exception to the need for large ships was in the fishery. Fishing boats could not be made too large since they had to make their way into small harbours and often had to be beached at night. Fishing boats had to be highly maneouvrable since they often worked very close to shore. The fishery relied on a number of different types of ships but among them were all the types used in the early voyages of exploration; *barcas, barinéis, caravelas* (Godinho, 1962: 187). It was exactly those types that made trips to the northwest coast of Africa for fishing long before Prince Henry or any Portuguese nobleman thought of exploring further to the south. The typical pattern throughout Europe for all the voyages of discovery was to use ordinary vessels, the type of ship that was familiar and easily at hand. Explorers preferred if anything vessels designed for the coastal trade. The caravelle fit all the criteria (Parry, 1963: 67-68, 130).

Up through the 1440's and the voyages down to Senegambia sailors used the caravelle for the traditional tasks of the type. When it came time to extend the voyages forther difficulties arose. Regular voyages to the Canaries had been going on since the mid-fourteenth century and in caravelles. In the fifteenth century not only the Canaries but Madeira, the Azores and the Cape Verde were colonized and so became regular goals of trading ships (Verlinden, 1980: 11; Mauny, 1970: 219-220). As distances travelled increased so too did the demands made on carrying capacity. Caravelles as a result became bigger. Crews were always relatively large on voyages of exploration. The men had to be fed so much of the hold was taken up with their food and especially water, a serious consideration along the Saharan coast. On typical voyages caravelles would carry four months worth of food and one month worth of water per man or about one-half ton. But for voyages of exploration that would more than double to over one ton per man. Given the size of the first caravelles and the size of the crews almost the entire carrying capacity would have been taken up with supplies (Denoix, 1966: 142-143; Chaunu, 1969: 287). That must have limitated reconnaissance along the northwest coast of Africa. By giving up oared vessels, barques and barinels, shippers decreased crew size. Improvements in the latten rig meant the caravelle could be built bigger without increasing the number of sailors. In the fifteenth century the caravelle also served effectively for many types of coastal navigation which explains its appearance in Bordeaux and in England. The growth in the size of the caravelle could have come from its role as a coastal trader along the Atlantic front as much as from the needs of explorers. Whatever the case by the mid-fifteenth century caravelles made trading voyages to Senegambia and to England, in both cases profitably.

There were two major changes in the equipment of Portuguese exploration. The first came about 1440, the second about 1460. In 1439-1440 the Portuguese first used caravelles and in 1441 they first traded in slaves along African coast (Chaunu, 1969: 141). The change in ship type was certainly related to the move south of Cape Bojador but it was also seeminly related to the need for greater carrying capacity and smaller crew to carry economically the new slave cargoes. While the caravelle dominated from 1440 to 1460 after that increasingly *naus* appeared in the fleets going to Africa. It is possible that *naus* were just as effective as caravelles on the west Africa route (Leite, 1958: 138-139). In 1452 an *urca* traded to Arguim (J. de Cortesão, 1950: I, 326). In the fleet that was sent out to set up the fortress at Mina there were 90 caravelles but 2 *urcas* of 400 tons each which carried much of the artillery and stone (Martinez-Hidalgo, 1966: 23; Haring, 1918: 264). By the late 1450's Portuguese traders were active in Senegambia which was perhaps the most highly developed part of black Africa. It was more urbanized and had better access to gold as well as other trade goods. Prince Henry not long before his death in 1460 called for a change in the character of African exploration. Rather than fighting the natives he urged seeking alliances with them, trying to trade with them and trying to convert them to Christianity (Scammell, 1981: 253; Mauny, 1970: 227). By the last quarter of the fifteenth century the *nau* was the type preferred for voyages to west Africa (Leite, 1958: 181, 183). In both instances as the Portuguese moved down the coast it was necessary to change ship type in response to the needs of expanding commercial opportunities.

When it came time for the voyages to India the ship, the *nau,* unequivocally replaced the caravelle. Once across the equator Portuguese navigators thought it might be possible to find a way to India. Bartholomeu Dias left Lisbon in 1487 on a voyage of exploration with three caravelles, doubled the Cape of Good Hope and showed the way to India. When Vasco da Gama set out for the Orient in 1498 principally to trade he took one caravelle and three *naus* for their greater durability and cargo capacity. The latter was certainly the most important consideration (Penrose, 1952: 270; Parry, 1975: 165-167, 171; Godinho, 1969: 550-551). The ships carrying trade goods out to India and spices back in the early sixteenth century were typically full-rigged with capacities of 500 tons but to get the spices from Lisbon to their principal market in the North, Antwerp, Portuguese shippers used caravelles (Godinho, 1969: 673, 727).

As the fifteenth century progressed Portuguese shipbuilders developed more specialized types of caravelles. A 1478 document for example mentions a *caravela de descobrir,* (J. Cortesão, 1950: I, 490). By 1500 a kind of caravelle was built especially for trading voyages along the Atlantic front. Called the *caravela redonda* to distinguish it from the older and now traditional *caravelas latinas,* it showed connections with the fifteenth century caravelles of the Andalusian style (Martinez-Hidalgo, 1966: 25; Parry, 1975: 170; Fonseca, 1934: 608-609). The *Nina* and *Pinta* of Columbus's first expedition in 1492 were good examples of that Andalusian variation. Hull design was typical of caravelles with shallow draft and little upperworks. Sometimes the vessels were rigged with just lateen sails but captains, like Columbus at Las Palmas, would and did rerig them with square sails on the fore and mainmasts and a lateen mizzen when they thought it appropriate. Columbus' caravelles performed well especially in sailing close to the wind (Martinez-Hidalgo, 1966: 8-9, 25, 87-100; Haring, 1918: 261, 263; Gille, 1966: 175). The sixteenth century version had three or four masts. The foremast carried two or even three square sails while the other masts rigged lateens. The stern was square. The hull form was the same as that of earlier caravelles but with a lower ratio of length to breadth. It was the round caravelle that appeared in northern Europe in the sixteenth century bringing cargoes from Iberia (Unger, 1975: 110-112; David, 1966: 155). The type also served to carry sugar from the Atlantic islands and black slaves from the Guinea coast. For the slave trade speed was important since delays at sea would increase the mortality of the valuable cargo. As the volume of Portuguese trade increased so too did the size of ships, the averaging doubling from the mid-fifteenth to the mid-sixteenth century. The round caravelle was part of that general growth. The *redondas* got to be as big as 200 tons with lengths of over 18 metres (Barata, 1965: 405; Denoix, 1966: 145; Godinho, 1969: 28; Prestage, 1933: 332). As time went on the greater carrying capacity of the round caravelle probably made it possible for it to dominate its ancestor.

Caravelles found a use by being smaller and therefore more versatile than the contemporary *nau*. Writing in the early sixteenth century a Venetian spy noted that for voyages to Brazil and the Guinea coast the Portuguese used small ships (Teixeira da Mota, 1966: 179). Shippers bringing sugar from Brazil to Lisbon preferred caravelles. Because of their size their holds could be filled more quickly so they had a chance of making two round trip voyages a year. Ships could not sit too long with the sugar since the cargo would deteriorate. The caravelle travelling alone was more vulnerable to pirates than the *nau* but despite that shippers still used round caravelles (Boxer, 1969: 220-221). The type also made its way into the Mediterranean. The first caravelles appeared there in the 1460's bringing sugar from the Atlantic islands. By the end of the century they were making long distance voyages all the way to the eastern shores of the Mediterranean. Trade goods in the Levant were increasingly luxuries, lighter goods and caravelles had no disadvantage in competing with large ships for the carriage of such goods (Heers, 1958: 116-118). Trade to the Antilles gave employment to both lateen and round caravelles, at least in the early decades of the sixteenth century. From 1506 to 1550 of the 285 ships which went from Spain to the New World and were identified by type 163 or over half were called caravelles. The 285 made up only a small proportion of the total of 4,385 ships making such voyages. In the seconde half of the sixteenth century, 1551-1600, the proportion of caravelles among the identified ships fell to just 63 out of 1,086 which in turn was a larger proportion of the 7,110 ships which made the voyages. Throughout the greatest use of the caravelle was for carrying slaves from the Guinea coast to the newly colonized islands in the Caribbean (J. Cortesão, 1950: I, 328-329; H. and P. Chaunu, 1955: VI, table 12E; Teixeira da Mota, 1969: 15).

What started as voyages to catch fish became in the fifteenth century voyages of exploration. Starting in the 1450's the trips became trading ventures. As exploration continued so the volume and scope of trade expanded. In each case the changes led to different transportation needs and shipbuilders adjusted to try to meet those needs. Often they did. In some cases other technological developments let them avoid confronting the problem. Improvements in navigational techniques saved shipbuilders from dealing with even greater problems. The prevailing winds and currents on the northwest coast of Africa are northerly. It is very easy to get down the coast but extremely difficult to return. It was possible to claw back up the coast using the variation of shore and sea breezes. But it was a great deal of work and sailing like that could never prove economical. The solution was to stand out to sea away from the coast on the return journey and describe a great arc in the Atlantic going north to about the position of the Azores and then turn east to use the prevailing westerly wind to bring the ship back to Portugal. In other words it was necessary to practice oceanic navigation. Certainly by the 1440's Portuguese sailors were able to do that (Boxer, 1961: 9-10; Teixeira da Mota, 1958: 130-132). In fact it is likely that they knew how to use the stars to guide them at sea for a long time before the voyages of discovery. Portuguese sailors made it to the Canary Islands and back in the fourteenth century so presumably they had developed some method of celestial observation which could allow long trips out of sight of land (Verlinden, 1980: 9; Chaunu, 1969: 289-293). There is a long-standing controversey over the exact date for the start of celestial navigation but it seems clear that Iberian sailors had developed many of the necessary skills long before there is any written indication of their accomplishments. Maps made in Portugal show that certainly by the 1470's pilots were using latitude observations to mark their position so they must have already been using celestial observations before that time. The transfer of the traditional practice to paper led to a revolution in mapmaking (Teixeira da Mota, 1958: 140; Teixeira da Mota, 1966b: 229-232).

The caravelle was the ship capable of making both the coastal voyage down to west Africa and the long run on the high seas to the Azores and then back to Portugal. Their speed on those long reaches at sea combined with their maneouvrability along the coast made them ideal for work to Guinea. The caravelles could follow the prevailing current south to Cape Verde. They stopped to take on fresh water at Beziguiche and then picked up the Guinea current which runs steadily eastward. Once along the coast they had to be careful of the many shoals and hidden rocks. For the return voyage the caravelles sailed south until they picked up the equatorial current. That ran parallel to the Guinea current but in the opposite direction. Making a wide sweep they got back to the Cape Verde Islands and from there they made that wide arc which took them to the Azores and then home (Blake, 1977: 13-14; Mauny, 1970: 229-230). One chronicler claimed that the reason for using caravelles as opposed to round ships, *naus,* was the currents (J. Cortesão, 1950: I, 490). The caravelles carried relatively high value cargoes. They competed with the caravans which carried goods across the desert between west and north Africa. Ibn Khaldun of Tunis writing about 1370 mentions slaves shipped from the Canaries to Moroccan markets. Portuguese ships brought pepper and gold north too in exchange for cloth and copper goods (Verlinden, 1980: 8; Teixeira da Mota, 1969: 6, 9-15). By the 1530's century Portuguese shipping had come to completely dominate the Muslim caravan routes across the desert (Godinho, 1969: 217).

The Portuguese government maintained the myth that only caravelles could make a round trip to the Guinea coast. The king ordered the two *urcas* sent to carry supplies for setting up the fortress at Mina destroyed rather than allow them to return to Portugal. There is a story that one day over lunch a famous pilot told the king of Portugal that he was sure he could make a round voyage to west Africa with any good ship. The king said he was exaggerating but took him aside later telling him that though he was right he should keep it to himself for the good of the king's service (Martinez-Hidalgo, 1966: 23-24). Even if the government insisted that only caravelles could make it back that would not have prevented sailors from other countries from making voyages to west Africa. Caravelles could be and were built in a number of places outside of Portugal. In Bordeaux in 1493 a builder contracted to supply a *caravela redonda* of 80 tons (Bernard, 1966: 217). Andalusia certainly produced caravelles in the last decade of the fifteenth century and probably for many years before that. Portugal forbade the export of caravelles. The date of the law is not certain. It could be from the first quarter of the sixteenth century but in 1454 a man was executed for having sold a caravelle to an English buyer. Even if the legislation is from the early fifteenth century and the time of Prince Henry it was not only caravelles that worried the lawmakers. The law forbade the export not only of carvelles but also of *naus*. It also prohibited Portuguese pilots and masters from serving on foreign ships (J. Cortesão, 1950: I, 328-300; Teixeira da Mota, 1966a: 181). The lawmakers were trying to keep all Portuguese maritime skills at home, not just the design of the caravelle. The type may have had unique characteristics but those characteristics were widely known certainly by the late fifteenth century if not well before. Since the caravelle found use in a variety of different trades travelling to many different parts of Europe it is doubtful that the Portuguese government could have kept others from building them.

The Portuguese protected their African trading monopoly not so much by denying technical information about shipbuilding to potential competitors as by the use of force. The caravelle proved itself capable of performing that task as well. The Portuguese navy used caravelles as warships against Castile in 1476. It was a year later that the Castilians sent their own caravelles against the Portuguese in Africa (J. Cortesão, 1950: I, 471-473). The ability to deploy armed caravelles, that is armed with artillery, was a critical factor in the voyages of discovery (Teixeira da Mota, 1966a: 180). It was also a critical factor in maintaining the Portuguese hold on African trade. In 1557 3 Portuguese caravelles sailing with 2 *naus* off the Guinea coast defeated a combined French and English force of sailing ships. The nameless battle set the pattern for naval warfare for the next two and one-half centuries. The Portuguese demonstrated the defensive advantages of lining ships up one behind the other, of firing broadside at the enemy and of getting to windward of the enemy. The fire from the caravelles proved devastating. Since the Portuguese were only interested in driving off their enemies they did not have to bother with trying to board the English and French ships. They took advantage of the essentially defensive nature of their armed sailing ships (Guilmartin, 1974: 84, 91-93). In actions like that against Europeans and non-Europeans alike Portugal asserted her dominance of navigation and trade in the south Atlantic and Indian Ocean.

It is abundantly clear that the small Portuguese shipbuilding industry of the late Middle Ages was a vibrant and active one. It was capable of meeting the needs of local shippers and fishermen while at the same time supplying or evolving new forms of vessels. By the early fifteenth century Portugal already had the capacity in its shipbuilding industry to deal with any of the desires or ambitions of her rulers. Shipbuilding benefitted from the exposure over the centuries to different techniques from all over Europe. Shipbuilding and shipping too for that matter benefitted from the active support of the government. Kings like Diniz (1279-1325) and Ferdinand I (1367-1383) gave a great impetus to naval shipbuilding. Dinis planted the royal pine forest at Leiria which later furnished large quantities of wood for building ships. He also in 1293 set up a unique insurance system for all ships leaving Portugal (David, 1966: 156-157; Duffy, 1955: 3; Diffie, 1960: 39, 44-45). A more extensive insurance plan was instituted by Ferdinand. A charge of 2% was made on the value of all cargoes of ships over 50 tons leaving Portugal. The funds so collected were used to pay for any loss at sea of goods or, if the loss was total, of the ship as well. It may well be that the insurance scheme was never fully implemented but other legislation of Ferdinand to aid shipping proclaimed in 1377 and 1380 does seem to have been effective. He authorised purchase of wood from the royal forests by native shipbuilders without paying tax. If they built a ship of 100 tons or more then they paid nothing for the right to cut and carry the wood. He gave those same men the rights to all duties on imported shipbuilding supplies like hemp, pitch and sails. Shipowners were exempted from military service on land or sea. They got freedom from any duties on the first cargo a new ship brought into the country (Boiteux, 1968: 40-41; Engelbrecht, 1940: 5-6; Prestage, 1933, 10-12; Diffie, 1960: 68-69). The king of Portugal was involved in trading and trading ventures (Diffie, 1960: 30). Both the royal example and extensive legislation served to promote the Portuguese merchant marine and in its train shipbuilding. It is not suprising that kings took a hand in the voyages to new lands in search of potential trade.

The caravelle was not just a small vessel with some features which made it well-suited to exploration. Sailors and landsmen used the term to describe a range of vessels built and used in Portugal from the thirteenth century through the seventeenth. In the fifteenth and sixteenth centuries Portuguese shipwrights produced a number of variations on the essential caravelle design. To concentrate on a single type, on a single precise set of attributes would be incorrect. It is excuseable given the limited nature of the sources but none the less wrong. The caravelle had a number of uses. The range of uses increased through the fifteenth century as explorers found their way to the Guinea coast and back. First used as a fishing vessel the caravelle became important and then critical to the altogether new and suddenly thriving trade between Portugal and western of Africa. On the first voyage to some new region the problem was to get down the coast and back but on the second and all subsequent voyages the ships had to get back with a cargo which would pay for the cost of the trip as well as have something left over for the investors. The caravelle in its different forms solved a number of economic problems for shippers.

It may never be exactly clear what the products of Portuguese shipbuilding yards looked like in the fifteenth century. Discussion will always be made up of some guesswork because of the shortage of accurate information from the period. It is somewhat presumptuous to suggest where, some answers might be found since Portuguese scholars have long been devoted to the quest to find out more about caravelles, how they were built and how they were sailed. Still experience with the study of shipbuilding elsewhere in Europe may offer some guide. At least it is possible to suggest what might be of interest to those outside of Portugal where little is known of contemporary Portuguese research. The search can and will go on for now unknown illustrations of late medieval Portuguese ships. After so much research has already been done though it is not likely that that avenue will be very fruitful.

It is always hard to work backward, that is to interpolate the design of ships from vessels built later, but Scandinavian scholars have had some success in learning about Viking shipbuilding techniques from traditional boatbuilding practice still in use in small yards in Norway and Sweden. There were without doubt descendants of the caravelle and barque in Portugal into the twentieth century. *Barco* means simply boat now and boatbuilders often do not bother to distinguish one type from another. Yet it may well be that in one of the older types still being constructed along the north coast of Portugal hints can be found for the basic design of the barques of the fifteenth century voyages (Johnstone and Tilley, 1976: 15-17). The caravelle seems to have spawned a number of different types, some where the connection with the original became in time unrecognizeable. The frigate was probably a descendant of the caravelle, at least in its early seventeenth century form when it was brought to northern Europe (Mauro, 1966: 192-193; van Beylen, 1970: 67). The xebec or chebec, a popular two-masted lateen rig coaster of the Mediterranean through to the early twentieth century, showed many similarities with the caravelle in both rig and in hull design. A mass of information has survived about that type (Martinez-Hidalgo, 1966: 96-97). Along the Tagus River before Lisbon a number of small sailing vessels still remain in use. Some of them are probably related to the caravelle. The *fragata* bears many similarities to the caravelle but then so do almost all round-bottomed boats on the Tagus. The most likely candidate for direct descent is the now extinct *falua*. This may have been a smaller version of the *felucca*, a lateen rigged coastal trader which frequented many Portuguese ports. The *falua* carried two lateen sails on two masts and also had 4 oars. It had finer lines than the *fragata* (Landström, 1961: 210; Kochiss, 1978: 169-171, 175-177). Information about the construction and sailing qualities of these later types may help to fill gaps in knowledge about the fifteenth century caravelle.

Archaeology for the last three decades has provided a wealth of new information by turning up the ships themselves from the past. While archaeological finds have meant a great deal to the study of Roman, medieval Mediterranean and Viking ships it is doubtful that archaeology will be able to produce much about ships used to trade to Africa. Not many caravelles were lost and the waters along the African coast are not the type to aid in the preservation of a sunken ship. The high temperatures and salinity of the water make ideal conditions for shipworms which devour wood. Certainly it would be worth looking for wrecks underwater but the chances of such a search producing a major source of new information is small.

While Portuguese observers knew about the variety of types produced in domestic shipyards in the fourteenth and fifteenth centuries modern readers, especially outside of Portugal hardly know about them. Knowledge of those other designs would show why certain types were chosen for voyages of discovery over others. Also knowledge of the variety of products which came from Portuguese yards would reinforce the idea of the vibrancy and versatility of the Portuguese industry in the period. It would show that it was not just one man, by his own will, but rather a society with a wide range of seafaring skills, an expanding commercial and shipping sector and a strong interest in trade and navigation which produced travel and trade to Africa. After all in 1439, when the Duke of Burgundy wanted to build ships of the latest design using the most modern methods he sent for Portuguese shipwrights to come to Brussels to build such ships (van Beylen, 1970: 7; J. Cortesão, 1950: I, 193). Contemporaries thought highly of the skills and resources of Portuguese shipbuilding. It would be wrong for historians in the twentieth century to have a lesser view. While only a small proportion of the Portuguese population in the fourteenth and fifteenth centuries was involved in seafaring still there were skilled sailors and shipwrights capable of making the voyage of exploration and of building ships suited for such voyages. It was from the vigourous and innovative shipbuilding industry that the caravelle and many other types came.

The organization of shipbuilding yards is another part of the history of Portuguese industry which seems absent and which might add something to the study of ships and the early voyages to the Guinea coast. It was common in the High Middle Ages for governments in the Mediterraean to set up their own shipyards, arsenals like those in Venice and Barcelona. Castilians followed a similar practice when they conquered Seville in 1246. There shipbuilding, naval and commercial, was restricted to one part of the city (Godinho, 1969: 35; Dufourcq, 1966: 88; Galdacano, 1920: 18). Portugal followed suit with royal shipyards in existence in Lisbon by at least 1237. There were royal officials responsible for the proper operation of those yards (Diffie, 1960: 33-34; Godinho, 1962: 33). The centralization of naval construction must have helped to direct production of ships, at least of naval vessels (David, 1966: 165). The number of people involved in shipbuilding in Lisbon and other major centres was small in 1400 but over time it grew dramatically. In the mid-fifteenth century Lisbon boasted some 50-60 caulkers. In 1551 there were 100 caulkers along with 227, shipcarpenters and 18 riggers in the Lisbon yards alone. By the seventeenth century the royal dockyards in Lisbon employed 1500 men (Boxer, 1969: 11, 210; Duffy, 1955: 50-51). The problem of lack of sources is perhaps most critical in the case of study of the shipbuilding yards. Yet the organization of the yards in Portugal and in different parts of the Portuguese empire should reveal something about how the ships were built, about how the men who built them acted and were rewarded and most important about the ability of the Portuguese shipbuilding industry to adjust designs to perceived needs. Portuguese shipbuilding apparently bears important similarities to industries in other parts of Europe. Comparison with organization elsewhere might prove productive.

It is no longer necessary it seems for scholars either in or outside Portugal to worry about the invention of the caravelle. It is clear that a traditional type, a fishing boat with a long and probably Mediterranean ancestry became the subject of modification for Portuguese shipwrights. The builders took that ship type through at least four stages from fishing boat through exploration vessel to cargo ship and finally to the large round cargo ship. The first of those adjustments seems to have come around 1440 while the second came in the late 1450's. The final one appears to have been made at the close of the fifteenth century. Of course, shipbuilders while they might develop modified versions never gave up the established designs. Nor did they ever stop modifying the new variation on the traditional type. So many forms of caravelles always existed side by side. The same was true of all ship types. In general it is no longer necessary to unearth specific great Portuguese inventions to explain the successful voyages of exploration in the fifteenth century. This is true as much of the history of navigation as in other fields of the history of technology. As Teixeira da Mota and many other scholars have shown over the last generation there was a wide variety of scientific and technological activity in fifteenth century Portugal which laid the basis for expansion. The voyages of discovery had many different sources, political, economic, religious and they did not grow out of a single invention nor did any one single development make them possible. It was the choice of the right ship types combined with courage, obstinance and advanced cartographic and astronomical knowledge which assured Portuguese success (Gille, 1966: 176). Portugal was already a maritime nation in the fourteenth century well before embarking on the adventure of expansion.

It is not novel to suggest that the internal history of Portuguese shipbuilding, its organization, the variety of its products and the vestiges of those products in the nineteenth and twentieth century is worthy of investigation. However, it is audacious for an observer far away, limited in the works available and in the ability to use many of them, to explain what is needed to scholars already very familiar with such study. The caravelle finally virtually disappeared in the seventeenth century. The last mention of it was in Portugal in 1738 where it stayed in use longer than anywhere else in Europe (Martinez-Hidalgo, 1966: 27). The modification of the caravelle and its evolution was and is important as part of the general technical advance made in Europe in closing years of the Middle Ages. The results of those advances were massive and still too great for historians to properly describe or even comprehend. In the case of Portuguese shipbuilding, technical advance meant the creation of vessels capable of handling high value cargoes and thus of making economically profitable voyages to the Guinea coast. It was the caravelle in the fifteenth century which served to integrate Guinea geographically and commercially with Europe. So the products of Portuguese shipbuilding yards an immediate impact on interregional trade. The discovery of the sea route to India and the New World by Europeans meant the ultimate integration of the regional economy of west Africa with the economies and societies of other continents (Teixeira da Mota, 1969: 4, 16-18). The value and importance to the economic and cultural history of west Africa of that integration makes the study of Portuguese shipbuilding both more interesting and more important. But ultimately it is from the study of the internal history of the industry that the most valuable results for all historical study will come.

SOURCES

ANDERSON, R. C. (1932) — «'Carvel' and 'caravel',» *The Mariner's Mirror*, 28: 189.

BARATA, João da Gama Pimentel (1975) — «Note: Four Dordrecht Ships of the Sixteenth Century», *The Mariner's Mirror*, 61: 404-405.

BEAZLEY, Charles R. and Edgar Prestage, trans. and eds. (1896 and 1899) — *Gomes Eannes de Azurara The Chronicle of the Discovery and Conquest of Guinea*, 2 volumes, London.

BELL, Christopher (1974) — *Portugal and the Quest for the Indies*, London.

BERNARD, Jacques (1968) — *Navires et Gens de Mer à Bordeaux (vers 1400-vers 1550)*, 3 volumes, Paris.

——, (1965) — «Les Types de Navires Ibériques et Leur Influence sur la Construction Navale dans les Ports du Sud-Ouest de la France (XVe-XVIe Siècles),» *Actes du Cinquième Colloque International D'Histoire Maritime*, Michel Mollat, ed., Paris: 195-217.

VAN BEYLEN, Jan (1970) — *Schepen van de Nederlanden*, Amsterdam.

BLAKE, John W., trans and ed. (1942) — *Europeans in West Africa, 1450-1560*, 2 volumes, London.

—— (1977) — *West Africa Quest for God and Gold 1454-1578*, London.

BOITEUX, L. A. (1968) — *La Fortune de Mer, le Besoin de Securité et les Débuts de l'Assurance Maritime*, Paris.

BOXER, Charles R. (1961) — *Four Centuries of Portuguese Expansion, 1415-1825: A Succinct Survey*, Johannesburg.

—— (1969) — *The Portuguese Seaborne Empire*, London.

BRESC, Henri (1975) — «Una flotta mercantile periferica: la marina siciliana medievale,» Henri Bresc *et al.*, *Studi di storia navale*, 4: 7-24.

BURWASH, Dorothy (1947) — *English Merchant Shipping, 1460-1540*, Toronto.

DE CAPMANY, Antonio (1779-1792) — *Memorias Historicas sobre la marina commercio y artes de la antigua ciudad de Barcelona*, 5 volumes, Madrid.

CASSON, Lionel (1959) — *The Ancient Mariners Seafarers and Sea Fighters of the Mediterranean in Ancient Times*, London.

CHAUNU, Pierre (1969) — *L'Expansion Européenne du XIIIe au XVe Siècle*, Paris.

—— and Huguette (1955) — *Séville et l'Atlantique 1540-1650*, 8 volumes, Paris.

CHILDS, Wendy R. (1977) — *Anglo-Castilian Trade in the Later Middle Ages*, Manchester.

CORTESÃO, Armando and Avelino Teixeira da Mota (1960) — *Portugaliae Monumenta Carthographica*, Lisbon.

CORTESÃO, Jaime [1950] — *Os Descobrimentos Portugueses*, 2 volumes, Lisbon.

—— (1961) — *Os Descobrimentos Pré-Colombinos Dos Portugueses*, Lisbon.

CRONE, G. R., trans. and ed. (1937) — *The Voyages of Cadamosto*, London.

DAVID, J. Ferreira (1966) — «Aperçu Historique des Types des Navires à Travers les Siècles et Influence des Navires Ibériques sur la Construction Navale Européenne,» *Actes du Cinquième Colloque International D'Histoire Maritime*, Michel Mollat, ed., Paris: 149-169.

DENOIX, L. (1966) — «Charactéristiques des Navires de l'Epoque des Grandes Découvertes,» *Actes du Cinquième Colloque International D'Histoire Maritime*, Michel Mollat, ed., Paris: 137-147.

DIFFIE, Bailey W. (1960) — *Prelude to Empire Portugal Overseas before Henry the Navigator*, Lincoln.

—— (1977) — *Foundations of the Portuguese Empire 1415-1580*, Minneapolis.

DUFFY, James (1955) — *Shipwreck and Empire*, Cambridge, Mass.

DUFOURCQ, Charles-Emmanuel (1966) — *L'Espagne Catalane et Le Maghrib aux XIIIe et XIVe Siècles*, Paris.

ENGELBRECHT, W. A. (1940) — *Schets der Historische Betrekkingen Portugal-Nederland*, The Hague.

EWE, Herbert (1972) — *Schiffe Auf Siegeln*, Rostock.

DA FONSECA, Quirino (1934) — *A Caravela Portuguesa e a prioridade téchnica das Navegações Henriquinas*, Coimbra.

GALDÁCANO, Gervasio de Artiñano y (1920) — *Arquitectura Naval Española (en Madera)*, Madrid.

GATTI, Luciana (1975a) — «Construzioni Navali in Liguria Fra XV e XVI Secolo,» Henri Bresc *et al.*, *Studi di storia navale*, 4: 25-72.

—— (1975b) — «Imbarcazioni Ragusee Nel Secolo XVI,» Henri Bresc *et al.*, *Studi di storia navale*, 4: 73-96.

GILLE, Paul (1966) — «Navires Lourds et Navires Rapides avant et après les Caravelles», *Actes du Cinquième Colloque International D'Histoire Maritime*, Michel Mollat, ed., Paris: 171-178.

GODINHO, Vitorino Magalhães (1969) — *L'Economie de L'Empire Portugais aux XVe et XVIe Siècles*, Paris.

—— (1962) — *A Economia Os Descobrimentos Henriquinos*, Lisbon.

GUILMARTIN, John F. (1974) — *Gunpowder and Galleys*, Cambridge, Mass.

HAGERDON, Bernhard (1914) — *Die Entwicklung der wichtigsten Schiffstypen bis ins 19. Jahrhundert*, Berlin.

HARING, Charles H. (1918) — *Trade and Navigation between Spain and the Indies in the Times of the Hapsburgs*, Cambridge, Mass.

HEERS, Jacques (1958) — «Type de Navires et Spécialization des Trafics en Méditerranée à la Fin du Moyen Age,» *Actes du Deuxième Colloque International d'Histoire Maritime*, Michel Mollat, ed., Paris: 107-117.

JOHNSTONE, Paul and A. F. TILLEY (1976) — «An Unusual Portuguese Fishing Boat», *The Mariner's Mirror*, 62: 15-21.

KOCHISS, John M. (1978) — «Tagus River Barges and Passenger Carriers», *The Mariner's Mirror*, 64: 169-185.

LANDSTRÖM, Bjorn (1961) — *The Ship*, London.

LEITE, Duarte (1958) — *Historia dos Descobrimentos Colectânea de esparsos*, Lisbon.

LEWIS, Archibald R. (1975) — «The Medieval Background of American Atlantic Development,» *The Atlantic World of Robert G. Albion*, Benjamin W. Labaree, ed., Middletown, Conn.: 18-39.

—— (1976) — «Northern European Sea Power and the Straits of Gibralter, 1031-1350 A. D.,» *Order and Innovation in the Middle Ages: Essays in Honor of Joseph R. Strayer*, Wiliam C. Jordan *et al.*, eds., Princeton: 139-164.

MARQUES, A. H. de Oliveira (1976) — *History of Portugal*, 2 volumes, New York.

—— (1960) — «Navegação Prussiana Para Portugal Nos Principios Do Século XV,» *Revista da Universidade de Coimbra*: 19, 329-349.

—— (1959) — «Navigation entre la Prusse et le Portugal au début du XVe Siècle,» *Viertel-jahrschrift für Sozial -und Wirtschaftsgeschichte*, 46: 477-490.

MARTINEZ-HIDALGO, José M. (1966) — *Columbus' Ships*, Barre, Mass.

MAUNY, Raymond (1970) — *Les Siècles Obscurs de l'Afrique Noire Histoire et Archéologie*, Paris.

MAURO, Frédéric (1966) — «Navires et Constructions Navales en Europe Occidentale aux XVIe et XVIIe Siècles Points de Départ pour une Etude Comparée», *Actes du Cinquième Colloque International D'Histoire Maritime*, Michel Mollat, ed., Paris: 183-193.

MORISON, Samuel E. (1940) — *Portuguese Voyages to America in the Fifteenth Century*, Cambridge, Mass.

PARRY, John H. (1963) — *The Age of Reconnaisance*, New York.

—— (1975) — *Discovery of the Sea*, London.

—— (1961) — *The Establishment of the European Hegemony, 1415-1715*, New York.

PENROSE, Boies (1952) — *Travel and Discovery in the Renaissance, 1420-1620*, Cambridge, Mass.

PRESTAGE, Edgar (1933) — *The Portuguese Pioneers*, London.

SCAMMELL, Geoffrey V. (1981) — *The World Encompassed The First European Maritime Empire c. 800-1650*, London.

SOTO, José L. Casado (1975) — «Arquitectura naval en el Cantabrico durante el siglo XIII», *Altamira* (Santander), 23-56.

TEIXEIRA DA MOTA, Avelino (1958) — «L'Art de Naviguer en Méditerranée du XIIIe au XVIIe Siècle et la Création de la Navigation Astronomique dans les Océans», *Actes du Deuxième Colloque International d'Histoire Maritime*, Michel Mollat, ed., Paris: 127-148.

—— (1966a) — «Discussion des communications de MM. Denoix, Ferreira David et Gille», *Actes du Cinquième Colloque International D'Histoire Maritime*, Michel Mollat, ed., Paris: 179-181.

—— (1966b) — «Influence de la Cartographie Portugaise sur la Cartographie Européenne à l'Epoque des Découvertes,» *Actes du Cinquième Colloque International D'Histoire Maritime*, Michel Mollat, ed., Paris: 223-248.

—— (1969) — *Der Portugiesische Seehandel in Westafrika im 15. und 16. Jahrhundert und seine Bedeutung für die Entwicklung überregionalen Handelsverkehrs*, Köln.

UNGER, Richard W. (1975) — «Four Dordrecht Ships of the Sixteenth Century,» *The Mariner's Mirror*, 61: 109-116.

VERLINDEN, VERLINDEN, Charles (1980) — «Perspectief-Verschuivingen in de Vroege Geschiedenis der Europese Expansie», *Mededelingen van de Koninklijkke Academie voor Wetenschappen, Letteren en Schone Kunsten van België, Klasse der Letteren*, 42: 3-22.

VOGEL, Walther (1915) — *Geschichte der deutschen Seeschiffahrt*, Volume I, Berlin.

XIII

Northern Ships and the Late Medieval Economy: Columbus and the Medieval Maritime Tradition

The discussion of Columbus and the important events of 1492 does not normally include any examination of the maritime history of northern Europe and specifically of ships in that part of Europe. Concentration is invariably on developments in the Mediterranean and in Iberia in the later Middle Ages, where the navigational techniques of those oceanic sailors were developed and where those men gained their early training in sailing. Northern Europe is usually and correctly seen as peripheral to advances in navigation as well as in the design and construction of seagoing ships.

The usual reason for studying the economy and shipping of northern Europe, however, is that the economic, commercial, and demographic focus of the continent shifted from the Mediterranean to northwestern Europe in the sixteenth and especially in the seventeenth century. By 1700 Europe's largest cities, fastest growing economies, and greatest shipping fleets were to be found along the shores of the North Sea and English Channel.[1] During Columbus's lifetime, however, England, the Low Countries, northern Germany, and northern France were still very much on the edge of Europe, both economically and politically.

Despite those facts, there are at least three reasons why any discussion of the maritime history of the fifteenth century should include some discussion of northern European ships. First, voyages made from the British Isles and Scandinavia in the Middle Ages made a contribution to the European conquest of the sea. How much of a contribution the Irish and Vikings of the early Middle Ages made is very much a subject of controversy, but the discussion must be about extent or quality of their influence rather than about the fact of influence. Second, northern Europeans proved themselves capable of taking advantage of Columbus's voyages of exploration very quickly. The trips across the North Atlantic by John Cabot and his son, Sebastian, were matched, and in large numbers, almost immediately by many Norman fishermen going after cod near the coast of Newfoundland. English fishermen and traders from Bristol were not far behind. Third, the technology of shipbuilding in northern Europe made a significant contribution to the development of designs and building methods used for the ships of Columbus and for other explorers and shippers making long-distance deep-sea voyages. As a result of regular and increasing contact and, along with that, the exchange of technical knowledge among all parts of Europe from the early fourteenth century, a series of common ship types emerged which in turn created the basis for the highly effective and efficient European seagoing sailing ships of the next three or four centuries. This contribution of northern European shipbuilding to the development of the distinctive and ultimately technically superior ship emerged, in turn, from the unique demands placed on shipping in the fourteenth and fifteenth centuries, demands created by major changes in the economy.

Historians have tended to blame the Black Death of 1346-1350 for virtually everything that happened in and to the economy of Europe in the fifteenth century. The pandemic which spread from the Black Sea littoral to Italy, Iberia, and England and then to France, the Low Countries, and Germany, generated a population loss in just four years of something on the order of 40 percent. It seems inconceivable that the economy was not disrupted in the short run and very possibly in the long run as well. The recurrence of plague outbreaks for the rest of the century and well into the fifteenth century tends to confirm the belief that the economy would not have escaped. The now-orthodox view of that disruption grew out of macroeconomic ideas of the 1950s and 1960s. It has often been challenged and, in some aspects, revised but remains largely intact.[2]

The readjustment after the Black Death and in the wake of subsequent periods of high mortality down to the mid-fifteenth century led to falling grain prices as the surviving farmers concentrated their efforts on only the most productive soil. The shift led in turn to the

1. For example, see Richard W. Unger, "The Tonnage of Europe's Merchant Fleets 1300-1800," *The American Neptune* 52, no. 4 (Fall 1992), 247-261.

2. Harry A. Miskimin, *The Economy of Early Renaissance Europe, 1300-1460* (Englewood Cliffs, NJ: Prentice-Hall, Inc., 1969).

Reprinted from *The American Neptune* 53 (Fall 1993). Peabody Museum of Salem, Essex Institute, East India Square, Salem, MA 01970, USA

desertion of many agricultural holdings. The decline in population decreased the number of mouths to feed and with that the demand for grain, the demand falling faster than the supply. Meanwhile, the inability to replace skilled workers quickly, if at all, led to rising prices for industrial goods. Industrial goods were almost completely made by hand — there was true manufacture — and so supply depended on having trained individuals to do the work. The education of workers took time and knowledgeable individuals to teach them. There were few of the latter left, especially since mortality rates were higher in towns than in the countryside. During the many years it took to train new workers, Europeans had to accept lower quality goods and higher prices for those goods.

On the other hand, agricultural raw materials cost less. The greater efficiency of farmers, the shift to less labor-intensive forms of production in the countryside in order to use land otherwise deserted, and the fall in the population all made goods like wool cheaper within a few decades. The combination of price changes in agricultural and industrial goods produced a swing in the balance of payments between the country and towns, to the advantage of the latter. Money drained from the countryside to the towns in the years after 1350.

There was also an increase in the consumption of luxuries per person because of increases in incomes per person and because of changes in income distribution which favored certain groups, including some skilled workers and some merchants. To enjoy their rising real incomes, such men and women turned to buying better and more varied foods and clothes. In some cases the difference was as simple as having another measure of beer now and again. In others the difference was the purchase of works of art or fine garments brought from afar. The increase in purchases of luxuries led to a rise in imports into northern Europe from Italy, generating a long term balance of payments deficit for the North. That in turn led to a chronic specie shortage in England, France, and the Low Countries until late in the fifteenth century.[3]

All those adjustments in the economy meant that after 1350 the market for shipping services changed as well. Rising real wages which were typical of the period for many workers made shippers and shipbuilders more conscious of costs, and especially labor costs, both in the construction and in the handling of ships. It proved advantageous to be able to carry goods over longer distances (that is, without increasing costs) in order to be able to reach more markets both within northern Europe and beyond. Speed, always a consideration, became even more important as the faster movement of goods, especially luxuries, enjoyed greater relative rewards. With the total volume of traded goods down and with novel trade opportunities emerging, shippers could gain real advantages from the diversification of shipping services. Just doing more of the same as in the past was not productive. In fact, often it was not even possible. The reorientation of many trades and the efforts to discover new ones to replace those diminished or lost gave shipbuilders reason to be concerned with innovation.

The starting point for any innovation was the ships already in use in 1300. Over the course of the late fourteenth and fifteenth centuries, shipbuilders generated a greater variety of vessels for the use of shippers and also for governments interested in naval power. Shipbuilders responded to the demand for new kinds of shipping services in three ways. First, they modified and improved what they had, in the process creating an evolution in the old designs. Second, they developed entirely new designs locally, departures from past practice. Third, they imported new construction methods from the South and modified those methods to suit their own skills and local needs. In time they exported or, in the case of designs brought in from southern Europe, reexported all those developments, in greater or lesser degree, to Iberia and to the Mediterranean. As a result, then, all Europeans, eventually, benefited from the accomplishments of the shipbuilders working on the shores of the North and Baltic Seas.

The dominant bulk carrier in northern waters had, since the twelfth century, been the cog. A product of the addition of a keel to an earlier Celtic coaster which bore the same name, this design was well established by 1350, and cogs could be found in most northern ports. The design evolved in the fourteenth century to generate an even more efficient vessel. Shipbuilders merged the cog design with that of the hulk, another type of Celtic ancestry and dating back to Roman or possibly pre-Roman times. Using the more rounded form, especially at the bow, of the hulk, shipbuilders produced a vessel which was easier to handle in the open sea and less likely to drift in a current or tide than the more square cog. On larger versions of the new merged type the bottom was not flat, or at least not as flat on the earlier versions of the cog. The new combined cog-hulk sacrificed little in carrying capacity relative to earlier cogs, so the better handling attributes were achieved at little cost.

3. For a balanced discussion of the problem of specie flows, see John H. A. Munro, "Bullion Flows and monetary contraction in late-medieval England and the Low Countries," in *Precious Metals in the Later Medieval and Early Modern Worlds*, John F. Richards, ed., (Durham, NC: Carolina Academic Press, 1983), 97-158.

Cog-hulk combination from the town seal of Gdansk, around 1400. From Siegfried Fliedner, ' "Kogge" und "Hulk" Ein Beitrag zur Schiffstypengeschichte,' in *Die Bremer Hanse-Kogge Fund Konservierung Forschung* (Bremen: Verlag Friedrich Rover, 1969), 99, no. 23.

The more efficient bulk carrier that emerged by 1400 in ports like Gdańsk, where it appeared on the new town seal of that year, proved a great asset to grain farmers along the Baltic coast from Pomerania through Estonia. After the sharp drop in the number of mouths to feed in the wake of the Black Plague, those farmers were looking for outlets for what had suddenly become surplus production. Their interest in more efficient movement of their grain was greater because transport costs were about half of the delivered price of grain in the Low Countries, and because consumers in the Baltic needed to increase exports to pay for rising imports of relatively more expensive manufactures.[4] The simple solution of increasing the size of bulk carriers was out of the question. Larger ships would be hard to handle. More important, with a single sail the maximum size of a sail that the crew could raise fixed the maximum size of the ship. Even with the help of a capstan, usually fitted on a deck at the stern on cogs as well as on other types of single-masted sailing ships, there were physical limits on what the crew could do. There were also economic limits on how big the crew might get. Each man added for his muscle power to raise the sail was a mouth that had to be fed out of stores carried on board. Those stores reduced payload. The solution in the course of the fifteenth and particularly in the sixteenth century would be to divide the sail plan. Shipbuilders increased the number of sails while decreasing the of each of the sails. The mainsail did not so m decrease in size, though there was some shrinkage the other sails grew. The greater division among sails allowed a smaller crew to handle the same or n total canvas. The crew simply dealt with the canva smaller units. The superior sail plan also gave captain greater ability to control his ship.

When northern European shipbuilders develo new designs locally, in almost every instance it wa meet some specific need. There was a numbe different types of smaller vessels, such as the bali and various types of barges, which combined sail oar for propulsion.[5] They were popular for fighting for amphibious operations and were used, for exan in the wars between England and France, but they served to carry goods along the coast and as fis vessels. Not all of the different types developed lo found only geographically limited use, though. One type proved to be a major asset in the developme an industry which depended on technical advance the changing consumption patterns in the wake of Black Death.

The buss, which emerged in the course of fifteenth century, was designed for use with large nets in the herring fishery. Novel methods of trea the fish at sea created a better and more acces luxury foodstuff. The better and more abundant su of cured herring met a rising demand, one create falling grain prices which freed more of consur food budgets for things like fish.[6] The buss may l owed something to a high medieval Scandina design, a vessel called a *kongebuss*. In its late medi form in the Low Countries the buss was long relati its width, a ratio of 3 or 4 to 1, and carried three m each with a square sail. During fishing, two of t masts were struck. The fishermen set the sails and on racks on deck. The third stayed rigged and was to keep some pressure on the long net. The combina of the new net, introduced in 1415, the superior me of treating fish, and the buss all created a thri fishery and supplied a trade good for shipment along coast of the North and Baltic Seas as well as the c of France.

4. Richard W. Unger, *The Ship in the Medieval Economy, 600-1600* (London: Croom Helm, 1980), 163-171.

5. Smaller coastal vessels of all types have received attention because of the difficulty of finding records of their use. some suggestions on the importance of such vessels see J Sherborne, "English Barges and Balingers in the Late Fourt Century," *The Mariner's Mirror* 63 (1977): 109-114.

6. Richard W. Unger, "The Netherlands Herring Fishery i Late Middle Ages: The False Legend of William Beukel Biervliet," *Viator* 9 (1978): 335-356.

A herring buss from around 1570: Gemeente Archief Dordrecht, Archief der Gemeente Dordrecht, I, De Grafelijke Tijd, 1200-1572, X, Keur-en Handvestboeken, 12.

It was another new design, the boyer, which played a role in the more efficient distribution of goods such as fish but also more traditional trade goods such as grain. That type combined a relatively high length-to-beam ratio and a flat-bottom or modified flat-bottom and squared-off cross section with a novel rig. By the sixteenth century at the latest, the boyer had a spritsail, a forestaysail, a square main and maintopsail, and a lateen sail on a small mizzen mast. The somewhat complex, highly divided sail plan and the variety of sails allowed crews to be kept small with no loss of handling ability. Boyers required few men to handle them, registering figures as high as twenty tons served for each crew member. Since warships would have one crew member for each ton and bulk carriers would be very fortunate to have ten tons per crew member, the boyer was by standards of the day very efficient and well-suited to use on coastal and short distance trading voyages. At lengths of twenty meters or more by the late sixteenth century, boyers could carry sizeable quantities of goods at relatively low cost.[7]

Of greater importance to the development of ocean sailing for all of Europe, though, were designs brought in from the South, designs which were then adapted to local needs and local knowledge. Not all designs imported from the South succeeded. Galleys in general and the great galley — a Venetian cargo ship with three lateen-rigged masts and a large complement of rowers — never succeeded in finding a home in the North even though the type was known and used in northern waters. Venetian, and after them Florentine, shippers sent great galleys to Southampton and Antwerp for economic reasons. Those vessels brought north from the Mediterranean luxuries, including tropical and sub-tropical fruits, goods which needed rapid shipment to prevent spoilage. In return they took back high quality wool from England and Flanders, a necessary raw material for the luxury textile industries which developed in late fourteenth- and fifteenth-century Italy. Galleys, whether large trading galleys or naval vessels with a single deck, failed to gain anything more than a foothold in the North, primarily for technical reasons. The waters of the Atlantic and even the North Sea proved to be too rough for vessels with such low freeboard. To combat the rough seas, the answer was to raise the sides. But such an increase in height would either raise the angle at which oars entered the water, decreasing their effectiveness, or force builders to find some way of piercing the sides of the hull for the oars to pass through while keeping water from coming in those holes. Neither solution was satisfactory. Even in what was considered good weather in the North, rowers found it hard to move their galleys against the winds, tides, and generally rough waters. In the fifteenth

7. Unger, *The Ship in the Medieval Economy*, 262.

A boyer from about 1570. Gemeente Archief Dordrecht, Archief der Gemeente Dordrecht, I, De Grafelijke Tijd, 1200-1572, X, Keur-en Handvestboeken, 12

century, Italian maritime republics found it more and more difficult to recruit rowers for galleys. Eventually the solution was to use convicts to row. In the more thinly populated North, with less of a tradition of oared vessels, finding men to man the oars must have been difficult; in fact, it seems to have been virtually impossible. The problems of propulsion, of manning, and of security of the vessels proved insurmountable, and other than a few government experiments with rowing barges for naval and ceremonial purposes, galleys were not adopted in northern Europe.

The full-rigged ship combined square-rigged main and foremasts with a mizzenmast carrying a lateen sail. That type first appeared around 1400. It was a modification and alteration of the northern cog made in the Mediterranean or along the Biscay coast of Spain or France. Sailors had taken the northern cog into the Mediterranean as early as the end of the twelfth century. Southern shipwrights made a range of changes in that cog and converted it into a carrack with two masts, one with a square sail and the second with a lateen sail. It was in that form, carrying bulky goods from the Mediterranean or archers to fight naval battles in the English Channel, that the modified cog came back to the North.

The biggest change Mediterranean shipbuilders had made in the cog was in hull construction. They replaced northern European clinker or lapstrake building with smooth-surface or carvel-building. The former, in having each plank fitted meticulously and precisely to the previous one partially overlapping it, created an extremely sturdy and durable vessel. The latter was a more flexible method of construction, strength coming from the internal framing. It had been in use in the Mediterranean since the end of the first millennium. The change in building made the cog easier to handle, simpler and less expensive to construct, and just as reliable if not more so. The first change in the rigging came with the addition of that second mast with a lateen sail. It did not take long to realize that a third mast, at the bow, would balance effectively the sail at the stern and make it easier to control the vessel. First in the form of the two-masted Genoese carrack and then in the form of the full-rigged version, northern European shippers and shipbuilders realized the commercial and naval advantages of the new type. They had in the bulk carriage of grain and in the movement of a variety of goods between Iberia and the Low Countries an economic use for the new carrack.

In order to produce the new type, though, shipwrights had to adopt an entirely new and unfamiliar building technique. It is clear that northern shipbuilders had troubles in taking on and developing the new method of construction. As late as the 1670s, a Dutch writer on shipbuilding was still copying a description of how to go about building in the skeleton-first way that

A full-rigged ship with divided sail plan under full sail, from *The Fall of Icarus* by Peter Breugel the Elder, mid-sixteenth century. From Rijksmusueum Nederlands Historisch Scheepvaart Museum, K 1215a.

had been written in Portugal a hundred years earlier.[8] In the 1430s in the Low Countries, the Duke of Burgundy had to bring in Portuguese shipwrights to build in the new style. In the province of Zeeland, at the town of Zierikzee where skeleton-building was first tried, it was a Breton shipbuilder who taught local men how to construct a ship in the new style. In the Baltic at Gdańsk in 1462, a 700-ton sample of the new type was deserted by its captain. With that example at hand, local shipwrights, but only over the following two decades, were able to imitate successfully the new type of shell-first building.[9] It was not so much the placing of planks end-to-end that bothered northern builders. They had done that on the flat bottoms of cogs for centuries. Rather it was probably having to set up the frames and the ribs first, that is, before putting on the hull planking, which was the greatest obstacle to innovation. The builder had to know exactly what the ship would look like when it was finished before construction started. He had to imagine all the implications of the shape of the ribs and, in fact, just the shape of the single central rib before he drew it and gave direction to the carpenters who would put them together. Shipbuilders in the Mediterranean had perfected that art over the previous half-millennium and more. It was some time before northern Europeans became adapted to the changed requirements of shipbuilding.

A smaller cargo ship, full-rigged, from northern Europe around 1470, engraved by WA. From Rijksmuseum Ned. Hist. Scheepvaart Museum, K 1244 6.

In this case as in others it appears that, rather than absorb immediately and completely a new technology, shipbuilders found a halfway house, adopting the technique only partially and in modified form. Such mixtures of techniques are a common feature of technical change and most particularly of shipbuilding.[10] As late as the eighteenth century, Dutch shipbuilders, known to be among the best in Europe, were still beginning the process of construction just the way their predecessors from the high Middle Ages had, but then going over to Mediterranean building. They started the hull by putting in place the first four or five or more planks on either side of the keel, and only when that process was finished did they design, make, and put in

8. Nicolaes Witsen, *Architectura Navalis et Regimen Nauticum ofte Aaloude en Hedendaagsche Scheeps-bouw en Bestier* (Amsterdam: Pieter en Joan Blaeu, 1690), 88-90; Walther Vogel, "Ein Neuentdecktes Lehrbuch der Navigation und des Schiffbaues aus der Mitte des 16. Jahrhunderts," *Hansisches Geschichtsblätter* 17 (1911): 370-374.

9. Unger, *The Ship in the Medieval Economy*, 220-223.

10. See, for example, O. Hasslöf, "Main Principles in the Technology of Shipbuilding," *Ships and Shipyards, Sailors and Fishermen, Introduction to Maritime Ethnology*, O. Hasslöf, H. Henningsen and A. E. Christensen Jr., eds. (Copenhagen: Rosenkilde and Bagger, 1972), 27-72.

place the ribs. Mediterranean shipwrights, after setting up the keel and posts, went straight to forming and fixing the ribs. That radical change from the earlier practice of fitting each plank to the one below it, the traditional way of building using northern lapstrake construction, was obviously too much for Dutch shipbuilders. In the second half of the seventeenth century, when the French minister of finance J.-B. Colbert sent spies to England and Holland to find out about shipbuilding methods, he got reports of very different practices in the two countries. English builders followed the Mediterranean fashion, but Dutch builders used a modification of the traditional method, he was told. A Swedish writer on shipbuilding in 1691 illustrated two different kinds of construction, the Dutch method being different from the norm because shipbuilders put the first planks in place before setting up the ribs. The dogged retention of a then-outmoded technique showed not only the conservatism of shipbuilders, but also the ability to combine different traditions to the greatest advantage.[11] The products of those Dutch shipyards were not inferior to their English or their Mediterranean counterparts, but they were built differently. The hybrid method continued in use despite the recognition, as early as the fifteenth century, of the value of the design which came north in the form of the carrack.

Once mastered, or rather accommodated, the skeleton-first building method in the hands of northern Europeans led to the production of vessels of the highest quality. Northern ships might have lagged somewhat behind in complexity or in quality of their designs compared to their southern European counterparts in the fifteenth century, but by the mid-sixteenth century that was certainly no longer the case. Designs from the Baltic had formed the base for the development of the earliest ships of exploration. Portuguese barks that sailed down the coast of Africa and clawed their way back to Lisbon in the fourteenth century were probably just like Scandinavian ships of the period or even earlier. The vessels of Columbus and Vasco da Gama at the end of the fifteenth century owed their original form to the cog. Shipbuilders in northern Europe, through no choice of their own in the interchange of technical information, had contributed to the tools of reliable European long-distance ocean travel.

In northern Europe shipbuilding was better in 1492 than it was in 1300. That was true if for no other reason than it was more varied. The same can be said of southern Europe. There were more types of ships, with different hull forms and rigging, that were built in different ways. The economic changes of the late Middle Ages created greater potential for gain for merchants and for shippers, gain which could especially be found in long-distance trade. That certainly and most emphatically included trade between the two regions of the Continent. The interchange of all kinds — economic, political, personal — which resulted from those economic opportunities made technical knowledge more easily available along all the coasts of Europe from the eastern Mediterranean to the eastern Baltic. The greater variety of options known in part or in full to shipbuilders made northern Europeans, in turn, better able to exploit the opportunities which the economy presented to them in the fifteenth and even more so in the sixteenth century. The different types of ships available to shippers in northern waters by 1492 created more options and more efficient options, no matter if the designs were modifications of traditional, local ones or modifications and adaptations of ones brought in from the South. Northern Europeans, in part thanks to developments in the economy, also contributed to the development of ships capable of sailing to India and back and capable of sailing to America and back, and they produced such ships in large numbers very soon after Columbus returned from his first voyage. Furthermore, with their expanding range of ship types and nautical skills, northern Europeans also proved able to distribute quickly and economically the goods that came back from all those lands around the world, goods which improvements in ship design had added to the international trading network of Europe.

11. Richard W. Unger, "Dutch Design Specialization and Building Methods in the Seventeenth Century," *Postmedieval Boat and Ship Archaeology*, ed. Carl Olof Cederlund (Oxford: British Archaeology Report, 1985), 153-164.

XIV

THE TECHNICAL DEVELOPMENT OF SHIPBUILDING AND GOVERNMENT POLICIES IN THE FIFTEENTH AND SIXTEENTH CENTURIES

Since better ways of doing things are critical to modern economic growth it is not surprising that there are several theories about why technical advance takes place [1]. Scholars commonly make a distinction between the invention and its application. They have applied more effort to why and how ideas are applied. Diffusion of inventions has drawn the interest of economic historians, anthropologists, geographers, economists and students of business among others. They have developed a variety of different explanations for the patterns and processes of diffusion [2]. The perspective of historians is somewhat different because they deal in the past, because they focus heavily on technological aspects of innnovations and because they concentrate more on preconditions for diffusion than on the diffusion process itself [3]. There are difficulties with studying the diffusion of inventions [4]. In economic terms and probably in sociological terms adaptation, modification and improvement in design can be equally or even more important than the original invention [5]. Geographers point to barriers to adoption such as, first, problems of communication; second, problems of social resistance to changing ways of doing things; and, third, to problems of economic resistance generated by practical considera-

[1] C.A. DAVIDS, *Zeewezen en Wetenschap De wetenschap en de ontwikkeling van de navigatietechniek in Nederland tussen 1585 en 1815*, Amsterdam, 1986, pp. 25-31.

[2] L.A. BROWN, *Diffusion Processes and Location. A Conceptual Framework and Bibliography*, Philadelphia, 1968, pp. 2-6 for example; ID., *Innovation Diffusion. A New Perspective*, London, 1981, pp. 3-9.

[3] L.A. BROWN, *Innovation Diffusion*, cit., p. 190.

[4] *Ibid.*, pp. 179-182.

[5] N. ROSENBERG, *Factors Affecting the Diffusion of Technology*, « Explorations in Economic History », X (1972), pp. 7-8.

tions[6]. For example, the adoption of one invention may change relative costs of doing something so that a second innovation becomes more desireable. One change may generate a compulsive sequence with each subsequent adoption in turn breaking down new economic and social barriers to innovation[7]. Individuals may also be subject to a type of contagion where the idea travels around quickly that an innovation is a good thing[8]. Institutions can affect the pace and the direction of technical advance. The institutions can be informal ones which, for example, set community norms of behavior. They are more typically formal institutions such as schools and producers' organizations. The most important of such institutions historically have been governments. They can and did promote or deter innovation without or with overt coercion. They also often had a role in determining the distribution of the rewards from innovation and so could affect the level of economic resistance.

The reasons for the many inventions which made ships better through the fifteenth and sixteenth centuries are difficult to isolate. The general development in ship design in the period is well known and the same is true, perhaps to a lesser degree, of the evolution of shipbuilding. The great breakthrough of the late Middle Ages embodied in the development of the full-rigged ship of three masts with square sails on two and a triangular sail on the third dwarfs all subsequent improvements, that is down to the introduction of steam power in the nineteenth century. The exploitation of the implications of that early fifteenth century development was the hallmark of shipbuilding in Renaissance Europe. Once the great invention was made there were many small improvements which made ships more efficient. All those small changes added up to three prin-

[6] L.A. BROWN, *Innovation Diffusion*, cit., pp. 19-31; T. VALKONEN, *On the Theory of Diffusion of Innovations*, « Sociologia Ruralis », X (1970), pp. 166-167.

[7] N. ROSENBERG, *The Direction of Technological Change: Inducement Mechanisms and Focusing Devices*, « Economic Development and Cultural Change », XVIII (1969), pp. 4-7, 11.

[8] T. VALKONEN, *op. cit.*, p. 171.

cipal long term trends: first, a tendency toward the standardization of the design of vessels so that ships throughout Europe tended to be more alike; second, an ever increasing specialization in the design of ships most noticeable in the greater differentiation of cargo ships from warships; third, changes in scale with the ships, the shipyards and the merchant marines of different states getting bigger. Inventions led to these changes but so did the choices of many different shipbuilders to adopt those inventions, choices made easier and often even promoted vigorously by governments.

Standardization of design had to some very limited degree begun in the Middle Ages. By the eleventh century skeleton-first construction was known in the Mediterranean[9]. In the course of the fifteenth century that method of building spread to northern Europe. Northern shipcarpenters had been visiting the Mediterranean since the eleventh century. Mediterranean vessels of skeleton-first construction were certainly visiting northern ports in the second half of the fourteenth century. But it was not until the fifteenth century, and then only under some pressure, that shipcarpenters in northern Europe made the first tentative attempts to build in the new style. Northern builders did not fully absorb Mediterranean practice but adjusted it to make it more like their traditional ways of doing things. In the Netherlands in 1439 when the Duke of Burgundy wanted a ship built in the southern style he had to import Portuguese shipwrights to do the job[10]. At Zierikzee in 1460 local builders imported a Breton to show them how to build in the new way. When in the sixteenth century Dutch builders had already proven that they could build skeleton - first, they still built inland vessels and small boats shell - first, even retaining the method for some seagoing ships of medium size. Overlapping planking, a typical feature of traditional northern clinker build-

[9] G. BASS, F. van DOORNINCK, *An Eleventh Century Shipwreck at Serçe Liman, Turkey*, « The Intern. Journal of Nautical Archaeology », VII (1978), pp. 119-132.

[10] R.W. UNGER, *Dutch Shipbuilding before 1800. Ships and Guilds*, Assen, 1978, p. 189, n. 16; J. PAVIOT, *Les navires du duc de Bourgogne Philippe le Bon (vers 1440 - 1465)*, in this volume, pp. 167-195.

ing, appeared on the upperworks especially at the stern of large warships down even to the end of the seventeenth century [11]. Retention of earlier methods went even further. Though Dutch ships were built so that the skeleton was the source of strength it was not the skeleton which went up first. Dutch builders put up keel and posts and then, instead of putting the ribs in place as Mediterranean builders did, they laid down four or five hull planks on either side of the keel before setting up any of the ribs. They did that in part because the bottoms of Dutch ships were flatter than the typical vessels of the Mediterranean or for that matter the rest of northern Europe. But more than that they did it because it was consistent with how they had built ships for generations [12]. There may well have been other transitional methods of building which remained, that is where the advantages of the Mediterranean method were not so obvious.

In the fifteenth century Portuguese shipbuilders took the caravel, a coastal fishing boat first known in the Mediterranean, and made it into a versatile and reliable vessel for exploration down the African coast. Caravels were typically two-masters carrying a lateen sail on each mast. They proved highly valuable not only in voyages to unknown waters but also in trade from Portugal to the Atlantic islands. The two-masted sail plan created some problems with steering and was not as efficient as having a combination of lateen and square sails. So for long distance voyages, for example from Iberia to the Orient or across the Atlantic, sixteenth century sailors preferred vessels with the soon-to-be standard balanced rig with a large square sail on the mainmast for propulsion and small sails on fore- and mizzenmasts for control. In the sixteenth century shipbuilders thought it would make more sense to have four masts; square - rigged fore - and main-, and lateen - rigged mizzen - and bonaventure-mizzen-masts. They found by about 1600 that they

[11] R.W. UNGER, *Dutch Shipbuilding*, cit., pp. 28, 32-33.

[12] ID., *Dutch Design Specialization and Building Methods in the Seventeenth Century*, in *Postmedieval Boat and Ship Archaeology*, Oxford, 1985, pp. 155-156.

could gain the same balance and with much less difficulty using only one mizzen mast. The three-masted ship tended in the end to supercede all others. Over time the number of sails increased as captains pressed shipbuilders for a more divided sail plan. The same sail area but in smaller units gave captains more flexibility in deploying their canvas and also allowed them to work with smaller crews, having the men work on each sail in turn.

Standardization of design was certainly the tendency, a tendency most noticeable in the ships of war of European states. In northern Europe in the high Middle Ages warships were cargo ships. The cog, the dominant type, proved to be the best ship for fighting as well as the most efficient bulk carrier [13]. In southern Europe there were three different types: light galleys, great galleys and the predominant Mediterranean cargo ship which was a pure sailing ship, not narrow like the galleys but round with a relatively low length-to-beam ratio [14]. In the sixteenth century the differences in the Mediterranean began to boil down to one difference while in the North a similar distinction emerged between ships of war and ships for carrying goods. « The introduction and generalization of effective artillery armament at sea was the central fact of life of sixteenth-century naval warfare » [15]. Warships, as a result became specialized as gun carriers. The sailing ship carrying broadside armament proved its superiority in battle not just in the Mediterranean but also in the Atlantic and the Indian Ocean [16]. The guns and the crews to handle them as well as supplies of gunpowder and shot made armed ships of little use for carrying cargo. On the other hand specialized warships could and

[13] ID., *Warships and Cargo Ships in Medieval Europe*, « Technology and Culture », XXII (1981), pp. 233-252.

[14] J.F. GUILMARTIN, *Gunpowder and Galleys Changing Technology and Mediterranean Warfare in the Sixteenth Century*, Cambridge, 1974, p. 233; R.W. UNGER, *The Ship in the Medieval Economy 600-1600*, London, 1980, pp. 176-182.

[15] J.F. GUILMARTIN, *op. cit.*, p. 121.

[16] *Ibid.*, pp. 267-273.

did serve to protect cargo ships. Convoys became more common so trading vessels could and did drop armament. That made them more efficient with a significant decrease in the cost of operation because of the smaller crews and greater available cargo space. Dutch shipbuilders were the first to explore the possibilities with their "fluiten", but other European shipbuilders soon followed[17].

Governments had to take on the task of building and maintaining specialized warships. No individual and no other institution was willing or able to do it. As late as the second half of the sixteenth century larger cargo ships were impressed into the navy in times of war. Those vessels were little different from warships, especially if used in dangerous waters like the pirate infested Mediterranean[18]. The trend though was clearly away from that practice. Warships became more alike within navies but also more like ships in other navies. The galleon, for example, appeared in a number of variations in the naval forces of Venice, Spain, England, the Netherlands and a number of other states.

The specialized warships got bigger in the sixteenth century. Though in some cases specific navies did drop vessels of the largest types from their forces, especially in the Mediterranean, the tendency was for fighting ships of all classes to increase in size. That was only one of the many indications of the general increase in the scale of shipbuilding in the period. Construction yards grew. After 1500 the areas of towns where shipbuilders worked took on a more permanent character. More and heavier equipment appeared on the wharfs. Slipways and cranes and jacks all increased the ability of shipbuilders to produce bigger ships. Naval shipyards appeared everywhere in Europe, no longer confined to the Mediterranean arsenals for

[17] Ch. KONINCKX, *Convoyeurs et convoyés au XVIe siècle aux Pays-Bas espagnols. Quelques aspects de la vie à bord en mer*, in this volume, pp. 61-88; R.W. UNGER, *The Ship*, cit., pp. 262-264; N. ROSENBERG, *Factors Affecting the Diffusion*, cit., pp. 30-33.

[18] For example, G. MATTINGLY, *The Defeat of the Spanish Armada*, London, 1959, pp. 94-96, 195.

building and repairing galleys. The number of workers on a medieval wharf had always been small with perhaps two dozen being near the maximum. By 1600 private shipbuilding yards certainly had more, many of those being virtually permanent employees or they would at least expect to work at the same yard whenever there was work to be done.

The larger scale of shipbuilding yards, both naval and private, was dictated by the large size of ships. It is extremely difficult to measure changes in the average size of European vessels. There were at least two changes going on at the same time; the size of different types was changing and the mix of types was changing. Larger ships tended to replace medium size and smaller ships. Portuguese shipwrights proved that they could build wooden ships of the largest type. Their carracks for the India trade approached on occasion 2000 tons making them the largest wooden ships every built. But those were exceptional. Shipwrights and shippers found that the largest size was not the optimal size. The most efficient vessel proved to be one of under 500 tons, that discovery made in the earlier years of the seventeenth century. The Dutch were the first to realize the advantages of such ships, that in the Baltic trades where the average size of ships doubled to 280 tons between 1570 and 1640 [19].

The optimal size depended not on the technology of shipbuilding but on the time it took to gather together a cargo large enough to make a voyage worthwhile. Improvements in communication and the general rise in the volume of trade served to decrease, if only slowly, the time ships had to wait in port before enough paying cargo was assembled to warrant going on to the next port. The gradual fall in turnaround time paved the way for the rise in the optimal size of cargo ships. One result of the growth in the scale of ships and of shipbuilding was a tendency toward the concentration of the industry into one area or district, a place with access to large quantities of raw materials and a pool of skilled labour. The

[19] R.W. UNGER, *Dutch Shipbuilding*, cit., pp. 11, 38.

Zaanstreek outside of Amsterdam became the greatest centre of shipbuilding in Europe after 1600. In England the industry came to be concentrated along the Thames downstream from London[20] and Portuguese shipbuilding too became concentrated, in that case along the Tagus near Lisbon.

The growth in the size of the European merchant marines contributed to that tendency. The total number of ships in Europe's fleets, naval and merchant, grew in the fifteenth and sixteenth centuries. Not all fleets grew at the same pace nor did all classes of ship increase in number at the same pace. Sources are poor so the total size of the merchant marine is hard to measure. The best information comes from surveys done by the governments. They showed an increasing interest in just how much tonnage was owned in and operated from their own ports. A series of surveys done in the reign of Elizabeth I of England is only the best example. In 1560 when the crown made its first investigation the total English merchant marine was probably not more than 45,000 tons. By 1583 and the end of that unique set of surveys it was up to 90,000 tons, about the level in the early forteenth century at the height of the Anglo-Gascon wine trade. An estimate based on a survey done in 1629 showed that the growth trend had continued. By then the merchant marine had reached 115,000 tons[21]. Dutch growth was even more spectacular. In 1532 there were 400 seagoing ships. A 1567 estimate claimed 800 ships of over 200 tons. In 1636 Holland had at least 2,500 seagoing ships. By that last year the merchant fleet was between 450,000 and

[20] P. BANBURY, *Shipbuilders of the Thames and Medway*, Newton Abbot, 1971, pp. 24-25.

[21] M.K. JAMES, *The Fluctuations of the Anglo-Gascon Wine Trade during the Fourteenth Century*, in *Essays in Economic History*, New York, 1962, vol. II, pp. 130, 146-147; M. OPPENHEIM, *A History of the Administration of the Royal Navy and of Merchant Shipping in Relation to the Navy*, London, 1896, pp. 172-175; A.P. USHER, *The Growth of English Shipping 1572-1922*, « Quarterly Journal of Economics », XLII (1927), p. 467; S.M. JACK, *Trade and Industry in Tudor and Stuart England*, London, 1977, p. 98 and Table 1. For example BRITISH LIBRARY, *Harleian Mss.*, 168, fol. 248 *v* - 249 *v*, *Harleian Mss.*, 4228, fol. 45 *r* - 46 *v*, *Cottonian Mss.*, Otho, E. IX, fol. 144.

550,000 tons[22]. The Hanse ports of northern Germany saw their fleets grow from about 60,000 tons at the end of the fifteenth century to some 110,000 tons by about 1570. By that latter date the Scottish fleet was around 10,000 tons and the French at 80,000 tons. Those levels stayed much the same, increasing only in the later years of the seventeenth century[23]. While it took some time to reach medieval tonnage figures they were surpassed and European fleets were far beyond earlier levels by 1600. The growth of the fleets was a result of the increased volume of trade within Europe and between Europe and the rest of the World. That trade was made possible by, among other things, the increasing quality of the ships, the improvements in their design.

Technical improvements came from many different sources. Shipbuilders invented new designs and ways of doing things based on their own experience. There was never a problem, it seems, with generating new possibilities. If a shipwright had no new ideas himself he could, it appears, easily borrow inventions. He could and did borrow from fellow builders in the same place and from practice on different types of vessels. Small vessels proved a fertile source of ideas for big ships. If finding new ideas was no problem, the communication of those ideas was no problem either. Ships travelled over long distances, visited other ports, were seen and examined by shipwrights. There were few if any barriers — social, economic or geographic — to the communication of new methods of building ships. There was resistance though to the adoption of new methods. The latest designs were not always taken up immediately. The biggest deterrent was probably uncertainty. It was not possible to predict the effects of a specific innovation. Even if it had been tried a number of times there was still no guarantee it

[22] C.A. DAVIDS, *op. cit.*, p. 43; R.W. UNGER, *Dutch Shipbuilding*, cit., p. 11.

[23] W. VOGEL, *Zur Grösse der europäischen Handelsflotten im 15., 16. und 17. Jahrhundert. Ein historisch-statistischer Versuch*, in *Forschungen und Versuche zur Geschichte des Mittelalters und der Neuzeit. Festschrift Dietrich Schäfer*, Jena, 1915, pp. 331-333.

would work for the specific vessel being built. There could be some social resistance, usually typified as the inherent conservatism of the shipbuilder. There is little evidence for conservatism but some evidence for a reluctance to experiment for the sake of experiment and a reluctance to try something new without a clear or pressing reason to do so. Old designs had proven effective and could often be improved with small rather than revolutionary changes.

Institutions for promoting technical change increased in the sixteenth century. The production of treatises on shipbuilding was the sign of a growing informal network of theoreticians and practitioners in Europe and outside of Europe interested in exchanging and disseminating information about the practice of building ships [24]. Shipbuilders began to write descriptions of their products in a systematic way. Matthew Baker, the shipwright of Elizabeth I of England, showed that the practice of writing such books of notes had migrated from Italy to northern Europe. Even printed books began to appear, the first being published in Mexico City in 1587. It was a simple first step but it was also a beginning of what would be a flood of such works in the following centuries [25].

For warships governments in general proved to be a source for the dissemination of knowledge. Shipcarpenters imported from other countries brought the latest practices with them. Ships of other states were seized during war as prizes. The first goal was to cripple the enemy, the second to expand the size of the domestic merchant fleet. But one other advantage was

[24] R.C. ANDERSON, *Early Books on Shipbuilding and Rigging*, « The Mariner's Mirror », X (1924), pp. 53-54; ID., *Italian Naval Architecture about 1445*, « The Mariner's Mirror », XI (1925), pp. 135-163; D.J. PROCTOR, *A Late 15th Century Venetian Galley Builder and his Beliefs*, in this volume, pp. 229-243.

[25] V.D. TATE, *The Instruction Nauthica of 1587*, « The Amer. Neptune », I (1941), pp. 191-195; R.W. UNGER, *Design and Construction of European Warships in the Seventeenth and Eighteenth Centuries*, in *Les Marines de Guerre Européennes XVII-XVIII siècles*, Paris, 1985, pp. 21-23; ID., *The Ship*, cit., pp. 271-272.

the addition of the technology of another state, technology which could be adopted or rejected as suited local needs. Governments gave subsidies, direct and indirect, to shipbuilders and generally those subsidies implied a technical bias. In England for example the goal was to increase the size of merchant ships so they could be impressed into naval service in wartime. The Venetian government did the same thing in the years around 1500 as had Genoa. Spain and France had subsidy programs in the sixteenth century as well[26].

Governments were ultimately interested in national strength. A large merchant marine in the fifteenth and sixteenth centuries meant a pool of vessels to draw on in time of war. A large merchant marine was also a source of state income which could be used to build and maintain a navy. It was also a pool of shipbuilding and sailing talent which could be impressed in time of war. The desires of the governments dictated the character of their support for technical change and the dissemination of knowledge. At the same time the improvements in shipbuilding technology formed and shaped their needs and desires. There was a reciprocal relationship between the government demand for technical change and its ability to supply that technical change.

The fifteenth and sixteenth centuries saw a long term improvement in shipbuilding. Shipcarpenters came to build better ships, better both in their capabilities and in their efficiency. Shipwrights made an invaluable economic contribution to Europe making shipping one of the leading sectors of the economy and making possible growth in agriculture and industry not just in Europe but around the World. The population and the output of the economy reached their nadir around 1400. Technical improvements in shipbuilding made a significant contribution to and were also dependent in the long term sustained

[26] R. MEEKER, *History of Shipping Subsidies*, New York, 1905, pp. 1-4; A.P. USHER, *Spanish Ships and Shipping in the Sixteenth and Seventeenth Centuries*, in *Facts and Factors in Economic History*, Cambridge, Mass., 1932, pp. 192-194; F.C. LANE, *Navires et constructeurs à Venise pendant la Renaissance*, Paris, 1965, pp. 99-105.

recovery which followed over the next two hundred years. The improvements in ship design and shipbuilding came from a number of sources. On the side of supply builders themselves proved inventive. A number of conditions aided them but even more conditions aided in the dissemination of their inventions. Pressures to improve came from the demand side, from buyers of ships. The rise in total production in Europe made more goods available to be moved and so created a need for bigger and better ships. Equally or more important was the demand created by the voyages of discovery. The explorations which began in the fourteenth century and enjoyed their greatest successes in the last decade of the fifteenth century with the voyage of Columbus created a need for vessels of superior design able to deal with unknown and barely predictable problems. Governments, by promoting exploration, also indirectly promoted technical advance. As exploration increasingly gave way to settlement and the economic potential of lands far from Europe began to be exploited the demand for better ships and more ships increased. The discoveries of Columbus and others also created a climate where social resistance decreased, where trying new things seemed sensible.

Governments were among the principal institutions for diffusing and adopting new technology in shipbuilding. Over the long term pressures on shipbuilders came increasingly not from private buyers but from public institutions. Renaissance governments were invasive. That surfaces in many different ways and had a firm basis in contemporary political thought. Nicolò Machiavelli did not tell the Prince that he should promote technical advance in shipbuilding. But he did tell the Prince to do what was necessary to consolidate his power and increase his strength. The combination of the basic changes in ship design, the economic and military opportunities created by those changes along with other economic forces made it advantageous for states to promote improvements in shipbuilding technology. The economic, social and technical climate created the chances to do exactly that. Governments showed their interest in promoting technical advances in shipbuilding in a number of ways. They even went so far as to carry out the surveys and investigations and produce the reports generating the sources which make it possible to discuss, at least to a limited degree, the role of government in the diffusion of shipbuilding technology in the fifteenth and sixteenth centuries.

XV

DESIGN AND CONSTRUCTION OF EUROPEAN WARSHIPS IN THE SEVENTEENTH AND EIGHTEENTH CENTURIES.

In the years from the Spanish Armada to the Wars of Napoleon the sailing warship reached its greatest refinement. The shipbuilders who perfected the design and construction of those vessels were engineers, crafts-men and artists all at the same time. Each had his own approach to solving the many problems to satisfying the often conflicting demands which arose in designing and building an efficient and effective fighting machine. A comparison of European warships and for that matter cargo ships would show the many solutions they found to the problem of constructing the most effective vessel. Such a comparison, though highly informative, is at this point virtually impossible. The necessary specific studies of naval construction and administration in each of a number of different jurisdictions have yet to be done. The difficulties stem in part from the large number of navies involved. Those of Britain, France and the Netherlands are the best known. But there were also significant naval forces in the kingdoms of Denmark, Sweden and Sicily as well as Portugal and Spain. The naval forces of Turkey and the Barbary pirates played a part in the development of European naval construction as did those of newcomers to the naval competition, Prussia and Russia. Just to amass knowledge of the ships of all those states is an overwhelming task, one beyond most historians and certainly this historian. Rather than trying to offer some conclusions at this stage it is only possible to explore the principal long term trends, the similarities and differences which existed among warships from different states and suggest some explanations for the survival of similarities and the appearance of differences.

Among all the long term changes in shipbuilding practice one trend dominates all others in the two centuries before the Industrial Revolution : the replacement of a handicraft or craftsman's approach by a "scientific" one. Technologists more and more relied on advances in basic knowledge about nature to make improvements in their own methods and tools. It was the start of a pattern now very familiar. The major scientific advances of the 17th century did not prove easily applicable to shipbuilding. Kepler's astronomy or Newton's laws of motion and gravitation or the development of logarithms and the calculus did find many applications. They were critically important to navigation for example. But to the shipcarpenter on land it was virtually impossible to find a direct application for this new knowledge. What did matter to shipbuilders in the long term was the belief in the existence of certain immutable 'laws of nature'. Many came to believe that there were first principles about the world which were simply waiting to be discovered. The principles could and would be described in mathematical terms. They could then be applied to practical problems. It was not shipwrights who were the first to explore the use of mathematics to describe their "mystery". Rather it was scholars in search of another area to apply the new scientific thinking (Hasslöf 1972, 65-67). But the work of such men gradually turned into required reading for designers in the shipbuilding yards.

The obvious evidence for the long term trend toward a scientific approach is the many printed books on shipbuilding produced beginning in 1587 and increasing in number and complexity down through the following centuries. The most complex theoretical statement from 16th century writers was that since a fish travelled through the water quickly and with little resistance then the form of the ship souhl follow the form of a fish (Baker 1585 ; Timmerman 1963, 11-12). The first printed book on shipbuilding appeared in Mexico City in 1587. It was simple both in content and presentation (Tate 1941). Works in the 17th century had much more material with a broader range of subjects. It was common to list different types of vessels with their dimensions (e.g. Furttenbach 1629). That was true of the great work by the Amsterdam businessman and diplomat, Nicolas Witsen. His book had the necessary sections on ancient ships and then went on to discuss contemporary construction. The work was descriptive, far from analytical and based entirely on second hand information (Witsen 1690). The reporting of proportions for a number of ships combined in some cases with drawings seems to have been the established form for Dutch writing on shipbuilding in the late 17th century (e.g. Allard 1696). Cornelis Van Yk, an experienced shipwright, added his personal experience making his book more complete and more accurate (Van Yk 1697). Books of lists were not exclusive to the Netherlands. A contemporary English writer, also an experienced shipbuilder, could produce a work filled with proportions and tables (Sutherland 1711). More and more the books opened with chapters on the basics of geometry and trigonometry. Bushnell even in the 17th century made his first chapter "Of Geometricall Problemes" (Buschnell 1664). It was necessary, authors increasingly thought, to instruct shipbuilders in mathematics so they would be equipped to do the job of designing a ship. The famous mathematician Leonhard Euler even turned his hand to the question. His book was overwhelmingly a treatise in mathematics (Euler 1749). His contemporary, Mungo Murray, produced a less theoretical tract but was still concerned with how to design a ship using geometry (Murray 1754).

The French botanist, physicist and inspector of the French Navy, H. L. Duhamel Du Monceau, took a majoor step toward placing ship design on a scientific base (Duhamel Du Monceau 1752). He filled his book with instructions on how to draw the plans from which a ship could be built. Other authors relied heavily on Duhamel Du Monceau even going so far as to copy and to translate his words and drawings (e.g. Murray 1765 ; Timmerman 1963, 25-26). He succeeded in applying the theoretical advances of others to the essential problems of ship design. He like other writers in the 18th century pressed designers to calculate the metacentric height of the ship from drawings of the hull form so that they could predict stability and make correct decisions about other design features such as size of the upperworks and sail plan (Banbury 1971, 56 ; Timmerman 1963, 15-22). Such ideas depended on the discovery of principles of buoyancy which had begun with the work of the Dutch scientist Simon Stevin early in the 17th century. Though one expert said in the first decade of the century that no builder could predict how much water his product would draw, because of advances in theory by late in the 1600's ship designers could estimate the displacement of their ships even before launching (Howard 1979, 95 ; Naish 1957, 488 ; Perrin 1918, lxx-lxxiv). Fredrick Chapman brought together all these advances in his **Architectura navalis mercatoria** published in 1768. He combined basic knowledge of mathematics including the calculus and of theoretical advances in the study of bodies in water with practical knowledge of how to build a ship, knowledge based on years of experience in building yards. He presented all this with drawings of specific types of vessels. His book proved to be a combination of the tradition of practical works like those of Van Yk with the newer and more rapidly developing scientific tradition (Hasslöf 1972, 68 ; Timmerman 1963, 29-43).

Not incidentally the drive toward a more theoretical or scientific approach to building a ship created a new discipline, naval architecture. In the course of the 18th century the tasks of designing and of building ships became separated. Greater income and social status came to the designer, the man with some theoretical knowledge (e. g. Unger 1978, 95-97). Governments believed that the understanding and application of theory would allow them to bridge a technological gap, to allow them to catch up with more advanced states. While the Netherlands and to a lesser extent England in the 17th century could rely on a handcraft tradition to produce the best ships of the day France had to create a naval shipbuilding industry from virtually nothing. That explains interest there in theory which extended through the 18th century and to other states of Europe. It was not coincidental that Euleur's book on shipbuilding was published in St. Petersburg or that Chapman's book appeared in Stockholm. In each case the governments wanted to introduce a more scientific approach both because it was, as they saw it, the way of the future and also because it would cut short the training needed for their own designers. Perhaps too the scientific approach based on rules might allow the government greater control over the building of warships.

Though government officials may have favoured a theoretical appproach for building ships they did not succeed in eradicating traditional methods. Experience remained an important teacher on the wharfs, institutionalized through apprenticeships. Old ways remained in use in part because the new methods were not always reliable. More important though was the difficulty for the 17th and much of the 18th century in translating ideas and theory into practice. It was not really until well into the 19th century with advances in physics, in naval architecture through, for example, the use of tank tests, and changes in materials to iron and steel that "science" came to dominate the design and construction of ships (Hasslöf 1972, 70).

Two other trends become clear when examining the history of naval shipbuilding over two centuries. First, there was ever increasing specialization with warships. The distinction between warships and cargo ships had become blurred in northern Europe during the High Middle Ages but by the end of the 16th century ships for war could be and were purpose-built (Unger 1981). Cargo ships were pressed into service during wartime in the 17th century but the numbers kept falling so that by 1715 the division was complete (Albion 1926, 76). Among warships design specialization was increasingly obvious with a variety of types emerging for differing types of work. That made the vessels more efficient but also taxed the skills of ship designers. Second, warships got bigger. The displacement of the largest warships continued to increase throughout the period as did the dimensions. The rate of growth in size and number of guns was even greater for small warships as they closed the gap with the biggest vessels (Howard 1979, 181 ; Naish 1957, 482-483). Builders increased the firepower of the ships while also improving their sailing performance. The ability to do both at once proved that the scientific or whatever methods designers used did yield better and better results.

Great similarities existed among the ships of European navies. All vessels were of the same basic design. Though there was specialization in warship design, often the same types appeared in a number of different navies. Even the names of types were similar in different languages. Experience in naval battles in the century after 1550 had shown that the old tactics of fighting ship-to-ship where victory depended on individual courage and seamanship were obsolete. Theoreticians as early as the 1540's urged naval commanders to use their guns more effectively. But captains still usually resorted to boarding to decide battles. By the second half of the 17th century, however, navies fought at a distance. They used their long range guns first to try to disable or defeat the enemy. Fleets began to fight in lines so navies had to get rid of the weak vessels, to achieve homogeneity of the ships-of-the-line. That term became common and over time so too did the standardization of the largest fighting ships (Hough 1969, 108 ; Nicolas 1961, 31 ; Padfiled 1973, 32-35, 51-52 ; Tramond 1916, 124). The classic ship-of-the-line had three masts and two or three decks. The form was already fixed by the 1670's. The only obvious change over the next 100 years were the lowering of the rake with the castles becoming even smaller than before or disappearing entirely. Also the bows became more rounded so more guns could be mounted there (Hough 1969, 134).

The number, variety and size of supplementary vessels in European navies grew in the 18th century. It was with those ships and boats that designers were most inventive. Navies found uses for sloops, schooners, cutters, brigs and corvettes. Yachts, first developed in the Netherlands in the late 1500's, were transferred to England with the Restoration of the Stuart monarchy in 1660 and then found a place as a type of sloop-of-war. Oared ships were also subject to change and improvement but not in the Mediterranean. There galleys remained much the same, being dropped from the French navy only in 1773. It was Swedish and Russian builders in the Baltic who produced new forms of hybrid sail-oar vessels (Hough 1969, 155-161 ; Howard 1979, 156, 216-217 ; Unger 1978, 50-51).

There were two virtually new warships developed by European designers in the period. The first addition was an altogether new type, purpose-built to exploit a different type of gun and to solve a specific problem. In 1682 for a French attack on Algiers Bernard Renau d'Eliçagaray designed and built light galiots equipped with two mortars, one on each side of the vessel forward of the mainmast. Armies had used mortars on land for some years but because of their recoil no navy had found a use for them before the French designer. He is also given credit for starting the use of mathematical formulae in designing Fench ships and for beginning the rationalization of building. The bomb vessel had extra reinforcement to take the weight and recoil of the mortar. In the French form there were two masts, a mainmast amidships and a mizzen. Renau's vessels proved a great success against the fortifications of Algiers and other navies quickly imitated the French precedent, the English navy getting bomb vessels in 1687 and the Danish soon afterward (Hough 1969, 162-164 ; Howard 1979, 159 ; Nicolas 1961, 29-30 ; La Roncière 1920, 716-717 ; Symcox 1974, 38-39). The second addition was a new cruiser which French builders first produced in the 1730's. The name frigate was borrowed from an earlier oared vessel. The new type though was essentially a small version of the predominant warship but with the lower deck at or below the lead waterline so that no guns could be carried on what was called the gun deck. It also had in many cases two turns to the bilge which helped in both speed and stability. The frigate was not able to take a place in the line of battle but was large enough to be an effective patrol and convoy vessel. The new frigate also became a common component of all European navies, the English adding them in 1747, the Dutch by 1750 and the Danes in the 1750's (Gardiner 1975, 163, 165, 167-168 ; Howard 1979, 219).

The sail plan chosen for European warships also proved highly consistent. The similarities increased over time. If anything the similarities in rigging were greater than those of hull design. The tendency, already established in the 16th century, to divide the sails continued throughout the period. The increase in the number of sails meant an even more rapid increase in the quantity of standing rigging and a need for greater skill among sailors to handle it. As warships grew in size so too did the sail area. The fourth mast or bonaventure mizzen, common in the early 16th century, disappeared after 1625. More sails were put on the remaining three masts where, by the 18th century, topgallants were normal and royals not uncommon. The spritsail-topsail, and addition to give greater control, proved dangerously exposed out at the end of the bowsprit and not very effective. The introduction of headsails, taken over from small boats, in the late 17th century, solved the problem more effectively. The forestaysail or jib was the first to appear but soon all masts had at least one staysail. Despite the superiority of the jib the spritsail topsail and the spritsail, slung under the bowsprit, remained in use until about 1750. The mizzen lateen sail was always difficult to handle but it was not until the first years of the 18th century that navies replaced those sails with simpler and more effective gaffsails, drivers, hung from a single yard with the leading edge attached to the mizzenmast. Another addition was studding sails, trapezoids of canvas hung from small yards attached at the ends of the standing yards. Studding sails were difficult to rig and only useable in fair weather. They were only worth having on ships with large crews where speed might prove critical so 18th century warships almost always carried them (Anderson 1927 ; Naish 1957, 485-486, 496-497 ; Hough 1969, 126, 134 ; Howard 1979, 125-126 ; Unger 1978, 55-58).

As design and rigging of warships became more similar it was necessary to have some other way of recognizing foreign vessels. The increasing use of national flags suggest how difficult it had become for sailors to tell where a ship was from. Decoration of warships helped recognition when it was at its height it the late 17th century. The ships of Louis XIV reflected the power and authority of the absolutist monarch as much as did his baroque palaces. Other European monarchs followed as did the staunch Calvinists of the Dutch Republic. Some of the finest artists of the day did the carving, gilding and painting. ornamentation was extensive with hundreds of different works on a large warship. Around 1700 the decoration race came to an end. It proved too costly and the quantity of ornament was so great that it was beginning to hinder sailing. The final blow was a change in taste toward a greater classicism and simplicity of design (Hough 1969, 117-118 ; 148 ; La Roncière 1920, 379-381 ; Naish 1957, 493-495).

The reasons for the similarities among Europe's warships were and are largely obvious. Europe enjoyed a common technological tradition in ship design. Since the vessels in European navies were descended from common ancestors, the full-rigged ship and the galley, it is no surprise that they were much the same. The warships operated against each other in the same waters so it was logical for designers to opt for similar solutions. Also navies used the same type of armament. While there might be some variations in weight and length of the barrels still all navies came in the 17th century to rely on heavy muzzle-loading cannon as the principal weapon of warships. It was the job of designers to generate the most stable and spacious gun platform possible. The ratio of weight to firepower that gunfounders could generate as well as the reliability of the cannon circumscribed many of the decisions of shipbuilders. Governments recognized the limitations created by armaments. They also recognized the potential savings in wood and in inventory costs that standardized ships would give them. Homogeneity of fleets was dictated by fighting in line-ahead so governments from the mid-16th century on laid down categories for warships, the establishments. Precise dimensions and numbers of guns were fixed for each rate. But getting the government directives to take effect was difficult at the outset since designers had their own ideas. Colbert in France by the 1680's succeeded in imposing six rates on the ships of the navy, that is not including the sloops, bomb ketches, cutters and other small vessels. In England it started with the division of the navy into four ranks in the reign of King James I but that was changed to six rates in 1655. There were periodic revisions of the establishments generally reflecting the increasing size of warships. By the late 18th century Britain was using six rates while other states had three with two divisions of each of the three rates. By imposing establishments navies could and did make their warships basically like those of other states (Albion 1926, 81-82 ; Baugh 1977, 198-200 ; Hough 1969, 132-133 ; Howard 1979, 92-95 ; Jenkins 1973, 43, 143 ; Röding 1793-1798, 342-343 ; Symcox 1974, 33-34).

The general trend toward a more scientific approach to shipbuilding also contributed to the similarity of European warships. The books which were the foundation of that scientific approach were translated and circulated throughout Europe. But shipwrights did not necessarily have to read books to learn about building in other countries. There were other more immediate and effective methods of transfering information, methods directly subsidized by governments. Colbert sent representatives to other countries to find out about how ships were built there. He laid down very specific instructions for his "spies". Britain too sent experts overseas to learn about the practices of foreigners (Clément 1864, 1, 199-200, 211, 234-237, 2, 296-300 ; Colenbrander, 2-44). The agents often stayed some time, the welcome guests of another country. An even more direct way to get information across borders was to move the shipbuilders permanently. Shipwrights like most skilled workmen were mobile in the 17th century. Wars, religious persecution and shifting economic fortunes all played a role in keeping some builders on the move. Dutch shipcarpenters could be found in many Baltic ports in the early 17th century. Dutch immigrants designed and did much of the work on the **Gustavas Vasa.** Colbert was notorious for importing foreign workers, especially Dutch but also Italian and Spanish. Bernard Renau d'Eliçagaray was a Basque and the best of the foreigner designers. Biagio Pangallo, or Maître Blaise as he was known, came from Naples. Chapman was born in Sweden because his father had been hired away by the king from the British navy. His position in Sweden did not prevent Chapman from spending some time working in England. There was a long list of builders brought into Russia by successive czars in the 18th century. Peter the Great established the precedent for contact in the last years of the previous century working in Dutch and English shipyards. Even Holland, once the source of shipbuilding talent, became the recipient of foreign technicians when in the 1720's Dutch Admiralties imported English ship designers to stem the decline in the quality of Dutch warships (Bruijn 1972, 18-24 ; Clément 1864, 1, 132-133 ; Jenkins 1973, 40 ; La Roncière 1920, 373-374, 378-379 ; Timmerman 1963, 30-31).

If the transfer of personnel was not enough to make European warships similar than the transfer of ships guaranteed common designs. States purchased ships from overseas when their own shipbuilding industries proved unable to provide needed vessels. Richelieu turned to the Dutch when he wanted to build up the French navy quickly. Colbert later in the century did the same while developing domestic shipyards and skilled personnel (Bamford 1956, 14 ; La Roncière 1920, 375 ; Tramond 1916, 155-156). Many governments, with the exception of the British which insisted on building at home, imported ships from the Baltic or overseas colonies such as Cuba in the case of Spain because they found such vessels better and cheaper (Albion 1926, 75 ; La Roncière 1920, 373-374). If navies found imported ships better then they ordered native shipwrights to copy what they saw. Modelling off the design of another ship was common practice. Richelieu and later French ministers told builders to copy imported Dutch warships. In 1672 a French 74 - gun ship so impressed the King of England that he ordered a copy built and in turn nine copies were made of that ship. The first frigates in the British navy were copies of a French prize (Gardiner 1975, 163-165 ; Hasslöf 1972, 61-62 ; Hough 1969, 109). It was also not uncommon for ships of one navy to end up in the fleets of their enemies. In peace time warships were sold and in war time often captured as prizes. Some of the most valued ships in the English navy were French-built while, after modification, the French navy often found English-built warships both useful and durable. Ultimately one of the reasons the warships of different states ware similar was that they were the same ships.

Despite all the reasons for European warships being the same there were differences among them. Most shipbuilders in Europe, after setting up the keel and posts, built up the ribs and then added the hull planking. Dutch shipbuilders before building the ribs put a few of the bottom planks in place, some five to ten on each side of the keel. After finishing the ribs they nailed the planks to the floor timbers. Writers such as Witsen and Van Yk mentioned the building method. So too did Colbert's spies (Colenbrander 1919, 12 ; Hasslöf 1972, 58-60 ;

Timmerman 1979, 26-27 ; Witsen 1690, 164-165 ; Van Yk 1697, 67-73). A Swedish writer at the end of the 17th century illustrated the Dutch method as being different from the way Swedish shipwrights built warships. There were many Dutch shipwrights working throughout Europe in the 17th century and it would be suprising if they did not take their method with them. The problems Scandinavian governments had in trying to convince native shipwrights to conform to typical European practice, problems which continued into the twentieth century, suggest that the Dutch method was not unknown outside of the Low Countries (Christensen 1873, 141-143 ; Ralamb 1691, 43-44). By the 18th century for large warships the Dutch method must have been old fashioned, a leftover from the handicraft tradition. Even using that method the Dutch were still so fast at producing their ships as to be the envy of other countries. English builders were extremely slow taking three years normally to produce a warship. Smaller ships could be done faster, a 40-gun ship taking a year for example. Colbert tried to decrease the time it took to put out fighting vessels. His model was the Dutch and he had his spies report on all the labour-saving devices used in Dutch yards. In the late 17th century, thanks to tight administrative control, standardization of parts and of design, French shipyards performed impressive feats finishing a complete warship in a day and a galley in half that time (Albion 1926, 77-78 ; Baugh 1965, 254 ; Clément 1864, 1, 295 ; Colenbrander 1919, 15).

The shape of the central rib which dictated the design of the hull varied from one navy to another. Dutch ships had a flatter cross-section, the botton being broader than with English or French ships. The greater beam of Dutch ships led to violent rolling action in heavy seas but yielded a more stable gun platform in calm waters. The smaller draught kept the lowest guns well out of the water so long as the weather was good. It also allowed Dutch ships to carry less canvas. The Dutch built lighter ships using lower quality wood, worrying more about economy than quality so it was said (Baugh 1965, 249-150 ; Baugh 1971, 196-197 ; Clément 1864, 1, 336-337, 2, 304-305). Dutch ships were capacious, with more room for stores. This may have been a hangover from the use of merchantmen in navies. The change to exclusive use of warships was a distinct disadvantage to the Dutch since they had so many merchantmen. In the 1670's the Dutch had to replace many of their warships to catch up with the superior fleet of the French. They succeeded by cutting down the tumble-home going to almost straight sides, a practice later followed by others, and increasing armament. They still retained some typically Dutch design features however (Colenbrander 12-14 ; Hough 1969, 107 ; Symcox 1974, 36, 40-41). They could not lengthen their ships much since drooping, caused by the heavy guns, fast became acute with the nearly flat bottoms. The Dutch became masters at squeezing in extra guns since length was limited. Sterns of Dutch ships tended to be flat, lessening draught, well into the 18th century. Other states went over to a round-tuck stern in the second half of the 17th century. Dutch builders were also slower in getting rid of all the upperworks which were common to 17th-century warships (Howard 1979, 95-96 ; Timmerman 1979, 25-26).

Swedish and Danish and German warships were much like Dutch ones in the early 17th century. Over time Scandinavian builders changed to following English or French practice more closely, always adding features suited to their own unique circumstances. Warships in the North were certainly technically on a par with the ships of the larger western European navies (Howard 1979, 91).

The French navy started the 17th century following Dutch design as well. Under Colbert the French developed new designs, the result of the promotion and application of new ideas. French ships were larger carrying more men and heavier guns but fewer guns for a specific length than ships of other navies. They were typically much sharper with deeper draught and carried more canvas. The result was faster and more maneouvrable warships. Vessel for vessel French ships were at the close of the 17th century superior to those of England or the Netherlands and probably superior to any warships in Europe. The difficulty was that France could not afford to maintain a fleet large enough to contest control of the seas with the combined forces of Britain and the Dutch Republic. As a result she resorted to commerce raiding keeping her battle fleet only as a potential threat (Symcox 1974, 32, 37, 42). That strategic choice defined the design of French ships. After 1700 they became even more sleek, less well-armed, able to get away from an enemy ship. In the middle of the 18th century French frigates had fewer guns, were more lightly built and carried more canvas than the English copies (Gardiner 1975, 168-170). French shipbuilding had great influence throughout Europe in the 18th century. In Spain where French theory and design were combined with English skill shipyards produced some of the best sailing warships built (Hough 1969, 130-131).

English ships were built to hold command of the seas. The British navy expected to fight battles and to look after the protection of trade. British warships were typically longer, lower, carried more guns and were more sturdily built than their French or Dutch counterparts. Traditionnaly English ships had deep draught. Only slowly in the 17th century was that reduced to increase beam giving an almost semicircular cross-section by around 1700. That improved buoyancy and decreased crankness. English builders were slow to move toward straight sides retaining considerable tumble-home well into the 18th century. They did lower the superstructure and made the bows more bluff (Baugh 1977, 200-201 ; Clément 1864, 2, 326, 329 ; Hough 1969, 110, 117 ; Howard 1979, 95-96, 177-180; La Roncière 1920, 377). English shipwrights of the 18th century suffered a great deal of criticism. There seems to be little question that the largest of English warships did not compared favourably with foreign vessels. On the other hand for smaller vessels, for example 74-gun ships, English vessels could match any others (Albion 1926, 78-79 ; Baugh 1965, 250-252 ; Baugh 1977, 195-198 ; Hough 1969, 129-130).

Just as there were good reasons for similarities in the design of European ships there were good reasons for the differences. Supplies of personnel varied tremendously and not just skilled shipbuilders but also sailors. Britain or the Netherlands because of their extensive shipping had a large pool of talent on which to draw. Russia was in the opposite position. Navies needed experienced sailors and gunners. Even galleys needed at least one knowledgeable rower at each bench. If there were shortages of talent then some accomodations had to be made in the design of warships. English and Dutch designers had great latitude but designers in Sweden or Prussia knew that they had to keep the vessels relatively simple and highly reliable. The variation in the ways builders were trained in the different states affected ship design. in the Netherlands and England shipwrights learned through apprenticeships from experienced masters. In France there was formal schooling as well. Not only shipwrights but from 1680 all naval officers had to attend schools on construction. England got a school for warship designers only at the end of the 18th century (Albion 1926, 79-80 ; Symcox 1974, 35 ; Timmerman 1963, 17). The more extensive system of formal education in France, set by the state, certainly sped up the process of disseminating information and of creating new knowledge about shipbuilding.

Increasingly the state took responsibility for all aspects of naval shipbuilding. As warships became more specialized construction went to government-owned yards. When the state did contract work out to private yards bureaucrats oversaw, even if inefficiently, the work of the non-government shipwrights. It was in the private builders' interest to build the ships larger since they were paid, at least in England, by the total amount of wood they had to buy and form. Even so private yards could produce smaller warships faster and cheaper than government yards. Control over contract builders for maintaining quality and homogeneity of design may have been incomplete but such control did become greater over time. In the early 17th century builders began presenting drawings of their ships to get design approval. Admiralties demanded more and more information so that by the late 18th century builders were required to make presentation models for their bureaucratic superiors (Albion 1926, 88-89 ; Baugh 1965, 255-260 ; Gardiner 1977, 66 ; Perrin 1918, xciv, xcvii). By placing naval shipbuilding increasingly under the control of government officials design of ships became a product of the special ideas or even whims of the navies' rulers. Government interest could have a deadening effect on innovation. The establishments did stadardize designs but ran the risk of stifling development. Officials were often extremely conservative, interested in easing their own adminstrative problems rather than in bettering warships. The officials rarely had any practical experience. The establishments did vary from state to state and governing bodies did change the establishments periodically. So tighter government control over design introduced another source of variation in the warships of the states of Europe. Despite the establishments though shipbuilders still found it possible to experiment and innovate, to make design changes even within the parameters laid down by their superiors (Baugh 1965, 251-253 ; Baugh 1977, 198-200 ; Gardiner 1975, 163-164).

Each government was subject to a different limit on funding. The size and wealth of the country made a great difference to the level of support for the navy as did the policies which governments chose. Sicily, small, impoverished and suject to centuries of poor government could not mount the same naval effort as Spain. Britain with its rapidly growing overseas trade and flourishing colonies had to protect and could afford to protect her merchant vessels during wartime around the world, something never thought of in Russia. Navies were subject to tremendous variation in the attention they received from their governments. Enthusiasm like that in France of Richelieu, Colbert and his son, Seignelay, or the Duc de Choiseul led to rapid growth in naval building and in the ability to develop new designs. Other states had similar experience with monarchs like Charles I in England or Christian IV in Denmark who lavished favour on their navies. In between the times of expansion there were long periods of neglect. The variations in government enthusiasm led to variations in the quality of ships (Bamford 1956, 8-9 ; Hough 1969, 116, 135-136 ; Jenkins 1973, 128, 142-144 ; La Roncière 1920, 334-335, 387, 396 ; Tramond 1916, 179-180, 213). Changes in government policy did at times have a direct and major effect on ship design. For example the need for cruisers to defend convoys led the British navy to adopt frigates rapidly in the mid-18th century. The early 17th century desire to dominate the North led Sweden to building big warships. Fiscal reasons led the French government at the close of the 17th century to change to building commerce raiders (Gardiner 1975, 168-170 ; Symcox 1974).

There were two other forces which lead to international differences in ship design, forces which were perhaps the most obvious even to contemporaries. The first was problems with supplies of raw materials. It took a great deal of wood to build an 18th century warship. The largest at 2000 tons reached the limits set by the character and size of the biggest trees. Builders needed unique sizes and shapes of specific woods, especially oak. Navies increasing use of iron for supports and copper for sheathing ships' bottoms compounded the difficulties of supplying raw materials (Albion 1926, 5, 11 ; Banbury 1971, 56 ; Colenbrander 1919, 11). Naval shipbuilding gradually consumed the available suitable timber in western Europe. The Netherlands always had to import all of its wood. France tried to rely on domestic supplies and from 1669 legislation gave the navy first choice of any tree in the kingdom. Even so by the end of the 17th century the French navy had to go further afield to Italy and to Albania for ship timber. Those supplies proved more expensive. France never developed timber resources in the New World as did Britain and Spain. For all states in western Europe the principal source of shipbuilding wood was the Baltic. Access to supplies of fir masts and oak timber was critical for all navies in the West. Control over the Sound, the strait between modern Denmark and Sweden, was therefore of strategic importance. Navies could not stockpile enough wood to repair and replace their ships during wartime. They needed massive new supplies during hostilities and had to get them out of the Baltic. The ability of France to carry on war at sea was severely restricted by the ability of the British navy to cut off timber supplies. French designers had to be extremely careful in the ways they used wood and could never produce ships as robust and sturdy as the English since they could not be sure they could get the necessary wood. On the other hand builders in Sweden or in Russia could be profligate in their use of wood since they had an unlimited supply (Albion 1926, 25-27, 61-65, 140-144, 153-158, 166, 169, 198-200 ; Bamford 1956, 19-21, 27-28, 96-112, 206-211 ; Clément 1864, 2, 305-307 ; Colenbrander 1919, 27-28).

The character of harbours and domestic waters was a second critical factor in making warships different. The most obvious and extreme case was the Netherlands. Designers suffered a severe handicap since waters leading to Dutch ports were typically shallow, the extreme case being Amsterdam. For political reasons one of the naval shipbuilding yards of the Republic was vested in Amsterdam and even after the change to big and heavily gunned ships-of-the-line the Admiralty yard remained there. Dutch designers built their warships with something close to a flat bottom so they could negotiate the way into port. At Amsterdam that was not enough so designers had to create a unique piece of equipment, a "camel". It fit under the warship and lifted it up so it could go "over Pampus", that is over the shallows at the harbour mouth (Clément 1864, 2, 304-305 ; Hough 1969, 106 ; Naish 1957, 485 ; Timmerman 1979, 26 : Van Yk 1697, 360-361). A similar device was also used in Venice to get warships into port. Swedish and Russian warships of the late 18th century had shallow draught too since they were operated extensively in the Finnish archipelago. On the other hand since the English and French governments usually put their naval dockyards along estuaries there were no limits on the shape of warships in those countries.

The actions of governments, physical conditions and levels of skill all contributed to differences in the design of European warships. Even with those varying forces acting on designers they produced very similar vessels. By 1650 builders and governments had made all of the major decisions about warship design. After that it was a matter of accomodating the general design to local conditions and experimenting with potential improvements. With each experiment the potential for change decreased. Designers came closer and closer to the limits of what could be done, given the requirements set by government policies, the character and volume of wood available and the character of the weaponry. By the mid-18th century knowledge traveled easily and quicly from one state to another. Improvements were taken up right away by all fleets. The pace of technical innovation was slow, all the major breakthroughs having been made. Although the technology mays appear to have been stagnant that does not mean there was no effort to improve. With warships great effort was put into making the design of existing types better. Until the development of cheaper iron, steam engines and reliable breech-loading guns warship design was almost fixed. Designers showed inventiveness in the variety of their solutions to the problems they faced, always within narrow limits. It would be valuable to explore further the various responses, especially among the apparently less important navies of Europe. A complete comparison of the major technical features of European warships would illustrate not only the inventiveness of designers and builders but also the process by which technology improved, the process by which Europeans came to discover how to build the best warships in the world. The image of a stagnant naval shipbuilding technolgy especially in the 18th century is neigher wholly accurate nor entirely suprising. After the first phase of enthusiastic advance in the years up to about 1650 as the ship-of-the-line became established the pace of innovation had to slow. Even with that slow pace though there were improvements. Ship designers proved themselves capable of making the most of new knowledge, of exploiting advances in mathematics and science, and they did that while still dealing effectively the specific problems posed by the unique circumstances each one of them faced.

SOURCES

ALBION, R.G., 1926,
Forests and Sea Power, Cambridge, Mass.

ALLARD, C., 1695,
Nieuwe Hollandse Shceepsbouw, Amsterdam.

ANDERSON, R.C., 1927,
The Rigging of Ships in the Days of the Spritsail Topmast, 1600-1700, Salem, Mass.

BAKER, M., 1585,
"Fragments of English Shipwrightry", Magdalene College, Cambridge University, Pepys Library, 2820.

BAMFORD, P.W., 1956,
Forests and French Sea Power, 1660-1789, Toronto.
BANBURY, P., 1971,
Shipbuilders of the Thames and Medway, Newton Abbot.
BAUGH, D.A., 1965,
British Naval Administration in the Age of Walpole, Princeton.
BAUGH, D.A., ed., 1977,
Naval Administration 1715-1750, London.
BRUIJN, J.R., 1972,
"Engelse scheepsbouwers op de Amsterdamse Admiraliteits werf in de achttiende eeuw : enige aspecten", **Mededelingen van de Nederlandse Vereniging voor Zeegeschiedenis**, 25, 18-24.
BUSHNELL, E., 1664,
The Compleat Ship-Wright. Plainly and Demonstratively Teaching the Proportions used by Experienced Ship-Wrights..., London.
CHRISTENSEN, A.E., Jr., 1973,
"Lucien Basch ; Ancient wrecks and the archaeology of ships A comment", **International Journal of Nautical Archaeology**, II, 137-145.
CLEMENT, P., ed., 1864,
Lettres, Instructions et Mémoires de Colbert, III, 1 and 2, Paris.
COLENBRANDER, H.T., ed., 1919,
Bescheiden uit Vreemde Archieven omtrent De Groote Nederlandsche Seeoorlogen 1652-1676, II, 's-Gravenhage.
DUHAMEL du MONCEAU, H.L., 1752,
Elémens de l'architecture navale..., Paris.
EULER, Leonhard, 1749,
Scientia navalis seu Tractatus de Constuendis ac Dirigendis Navibus..., St. Petersburg.
FURTTENBACH, J., 1629,
Architectura Navalis Das ist : Von dem Schiff Gebäu..., Ulm.
GARDINER, R., 1975,
"The First English Frigates", **The Mariner's Mirror**, LXI, 163-172.
GARDINER, R., 1977,
"The Frigate Designs of 1755-57", **The Mariner's Mirror**, LXIII, 51-70.
HASSLOF, Olaf, 1972,
"Main Principles in the Technology of Ship-Building", in **Ships and Shipyards Sailors and Fishermen, Introduction to Maritime Ethnology**, O. Hasslöf, H. Henningsen, A.E. Christensen Jr., eds., Copenhagen, 27-72.
HOUGH, R., 1969,
Fighting Ships, London.
HOWARD, F., 1979,
Sailing Ships of War 1400-1860, Greenwich.
JENKINS, E.H., 1973,
A History of the French Navy From Its Beginnings to the Present Day, London.
LA RONCIERE, C. de, 1920,
Histoire de la Marine Française, 2nd ed., V, Paris.
MURRAY, M., 1765,
A Treatise on Ship-Building and Navigation..., 2nd ed., London.

NAISH, G.P.B., 1957,
"Ships and Shipbuilding", in **A History of Technology**, C. Singer, E.J. Holmyard, A.R. Hall, T.I. Williams, eds., London, III, 471-500.
NICOLAS, L., 1961,
Histoire de la Marine Française, 2nd ed., Paris.
PADFIELD, P., 1983,
Guns at Sea : a history of naval gunnery, London.
PERRIN, W.G., 1918,
"Introduction", in **The Autobiography of Phineas Pett**, London, xv-civ.
RALAMB, A.C., 1691,
Skepps Byggerij eller Adelig Ofnings Tionde Tom, Stockholm.
RODING, H.R., 1793-1798,
Allgemeines Wörterbuch der Marine..., Hamburg.
SUTHERLAND, W., 1711,
Britain's Glory or, Ship-Building Unvail'd Being a General Director for Building and Compleating the said Machines, London.
SYMCOX, G., 1974,
The Crisis of French Sea Power 1688-1697, The Hague.
TATE, V.D., 1941,
"The Instruction nauthica of 1587", The American Neptune, I, 191-195.
TIMMERMAN, G., 1963,
"Das Eindringen der Naturwissenschaft in das Schiffbauhandwerk", **Deutsches Museum, Abhandlungen und Berichte**, 30, 3, 5-53.
TIMMERMAN, G., 1979,
Die Suche nach der günstigsten Schiffsform, Oldenburg/Hamburg.
TRAMOND, J., 1916,
Manuel d'Histoire Maritime de la France, Paris.
UNGER, R.W., 1978,
Dutch Shipbuilding Before 1800. Ships and Guilds, Assen.
UNGER, R.W., 1981,
"Warships and Cargo Ships in Medieval Europe", **Technology and Culture**, XXII, 233-252.
WITSEN, N., 1690
Architectura Navalis..., 2nd ed., Amsterdam.
VAN YK, C., 1697,
De Nederlandsche Scheeps-bouw-konst Open Gestelt, Amsterdam.

XVI

The Tonnage of Europe's Merchant Fleets 1300-1800

The relative size of European merchant fleets was an issue of some importance at the beginning of this century. However, historians have explored the question of total tonnage rarely, if at all, since World War I. There were no conclusive answers given back at the start of this century, nor is it possible to offer any final authoritative measure now. Nonetheless, some of the large body of work on maritime history done over the past eighty years does offer more information about the size of merchant fleets. That information can be used to generate, if not a final answer, at least a more accurate picture of the scale of European shipping from the late Middle Ages to the beginning of the age of steam.

Measuring the relative size of merchant fleets was an issue for nationalist historians around 1900. They were concerned with questions about the relative economic and political strength of different parts of Europe. They used data on shipping capacity along with many other sources to indicate changes in the structure of international power in early modern Europe. Given the state of technology in that period, the size of the merchant fleet did offer an indication of relative naval strength. The size of merchant fleets can still offer a barometer not only of political strength but also of economic strength. It is clear that economic growth did not start with the Industrial Revolution and that Europe went through a long-term increase in productive capacity, with variations certainly, from the late Middle Ages. That growth is very difficult to measure, but the size of merchant fleets may be a way to explore the pattern of that economic growth.

I am indebted to the late Ralph Davis, Lawrence Stone, Murray Tolmie, Jon S. Kepler, and Geoffrey Scammell for their comments and suggestions on an earlier version of a related paper. I am also indebted to J. R. Bruijn and to the participants in the seminar of the International Commission for Maritime History, King's College, University of London, for their criticisms and observations.

Shipping was an industry that enjoyed rapid expansion in the early modern period. It was also a modern industry by the standards of the day. Ships and their sailors used inanimate power with increasing effectiveness, converting it to work which could in turn yield direct benefits. It was a pattern typical of the later Industrial Revolution. The data on the size of merchant fleets can also serve to show the effects of government policies designed to direct the size and composition of merchant marines. Governments took seriously their role in deciding what types of ships and shipping were best, so their actions deserve careful consideration as an indication of the kinds of government policies that officials thought were most effective in generating the desired results.

Though it is possible, in light of the intervening work since the turn of the century, to say more about the scale of European shipping in the period, it is still a matter of counting. It is premature to expect much more than the most simple analysis of the data. Discussion must be rather more descriptive and introductory. Despite that, some rather stark results emerge from simple inspection of the most straightforward of the figures.

I

The magisterial effort to compare the size of fleets was that of Walther Vogel, the prominent writer on German maritime history, in his "Zur Grosse der europäischen Handelsflotten im 15., 16. und 17. Jahrhundert," which appeared in 1915 in the *festschrift* for Dietrich Schäfer, another historian devoted to recording the historical importance of maritime affairs to Germany.[1] In the

1. Walther Vogel, "Zur Grosse der europäischen Handelsflotten im 15., 16. und 17. Jahrhundert," in *Forschungen und Versuche zur Geschichte des Mittelalters und der Neuzeit, festschrift Dietrich Schäfer* (Jena, 1915), 268-333.

Reprinted from *The American Neptune* 52 (Fall, 1992). Peabody Museum of Salem, Essex Institute, East India Square, Salem, MA 01970, USA

aftermath of German unification, the naval arms race with Britain, and then the First World War, such work took on a special urgency. In 1896 Michael Oppenheim, the historian of the English navy, had already given some data on that navy, most notably for the sixteenth century.[2] But since the great summary by Vogel there has been no effort to take on the whole of Europe over an extended period. There have been some notable forays dealing with a shorter term or for one jurisdiction. Notable among them is the quite ambitious work by the English historian of navigation, David W. Waters, who committed himself in a graph to English and Scottish figures for the years 1580 to 1780, connecting points where there are some moderately reliable figures though avoiding issues of fluctuations between those points.[3] J. R. Bruijn, the historian of the Dutch navy, offers some estimates as well for the size of the Dutch merchant marine, but his figures are restricted largely to the seventeenth century.[4] Combined with other more isolated estimates of shipping capacity and with figures collected earlier, those data do serve to give, at the least, a general impression of the development of European shipping.

In all cases, historians have had to rely on essentially two types of sources for their figures. The first and best source by far is government surveys. In the reign of Queen Elizabeth I, for example, the English state embarked on a comprehensive effort to establish the number of ships in England, their tonnage, and in some cases, the number of sailors available for service in them. For England there are port records and returns from royal efforts to impress ships for naval service from the fourteenth and fifteenth centuries, as well as a few returns done for the Court of Admiralty for some districts beginning in the 1530's. But none of these was on a scale of the reports from the reign of Elizabeth. The figures continue with a survey done in the reign of Charles I. The next step was a system of official returns which began in 1686 and continued throughout the eighteenth century, although the method of measurement was changed in 1773, making comparison with earlier figures difficult.[5]

Similar quasi-public investigations in Italian merchant republics in the fifteenth and sixteenth centuries preceded the English surveys. In France, the greater interest in the navy in the seventeenth century as well as the expansion of the scope of government generated efforts to find out, if not how many ships might be available, at least how many sailors could be counted on for naval service. Manning the navies of Europe was a constant problem throughout the period, and eighteenth-century French rolls listing seamen available for service were an attempt to deal with that problem.[6] The surveys were, in general, a product of thinking about the economy and about the role of government which came to be called mercantilist. They were also a product of existing maritime technology which created the need for the information.

While the surveys are the best source of data on the size of the European merchant marine, they are not without their problems. Surveys were not common in many parts of Europe. States without a central government or even without a strong central government, like the Dutch Republic, did not have surveys. Even in those ports, republics, or kingdoms where surveys were made, not all ships were included. Some, especially smaller vessels, could and did elude enumeration. Coastal and inland vessels might or might not be included, yet some of the coasters were sizeable, and in some parts of Europe, such as England or Genoa, the coasting trades employed an important if not massive portion of total tonnage. Whether or not such vessels were included in the surveys, however, depended on the enthusiasm of the individual keeping the records and the decision of the compiler of all the local reports. In addition, flags of convenience are by no means a recent invention. Because of the frequent wars, the piracy, and the legal uncertainties to do with seizure at sea, captains could and did change their home port when it suited them. Some ships carried multiple papers, producing the appropriate ones when needed, and the changes in declared home port could skew surveys. For example, in 1572 the port of Emden boasted a total fleet of 42,000 tons. That figure was larger than the one for the whole English merchant fleet in that year.[7] Emden was next to the Dutch border and so offered a safe and, as it turned out, temporary haven for shipowners from Holland and

2. Michael Oppenheim, *A History of the Administration of the Royal Navy and of Merchant Shipping in Relation to the Navy* (London, 1896), especially 171-177.

3. David W. Waters, "'The English Pilot' and English Shipping: English Sailing Directions and Charts and the Rise of English Shipping, 16th to 18th Centuries," The Taylor Lecture in *The Journal of Navigation* 42, no. 3 (September 1989): 317-354.

4. J. R. Bruijn, "Recrutement et enrôlement dans la marine marchande dans les ports néerlandais (1600-1900)," in S. Cavaciocchi, ed., *I porti come impresa economica* (Prato, 1988), 551-567. The same or similar estimates can be found in his "De vaart in Europa," in L. M. Akveld, et al., eds., *Maritieme Geschiedenis der Nederlanden*, vol. 2 (Bussum, 1977), 200-241.

5. Oppenheim, *Administration of the Royal Navy*, 171-177; Ralph Davis, *The Rise of the English Shipping Industry in the 17th and 18th Centuries* (London, 1962), 6-7, 395-397.

6. James Pritchard, *Louis XV's Navy, 1748-1762: A Study of Organization and Administration* (Montreal, 1987), 71-74; Alain Cabantous, *La Vergue et les Fers: Mutins et déserteurs dans la marine de l'ancienne France* (Paris, 1984), 80-83.

7. Vogel, "Zur Grosse der europäischen Handelsflotten," 283. His source is B. Hagedorn, "Betriebsformen und Einrichtungen des Emder Seehandelsverkehrs in den letzten drei Jahrzehnten des 16. Jahrhunderts," *Hansische Geschichtsblätter* (1909): 359.

Ships along the Scheldt River, downstream from Antwerp, from *Zelandia Descriptio 1548-1550*. Courtesy of College Van Burgemeester en Schepenen, Antwerp.

Zeeland. The 1572 Emden figure, then, reflects nothing more than the political upheaval of the Dutch Revolt.

The second principal source for data on the size of merchant fleets comes from records of specific trades. Customs officials collecting duties on all goods or on certain goods at certain points sometimes directly, but usually inadvertently, recorded the volume of shipping. Using figures on the quantities of goods, it is possible to extrapolate to estimate the tonnage needed to carry that volume and from there to extrapolate to the size of the merchant fleet. Records of trades were the principal source for Vogel, among many others. The best of such records are those for the tolls collected by the kings of Denmark on ships and goods passing through the Sund. The Toll Registers run with few gaps from the second half of the sixteenth century to the end of the eighteenth. They are unique in that they cover such a long period and enumerate shipping in a number of different ports at a critical site for seaborne commerce in northern Europe. The records of other places and other ports are not as extensive but offer a similar basis for estimating shipping tonnage.

With toll records, however, there are problems as well. While they show the quantity of goods passing a particular point, they do not show how frequently the ships which carried those goods made a trip nor do they typically show the average size of the vessels. Consequently, extrapolation from the volume of trade to the volume of shipping is open to sizeable error. Duties fell on some goods but not on all goods, and since no ship carried just one thing there may have been more capacity, the ability to carry more goods, than is shown from the duty paid. It is also true that shippers did not always report accurately the volume of goods carried. Smuggling was common, and the higher the duty and the more diligent the customs collector the more profitable was dissimulation. Smuggling could lead to an underestimate of carrying capacity both because some goods were not reported at ports and because some ships avoided ports entirely and thus never appear in the records.

Efforts to calculate the size of European merchant marines are bounded by categories of the nineteenth and twentieth centuries. In an era of national states and of official registration, what is being measured is clear, but the concept of a national merchant marine was unknown in the sixteenth century. There are, at most, only the beginnings of such an idea at that time. Equally hazy in the sixteenth century, if not more so, was the concept of what constituted a ship. How large did a vessel have to be, for example, before enumerators considered it worth counting? Equally important, how big did it have to be to make it worth counting as part of the maritime strength of a state? Also unclear was which state or jurisdiction could claim a vessel. Contemporaries may have had loyalties to their own port or to some larger political unit. In addition, many political divisions appeared during or even after the era of sail. The English and Scottish merchant marines were counted separately, although the two kingdoms went through changing relationships in the period. Both Germany and Italy existed in the minds of Europeans in the age of sail, but as political units they only emerged in the 1860s. To attach a figure for the size of a merchant marine to, say, Germany in the seventeenth century would have been little more than a curiosity to contemporaries.

The standard of measure presents problems as well. The carrying capacity of ships is measured in tons, obviously. But tonnage is a difficult concept. It is perhaps some comfort that shipbuilders and officials in the fourteenth through the eighteenth century were also confused about what tonnage meant. Since 1854 the gross registered ton has been fixed at one hundred cubic feet or 2.83 cubic meters of enclosed space. It is a measure of volume, of the carrying capacity of a vessel. Unfortunately, there are at least four other measures of tonnage applied to ships.[8] The measure chosen to compare fleets in the age of sail is metric tons burden, which is equal to deadweight tonnage. Contemporary measures are then converted to tons burden. The principal reason for the choice of deadweight tonnage was that, by the sixteenth century, commonly used measures were roughly comparable to modern measures of deadweight tonnage.[9] Contemporary measurements were never highly accurate since methods for calculation were simple. The usual and most reliable way to find out the tonnage of a ship was to load it and see how much it could hold. Typically, tonnages were rounded off for the sake of simplicity, often at the expense of accuracy. Errors of measurement are made worse by the summing of reported figures and also by approximations made now on the basis of approximations made at the time.

II

Despite all the problems of uncertainty about what is being measured, about the units of measure and about the accuracy of the data, it is still possible to generate figures, even with a sizeable margin of error, which show a consistency and which reveal the broad general trends in European shipping in the period. To show the method for developing estimates and also to establish the basis for comparison with other places and other times, as well as to illustrate the problems in dealing with the interpretation of resulting data, the first estimates are generated for the reign of Queen Elizabeth I of England. That is the best-known period, where the documentation is the most complete and where the role of the state is most clear. Surveys carried out in Elizabeth's reign show that English shipowners became committed to large defensible ships, a choice in part dictated by economic circumstances, in part by existing technology, and in part by a government policy which promoted shipbuilding and especially the building of cargo ships which were also able to act as warships. It is virtually impossible to say which factor was the most important in determining the path followed by the English merchant marine.

---

8. The four others are: displacement tonnage, which is the amount of sea water displaced by the vessel either empty or loaded; deadweight tonnage, which is the difference between light displaced and displacement loaded and so is equal to the weight of cargo; freight tonnage, which is the volume a ship can carry, one measurement freight ton being equal to 1.13 cubic meters; and registered tonnage, which is tonnage based on some calculation using the principle measures of ships. Two examples are the English registered ton of the seventeenth century, changed in 1696 and again in 1773, which was a proportion of the measured tonnage, and the Spanish *tonelada* of the sixteenth century, which from 1570 on was equal to 2.6 cubic meters. All of those measures of tonnage and variations on them were known before, during, and after the sixteenth century.

9. Tons burden was derived from practice in the Bordeaux wine trade, where one cask or ton was equal to 960 kilograms plus the weight of the cask, or about 1,000 kilograms or 2,200 English pounds (one metric ton equals 2,240.6 pounds). The wine barrels of 2,200 pounds took up about forty-two cubic feet and a modern "freight ton" is equal to forty cubic feet. Similarly, in Italy the wine cask was used for estimating the size of ships: *botta*, or *botte*, or *boute*, the same as the English butt, though the size of the butt varied from town to town. Shippers in the North (that is, in England) equated the ton with forty cubic feet (1.13 cubic meters) while in the Mediterranean it tended to be fifty-one cubic feet (1.44 cubic meters) because builders in the South included more space for lighter cargo. In the Baltic and North seas, ships were rated by lasts, the last being a weight of grain equal to about 2,000 kilograms. By the seventeenth century the last was 112 cubic feet (3.2 cubic meters). If that was the case, then the specific gravity of the cargo was .63. Jean-Baptiste Colbert, the minister of Louis XIV, standardized the measure in France at the *tonneau* of 979 kilograms, equal to 2,158 English pounds. See F. C. Lane, "Tonnages, Medieval and Modern," *Economic History Review* 2d series, 17 (1964): 213-229; A. van Driel, *Tonnage Measurement: Historical and Critical Essay* (The Hague, 1925), 6-11; W. Salisbury, "Early Tonnage Measurement in England," *The Mariner's Mirror* 52 (1963): 41-51; and Christopher J. French, "Eighteenth-Century Shipping Tonnage Measurements," *Journal of Economic History*, 2d series, 33 (1973): 434-435.

This seventeenth-century engraving by Claes Jansz. Visscher shows the *Griffin*, a heavily armed fighting ship from the reign of Elizabeth I. Courtesy of the National Maritime Museum, London.

Economic circumstances changed with the changes in English trade during the second half of the sixteenth century. The different demands placed on shipping certainly had to affect the size and character of the merchant marine. In 1558 most English trade went through the port of London to the Low Countries, overwhelmingly to Antwerp. By 1603, however, English traders had expanded their range to a wider variety of Channel, North Sea, and Baltic ports. They also ventured to the New World and the Orient. On average, by 1603 the goods carried were of greater value per unit volume and were moved over longer distances than at the start of Elizabeth's reign.[10]

The shipbuilding technology of the sixteenth and seventeenth centuries created opportunities for governments to influence the character of ships constructed. The explanation for this development lies in the increasing differentiation between ships for violence and ships for peaceful trading. In the high Middle Ages, the cog was the dominant type of ship used for war in northern Europe. It was also the most efficient long-distance carrier of trade goods. But in the years around 1400 a new type, the full-rigged ship, appeared. It combined square sails and the triangular lateen sail which had long been in common use

10. G. D. Ramsay, *The City of London in international politics at the accession of Elizabeth Tudor* (Manchester, 1975), vii, and G. D. Ramsay, ed., *The Politics of a Tudor Merchant Adventurer: A letter to the Earls of East Friesland* (Manchester, 1979), 58-66; Ralph Davis, *English Overseas Trade 1500-1700* (London, 1973), 9; Jan K. Federowicz, "Anglo-Polish Commercial Relations in the First Half of the Seventeenth Century," *Journal of European Economic History* 5 (1976): 365-368; and Peter Ramsey, *Tudor Economic Problems* (London, 1963), 76-78. George Nedham, a London merchant writing in 1564, talks about the rise of Antwerp and how it has taken over the trade of all nations and funneled that trade through its port. Burghley, Lord Treasurer to Elizabeth I, writing ca. 1581 and perhaps overestimating the earlier history of English commerce, said: "The navy of englanc was hertofore occupyed by merchantes into Levant, to spayn, to portyngall, and now the commoditees which english shippes wer accustomec to bryng thence is for the most part fovnd in Antwerp, from which place on hoye will bryng as much in one yere as x merchantes shippes war wont to bryng from the other placees in ij yers." R. H. Tawney and Eileen Power, eds., *Tudor Economic Documents*, vol. 2 (London, 1924), 125.

in the Mediterranean. The new type of ship, because of the combined rig, was more maneuverable than its single-masted predecessors with no sacrifice in size. The new type also embodied a form of hull construction new to northern Europe but known in the south since the end of the first millennium. Skeleton building, with internal frames set up before the hull planks were tacked onto those frames, made possible both lighter and bigger ships with no loss of security. The new ship type created opportunities for developing a more effective naval vessel, since the full-rigged ship proved to be an effective vehicle for carrying guns on board. By the late fifteenth century, guns were made large enough, accurate enough, and reliable enough to take up an important place in naval warfare.[11] In the last years of the sixteenth century, the many small anti-personnel weapons used to sweep decks of the enemy in preparation for boarding were replaced, as were the few large bronze guns, by many relatively inexpensive and now reliable iron guns.

As a result, by the middle of the seventeenth century broadside sailing ships dominated naval activity and the navies of Europe. Guns, combined with the design of the full-rigged ship, proved the technical basis for the separation of warships from cargo ships in the course of the seventeenth century. Until that distinction became complete it was still possible, using the design of the full-rigged ship and guns of both bronze and iron, to get an economically viable sailing cargo ship which was at the same time defensible and an effective naval vessel.[12] So long as that was true, in the fifteenth, sixteenth, and early seventeenth centuries governments often took an active role in directing shipbuilding to get merchant vessels of greater size able to carry guns and so able to be pressed into naval service in wartime. In the reign of Elizabeth I, the English government recognized and seized the opportunity created by the state of maritime technology to direct shipbuilding. Other states followed the same practice both before and after the English embarked on such policies.

Elizabeth's government expected three major gains from the range of policies designed to promote growth of the merchant marine: first, that more ships would earn more foreign exchange, making it easier to pay for England's large and critical imports of military supplies;[13] second, that an increase in the number of skilled seamen would create a source of crews for naval vessels in wartime; and third, that properly built new merchant ships could be converted for naval use. That expectation could not have been fulfilled in the following century, since by then ships and shipping had changed, and cargo ships were very different from warships.

The primary concern of the English government in the second half of the sixteenth century was unquestionably defense. It was that overwhelming concern for protection of the realm which directed the government in most actions, both political and economic. Elizabeth's Privy Council ". . . worked under relentless pressure of war and fear of war. . . . Permeating practically all of its major economic considerations was the idea of how best to avoid problems that would threaten 'if warres should chance.'" A program of rearmament was at base the reason for the extensive effort to expand and improve English shipping.[14] There were acts to force goods onto English-built and English-operated ships. From 1563, foreigners were excluded from coastal trades, the most important being the shipping of coal from Newcastle to London. There were attempts to increase fish consumption — the so-called political Lent — and a general encouragement of the fishing industry. In 1540 the English merchant marine was not large enough to carry all of England's trade, and the government, using these and other measures, planned to change that.

One policy among the many designed to promote and direct merchant shipping and shipbuilding was a subsidy program for the construction of larger ships. Such payments were not new to the England of Elizabeth I nor to sixteenth-century Europe. In the reign of King Henry VIII (1509-1547) the English state had paid five shillings per ton for every ship built of one hundred tons or more. In 1540 that would have been equal to about one-sixth of the total building cost. The proportion was to fall over time in real terms, since the cost of construction rose while the subsidy per ton remained constant. Grants to builders continued until early in the eighteenth century, with increasing restrictions to ever larger vessels. Elizabethan bounties were the only general construction bounties ever granted in Britain.[15]

11. Richard W. Unger, *The Ship in the Medieval Economy 600-1600* (London, 1980), 231-234.

12. Ibid., 163-166; Richard W. Unger, "Warships and Cargo Ships in Medieval Europe," *Technology and Culture* 22 (1981): 233-252; Peter Padfield, *Guns at Sea: a history of naval gunnery* (London, 1973), 33-43.

13. Lawrence Stone, "State Control in Sixteenth-Century England," *Economic History Review* 17 (1947): 111; Edwin M. Bacon, *Manual of Navigation Laws* (Chicago, 1912), 12-15; Sybil M. Jack, *Trade and Industry in Tudor and Stuart England* (London, 1977), 97-8. The government in 1558 was concerned about reliance on imports of armaments from Germany through the Netherlands and so also set out to develop native mining and metallurgical industries to reduce their dependence on foreign suppliers. Ramsay, *The City of London*, 115-116; Vincent Ponko, Jr., "The Privy Council and the Spirit of Elizabethan Economic Management, 1558-1603," *Transactions of the American Philosophical Society*, New Series, 58, no. 4 (1968): 57.

14. Ponko, "The Privy Council and the Spirit of Elizabethan Economic Management," 56-57.

15. Oppenheim, *Administration of the Royal Navy*, 38, 88, 107, 167-168, 472; Royal Meeker, *History of Shipping Subsidies* (New York,

In the last years of the fifteenth and first years of the sixteenth century, the Venetian government had a similar policy goal — that is, to get big ships with naval potential. In Venice, direct subsidies proved an insufficient incentive to get the large sailing ships the Venetians needed to send against the Ottoman Turks. Instead, the government ended up building ships itself and then fixing freight rates to favor the large sailing ships so that they could pay their way in peacetime. Genoa granted similar construction subsidies for building large ships in the late fifteenth century. The Spanish government did the same in the sixteenth, though in the reign of the Emperor Charles V there was a break in the practice. King Philip II raised the subsidy in 1563 when he perceived a fall in domestic production of big ships. The government raised the subsidy again in 1597 for ships suitable for trade to the Indies. From 1584, the French crown as well made grants for the building of ships over three hundred tons.[16]

Thus, subsidies for construction to direct the character of merchant marines were common in Europe in the fifteenth and especially in the sixteenth and seventeenth centuries. Those policies squarely established the involvement of governments in shipbuilding. One result was surveys to see if the efforts to promote shipping and to direct shipbuilding toward the construction of big ships had been effective. The surveys, at least for England, suggest that subsidy programs were successful.

The first English shipping survey dates from 1560, the second from 1568, the next from 1571-1572, the next from either 1576 or 1577, the next from 1582, and the next, which is a limited survey of the Cinque Ports, from 1587. There is as well a survey from the reign of Charles I done in 1626. Most of the surveys of merchant ships were associated with some crisis, internal or external, in English politics. For example, the report of 1568 came one year after a Spanish army under the Duke of Alba had entered the Low Countries. The 1571-1572 survey coincided with the Ridolfi plot and the height of fears for the security of Elizabeth's throne. There was also another trade depression. The catalogue of merchant ships done in either 1576 or 1577 came as the temporary settlement in the Low Countries had broken down and Elizabeth found herself aiding the rebels against the Spanish monarch.

The Elizabethan surveys show, first, the total tonnage of the English merchant marine, and second, the proportion of big ships, vessels of one hundred tons or more, in the merchant marine. The most comprehensive was that of 1571-1572, which gave the tonnage of each ship trading from English ports. It is possible to use those figures to calculate the total tonnage in various size categories and the share of the total tonnage which can be attributed to each category. The 1571-1572 survey offers a good first approximation and a contemporary set of multipliers to use in estimating the total size of the English merchant marine based on the evidence in the other surveys. In 1571-1572, eighty-six of the 1,383 ships, or 6.2 percent, were of one hundred tons or more, but those same eighty-six formed 22.7 percent of total tonnage, which was 51,912 tons. Similarly, vessels of from six to forty-nine tons, 1,032 of them in all, were 74.6 percent of the total of ships but contributed only 45.5 percent of total tonnage.

The first report of 1560 covered only ships of one hundred tons and more. The government was obviously worried about the availability of large and powerful ships. Extrapolating, using the 1571-1572 proportions, and assuming ratios were about the same in 1560, the 1560 survey suggests that total tonnage of the merchant fleet was over 45,000 tons.[17] Similarly, the second survey of 1568, using 1571-1572 multipliers, yields a total for the English merchant fleet of about 65,000 tons. The 1577 return, a list of all English ships of one hundred tons and over, probably has a few gaps because certain ships were left out. By 1577, the total of large vessels was up to at least 135 from the eighty-six of 1571-1572, with an average tonnage of 131 tons. The same method yields an estimate for the entire merchant marine of almost 78,000 tons. The expansion in total carrying capacity reflected the improved fortunes of the cloth trade as well as the growth of trade to new areas: the Mediterranean, Russia, and the New World.[18]

1905), 1-4; Edwin M. Bacon, *Manual of Ship Subsidies* (Chicago, 1911), 11-12; Jack, *Trade and Industry*, 97; Brian Dietz, "The Royal Bounty and English Merchant Shipping in the Sixteenth and Seventeenth Centuries," *The Mariner's Mirror* 77 (1991): 5-20.

16. F. C. Lane, *Navires et Constructeurs à Venise pendant la Renaissance* (Paris, 1965), 99-105; Charles de la Roncière, *Histoire de la Marine Française*, vol. 1 (Paris, 1909-1920), 245; S. B. Clough and C. W. Cole, *Economic History of Europe* (Boston, 1946), 217; A. P. Usher, "Spanish Ships and Shipping in the Sixteenth and Seventeenth Centuries," in *Facts and Factors in Economic History, Articles by Former Students of Edwin Francis Gay* (Cambridge, MA, 1932), 192-194, and Earl J. Hamilton, "Spanish Mercantilism before 1700," in *Facts and Factors in Economic History*, 235-236.

17. Public Record Office, London, State Papers, Domestic, Addenda, Elizabeth I, 15/22/1 (1571-1572); Oppenheim, *Administration of the Royal Navy*, 172, and "The Royal Navy and Merchant Navy under Elizabeth," *English Historical Review* 6 (1891): 472.

18. British Library, Stowe MSS 570, fol. 151; Oppenheim, *Administration of the Royal Navy*, 173; F. J. Fisher, "Commercial Trends and Policy in Sixteenth-Century England," *Economic History Review* 10 (1940): 96-97. Fisher's figures come from the data reported by William Harrison in Frederick J. Furnivall, ed., *Harrison's Description of England in Shakspere's Youth*, New Shakspere Society, series 5, part 1, no. 5, (London, 1877), 290. It is also the estimate used by Vogel, "Zur Grosse der europäischen Handelsflotten," 321.

Table 1
Tonnage of the English Merchant Marine 1560-1629

Year	Ships over 100 tons: Number	Ships over 100 tons: Total tonnage	Estimated tonnage of all ships
1560	76	10,360	45,000
1562	73*		65,000
1571-2	86	11,784	51,912
1577	135	17,670	78,000
1582	177	24,150	80,000
1583	183	25,000**	90,000
—	—	—	—
1629	323		115,000

*Major ports not included
**This is a minimum

The 1582 survey showed 177 ships of one hundred tons and more, a marked rise from the seventy-six of 1560 or even the 135 of 1577. The total tonnage of big ships was at least 24,150 tons, an increase of 133 percent over the 10,360 tons of the first Elizabethan report. The total tonnage of the entire English merchant fleet in 1583 was almost 80,000 tons, if the method of estimating is accurate. A survey by county in 1583 demonstrated the same general trends. Extrapolating as with the other figures, we derive a total tonnage estimate for the entire fleet in 1583 at close to 90,000 tons. The proportion of tonnage attributable to ships of one hundred tons or more, a total of 183 ships, had risen from under 23 percent in 1571-2 to over 28 percent in 1583. The 1583 list is the last comprehensive one before 1629.[19] There is no doubt that the total size of the English merchant marine grew in the reign of Elizabeth I. There is also no doubt that the proportion of large ships grew. The presence of alternative uses for large ships in a number of trades, in naval operations, and in privateering made it more likely that such vessels could return a profit to their owners and so made it more likely that they would be built.

III

While at its height the Elizabethan merchant fleet probably reached 90,000 tons, it was still small by modern standards and, even more significant, it was not impressive compared to figures from the Middle Ages. It was also not impressive compared to contemporary Holland or major northern Italian port towns. However, the pattern of growth and the distribution of tonnage established during the second half of the sixteenth century in England was similar to the pattern followed elsewhere in Europe both before and after the time of Elizabeth I.

For medieval shipping there are no government surveys — there was no government interested in directing shipbuilding — so data must come from trade figures. In the Mediterranean, government yards, or arsenals, built warships. In the North, merchants had cogs built without government help. There were few tolls for which records survive and even fewer regulated trades, so data are scarce. The Bordeaux wine trade offers one of the few sources. In the early years of the fourteenth century the average annual export of wine from Bordeaux was over 80,000 tons. English ships probably carried more than 33,000 tons of that wine. If that was the case, then it is doubtful that the entire English merchant fleet, including coastal shipping, was less than twice that size. It was then probably over 70,000 tons. Moreover, the Bordeaux customs accounts suggest that the proportion of English ships of one hundred tons and more was greater in the early fourteenth century than in the late sixteenth, but that may be because the Bordeaux wine trade involved the use of relatively large ships. In the early fourteenth century, shippers called Normans by the customs collectors carried more than 18,000 tons of wine each year out of Bordeaux. Using a similar multiplier for them, the total merchant fleet for the area which would come to be called France

19. Jack, *Trade and Industry*, 98, reports a tonnage estimate for 1629 which is presumably based on the 1626 survey.

Scenes from a Venice panorama, from B. von Breydenbach, *Bevaerden tot dat heilghe grafft*, 1488. Courtesy of the National Maritime Museum, London.

was probably at least 40,000 tons.[20] The Bordeaux wine trade, a large-volume trade in a bulky commodity, collapsed in the second half of the fourteenth century. The collapse caused a marked fall in the total size of the English merchant marine and presumably in the average size of English ships as well.

In the fifteenth century it appears that there was a widespread movement toward smaller ships.[21] The Italian merchant republics tried to overcome that trend and to promote an increase in the size of their ships. In 1424, Genoa did a survey to find out the total tonnage of big ships. The government was concerned not with the size of the whole merchant fleet but only with how many large ships could be deployed. For Genoa, large meant ships of five hundred tons or more. Around 1450, Venice also took an interest in the proportion of large ships. A survey showed that some 19 percent of the total tonnage was in ships of 240 tons or above, but more than 50 percent of the tonnage came from ships of one hundred tons or better and only 9 percent of ships fell into that category.[22] By the late sixteenth century, the combination of economic and political pressure had reversed the earlier trend, and there were more and more big ships everywhere in Europe.

The Dutch had the largest merchant fleet in Europe in the seventeenth century. An estimate for 16?6, based on statements made in the Dutch States-General, put the total tonnage for the Dutch Republic between 600,000 and 700,000. The large figure included a sharp rise in the

20. Margery K. James, "The Fluctuations of the Anglo-Gascon Wine Trade during the Fourteenth Century," in E. M. Carus-Wilson, ed., *Essays in Economic History*, vol. 2 (New York, 1962), 130, 146-147; Yves Renouard, "Les transformations économiques et sociales," in Charles Higounet, ed., *Bordeaux sous les Rois d'Angleterre*, vol. 3 of *Histoire de Bordeaux* (Bordeaux, 1965), 423-424, 429, and Renouard, "L'exportation des vins gascons," in *Bordeaux sous les Rois d'Angleterre*, 254-257. For some additional data on medieval shipping, see the survey of 1,291 ships in Timothy J. Runyan, "Ships and Fleets in Anglo-French Warfare, 1337-1360," *The American Neptune* 46, no. 2 (Spring 1986): 91-99.

21. Geoffrey V. Scammell, "English Merchant Shipping at the End of the Middle Ages: Some East Coast Evidence," *Economic History Review*, 2d series, 13 (1961): 331-334.

22. Lane, *Navires et Constructeurs*, 95.

average size of Dutch ships from the 1570s to the 1630s.[23] Around 1600, then, when the English fleet was about 90,000 tons, the Dutch fleet was at least above 300,000 tons. The report of a survey from 1629 shows that while the English fleet had grown some 28 percent in the intervening years to 115,000 tons, the Dutch had grown 100 percent to 600,000 tons. In 1600 the Dutch fleet was three times the size of the English, but by 1630 it was six times the size of its English counterpart. England was not alone. The Dutch merchant fleet grew rapidly relative to all other fleets in Europe in the late sixteenth and early seventeenth centuries.

In 1636 the Dutch merchant fleet boasted 1,750 ships in service in commercial navigation in European waters, including the Mediterranean. Thirteen hundred of those ships were rated at over eighty tons. The average tonnage of all ships in the estimate was just under 177 tons, up dramatically from the average of thirty-nine tons for all English ships in the survey of 1571-1572. It appears that there was a dramatic change in the period from 1570 to 1636 in the size of ships. The Dutch specialized in a different kind of trade, carrying bulk goods, principally grain, from areas of surplus and low production costs to markets in western and southern Europe. At the same time, Dutch merchants and shippers moved into more distant trades in the Far East and the New World where larger ships were typical. By 1636 the Dutch fleet was highly diversified but still dominated by slow-moving, lightly built and lightly armed cargo ships designed for the carriage of bulk goods. It was a different kind of merchant marine, with vessels which could never be used as warships. The rapid growth of the Dutch fleet and Dutch success in exploiting rapidly growing trades helps to explain the English concern and confusion in the late sixteenth century about their own merchant marine. It was not until the second half of the eighteenth century that the English merchant fleet would reach the size of the Dutch of 150 years before.[24]

The same was true of Spain and Portugal. When Philip II had united the two crowns in the 1580s, the combined fleet was smaller than the Dutch but still more than twice the size of the English or the French; that is, it was in excess of 200,000 tons.[25] For one Spanish trade there is extensive data on ship size because it was so carefully and so closely regulated. By the late sixteenth century an annual average of some 150 to 200 ships were involved in the Indies trade, the trade to the New World. The total tonnage in the trade ran to some 15,000 to 20,000 tons each year, so the average size of ships was about one hundred tons. In the 1520s there were nearly one hundred ships per year, with a total carrying capacity of 4,500 tons, carrying goods across the Atlantic between Spain and her American outposts; thus, the average tonnage of the ships was forty-five tons.[26] In the course of the century, both the number of ships and the average size of the ships doubled, which led to a four-fold increase in carrying capacity in that trade alone. The vessels used in the Indies trade were highly defensible, so the Spanish crown got exactly the result desired.

The capacity of the combined fleets of Spain and Portugal appears to have been much the same in the late eighteenth century as it was in the late sixteenth. The 1786-1787 tonnage for Spain was 149,460 tons and for Portugal 84,843 tons. The two kingdoms had separated long before, of course, so combining the two figures for the later date is misleading. More misleading may be the similarity of the 1585 and the 1786-1787 figures, which probably masks a significant fall in the size of the merchant fleets of the two kingdoms in the course of the seventeenth century. There was discussion of a revival in shipping and shipbuilding, especially in Spain, in the course of the eighteenth century. The 1786-1787 figure could reflect a long-term recovery with Spain and Portugal finally, at the end of the eighteenth century, reaching the levels of shipping of two hundred years before.

23. On the size of the Dutch fleet, see Aksel E. Christensen, *Dutch Trade to the Baltic about 1600* (The Hague, 1941), 93-94, 444-447; Maria Bogucka, "Amsterdam and the Baltic in the First Half of the Seventeenth Century," *Economic History Review*, 2d series, 26 (1973): 437; S. C. van Kampen, *De Rotterdamse Particuliere Scheepsbouw in de Tijd van de Republiek* (Assen, 1953), 75-76; Richard W. Unger, *Dutch Shipbuilding before 1800* (Assen, 1978), 11.

24. Jonathan I. Israel, *Dutch Primacy in World Trade, 1585-1740* (Oxford, 1989), 6-10, 408-411. Though he argues that ". . . the view that bulk freightage was the key to general primacy in world commerce has never seemed wholly convincing" (p. 9), he does agree that one of the pillars and indeed a critical pillar of Dutch primacy in world trade was her massive shipping capacity (e.g., p. 411) which was largely in bulk carriers. See also A. P. Usher, "The Growth of English Shipping 1572-1927," *Quarterly Journal of Economics* 42 (1927): 467. He gives official tonnage figures which are probably low since coasters and small craft may well have been left out, but then the same problem exists for data from the sixteenth century. As late as 1701 the official tonnage of the English merchant fleet was still only 261,222, well below the estimate by Ralph Davis for 1702 which makes allowance for vessels not included in the returns. See table 2 on pages 260-261.

25. On the size of the Spanish fleet, see Usher, "Spanish Ships and Shipping," 211-212. The estimate is highly suspect on a number of grounds, including the conversion of the contemporary Spanish measure, the *tonelada*. Pierre Chaunu, "La Tonelada Espagnole aux XVIe et XVIIe Siècles," in Michel Mollat, ed., *Travaux du [Première] Colloque d'histoire maritime (1956)* (Paris, 1957), 71-81; Earl J. Hamilton, "The Decline of Spain," in E. M. Carus-Wilson, ed., *Essays in Economic History*, vol. 1 (London: Edward Arnold, 1954), 215.

26. Huguette and Pierre Chaunu, *Seville et l'Atlantique*, vol. 6, pt. 1 (Paris, 1953-1959), 168, 337.

Portuguese carracks off Sintra, Portugal, by Cornelius Anthoniszoon, ca. 1540. National Maritime Museum, London.

The fleets of Mediterranean states offer even less in the way of reliable estimates of their size. Fernand Braudel, the historian of the Mediterranean in the time of Philip II, put total shipping capacity there in the late sixteenth century at about 350,000 tons.[27] He also gave a figure of 600,000 to 700,000 tons for all maritime activities in the Atlantic, including fishing, for the same period. In light of other data, his estimate for the Mediterranean appears to be too high and the figure for the Atlantic too low. The latter seems especially low given Dutch and English data. The former may be more accurate, however, since the Italian merchant republics certainly deployed large fleets. Fifteenth-century Venice already had over 80,000 tons of shipping, and the fleet was probably as big or bigger in the early fourteenth century.

As to the ports of what would become the German Empire after 1871, the estimates made by Vogel are perhaps a bit too optimistic. In the thirteenth and early fourteenth centuries, the fleet of the towns which belonged to the Hanseatic League was undoubtedly very large. The political and naval power of the League as well as incidental comments suggest not only that the fleet was large but also that many of the ships in the fleet were large — that is, of two hundred tons or more. The beer trade from Hamburg to Amsterdam in the 1360s alone required a minimum of 1,500 tons of shipping, and by the late fifteenth century 4,000 tons were needed. That trade was only a small part of the total of Hanseatic commerce. It appears that in the 1590s League ports had somewhere around 100,000 tons of shipping in long-range ocean-going commerce alone.[28] All indications are of a rise and recovery in total tonnage in the sixteenth century followed by stagnation in German ports, and for that matter throughout Europe, in the seventeenth century.

IV

In the eighteenth century, the tonnage of the Dutch merchant fleet remained stable. A 1700 estimate put 1,600 Dutch ships in European commerce, and an estimate for 1750 was 1,450. Those approximations are based on contemporary guesses. The total number of ships had fallen from 1,750 in 1636, but it is possible that the average size of vessels had risen, possibly even enough to compensate for the fall in the number of ships. If the total tonnage of the Dutch fleet fell in the eighteenth century, it was not by much. Around 1750, 4,400 ships entered Dutch ports each year, and by 1790 that figure had reached 5,500. The total included vessels in European trades, in whaling, and in trade to the East and West Indies.[29]

27. Fernand Braudel, *The Mediterranean and the Mediterranean World in the Age of Philip II*, vol. 1, trans. Siân Reynolds (New York, 1972), 445-448.

28. On the German fleet, see Vogel, "Zur Grosse der europäischen Handelsflotten," 273-302.

29. Bruijn, "Recrutement et enrôlement dans la marine marchande," 553.

While there may have been little change in the total tonnage of the Dutch fleet, there was a great rise in tonnage of the English fleet. The date for the rapid expansion of the English merchant marine is rather late and, indeed, comes just on the eve of the introduction of steam power. It is true that the wars of the eighteenth century and especially the wars of the French Revolution and Napoleon skew the figures. Even more important was the change in the government method of measuring shipping tonnage made in 1773 and enforced from 1786. Even reducing the official post-1786 figures for registered tonnage by one-third — a likely multiplier — in order to make them comparable with earlier numbers, the British gains in the late eighteenth century are still impressive.[30] By any measure the fleet more than doubled between the 1760s and the 1790s. The rapid growth in the English merchant marine made the total figure for Europe rise rapidly as well. The proportion of shipping attributable to England in 1786, however, was still less than the Dutch share of the total in the mid-seventeenth century. That did change rapidly, though, in the 1790s.

By the eighteenth century, Denmark-Norway was also making a significant contribution to Europe's total tonnage with a merchant marine almost as large as that of the Dutch Republic. The 1786-1787 figure of 386,020 tons suggests rapid growth in the preceding century. While the English fleet grew in size at an impressive pace through the eighteenth century, the rate of growth for Denmark-Norway, though admittedly starting from a smaller base, was at least as high or higher. While France was generally keeping pace with English growth and German ports were falling behind the rate of expansion set by England, Norway was able to establish a position as a major supplier of shipping services internationally. The British colonies in the New World did much the same thing in the eighteenth century. In the years 1769, 1770, and 1771, the average annual output of shipbuilding yards in what was to become the United States was over 21,000 tons, much of the production being exported to Britain. On the eve of the American War of Independence, British North America boasted by one estimate a merchant marine of over 450,000 tons, 250,000 of those tons in a coastal fleet of 5,200 vessels.[31] Not only the size but also the geographical scope of Europe's shipping grew rapidly in the eighteenth century as other parts of the world were integrated into the complex of trade and navigation.

For the total tonnage of all of Europe's merchant fleets, only a figure from the survey carried out by French consular officials in 1786 and 1787 can be given any credence. That survey gave an estimate for the total capacity of all the fleets of Europe of 3.5 million French tons, a French ton being 979 kilograms.[32] The 1676 estimate of just two million tons given by Sir William Petty in his *Political Arithmetick* is certainly questionable. He contradicted his own estimate of the size of the English fleet in the same book, giving in one case 500,000 tons and in the other 600,000.[33] Fernand Braudel offers an estimate for the total for all of Europe in 1600 of 700,000 tons, which is certainly too low.[34] It seems likely that by 1600 the European fleet was already up to around one million tons, grew to 1.5 million tons by the 1670s, and then, after a long-term sustained rise, reached 3.5 million tons by the closing years of the eighteenth century.

The 1786-1787 survey also shows that another trend dating from the reign of Elizabeth I and before had continued. The share of large ships in the merchant fleet had continued to grow. The French consular survey showed 3 percent of all vessels being over two hundred tons with those vessels comprising 31 percent of the total tonnage. Ships of one hundred to five hundred tons formed a minimum of 67 percent of the total tonnage of Europe's merchant fleet. In the course of the four centuries after the development of the full-rigged ship, European shipbuilders and shippers had discovered, by experiment and with government direction, the optimum size for the sailing ship. That size was somewhere around four hundred tons.

Data on the size of European merchant marines from the late Middle Ages to the end of the eighteenth century yield many well-known and predictable results. The total tonnage of European shipping grew over the long term, but that growth was not continuous. Rather, there were wide swings, up and down, in the total tonnage. There was a sustained rise in the fifteenth and sixteenth centuries and then slower growth through the seventeenth and early eighteenth centuries. Sixteenth-century levels indicated a recovery from a medieval high, but how high that medieval level may have been is hard to tell. In 1600 Europe may have just gotten back to medieval tonnage. It is true that late sixteenth-century ships were technically

30. Davis, *The Rise of the English Shipping Industry*, 395-396; French, "Eighteenth-Century Shipping Tonnage Measurements," 434-437.

31. James F. Shepherd and Gary M. Walton, *Shipping, Maritime Trade, and the Economic Development of Colonial North America* (Cambridge, 1972), 243; Fernand Braudel, *Civilization and Capitalism 15th - 18th Century*, vol. 3, trans. Siân Reynolds (London, 1984), 462.

32. Ruggerio Romano, "Per una valutazione della flotta mercantile europea ala fine del secolo XVIII," in *Studi in onore di Amintore Fanfani*, 6 vols. (Milan, 1962), 5: 578-80.

33. Sir William Petty, *Political Arithmetick or a Discourse Concerning The Extent and Value of Lands, People, Buildings; Husbandry, Manufacture, Commerce. . .* (London, 1690), 5. On page 5 England has 500,000 tons of shipping and on page 77 it is 600,000.

34. Braudel, *Civilization and Capitalism 15th-18th Century*, vol. 1, trans. Siân Reynolds (London, 1981), 362.

Convoy of Dutch cargo ships escorted by a warship (center), from a painting by an unknown seventeenth-century Dutch artist. Courtesy of the Museum of Fine Arts, Budapest.

better and more efficient than their medieval counterparts, so each ton of shipping could give more transport service in a given period of time by 1600. Even if tonnage levels were the same in 1300 as in 1600, there were greater shipping services available in 1600.

Rapid expansion of tonnage began after 1750; that is, on the eve of the introduction of steam propulsion. By the late eighteenth century tonnage in Europe's merchant marine far exceeded anything seen before. By that time, the transfer to northwest Europe in shipping was obvious as well. That shift had been, first, from northeast and central Europe in the fifteenth and sixteenth centuries and, second, from the Mediterranean in the very late sixteenth and in the seventeenth and eighteenth centuries.

The long-term increase in the number of large ships is obvious from the data. The trend began in the Mediterranean, in Venice and Genoa in the fifteenth century with the introduction of the three-masted carrack, a large version of the full-rigged ship. Northern Europeans followed the example, imitating their counterparts in the south. The new big ships dragged up the average size of ships as well as increasing the total tonnage. In northern Europe, the Netherlands was a partial exception since, though Dutch shippers had defensible ships for dangerous trades in high value goods, the great majority of their vessels were designed and intended for bulk carriage. Dutch figures, however, show the same tendencies toward a rising total tonnage and higher average tonnage because the merchant fleet was large and because few vessels were either very large or very small. English shipowners, on the other hand, concentrated on a smaller number of big ships in the late sixteenth century.

The goal of English state policy was to direct shipping into trades which were relatively dangerous and so away from bulk trades. Government action in the 1560s and 1570s helped to commit England to a relatively small fleet, relatively large ships, and relatively high costs of operating those highly defensible ships. Other states followed similar courses. In the reign of Elizabeth I those in positions of power had an overwhelming interest in defense and in regulating the economy to promote self-sufficiency, which was understood to be essential to that defense.[35] Construction of large ships fit that overriding interest. The rearmament policy, of which treatment of the merchant marine was an integral part, left a legacy of high-cost carriers. It was not until the eighteenth century that the English merchant marine escaped the choices made, for reasons of defense, in the second half of the sixteenth. The Dutch merchant marine, on the other hand,

35. Christopher Hill, *Reformation to Industrial Revolution* (London, 1967), 73; Lawrence Stone, *The Crisis of the Aristocracy 1558-1641* (Oxford, 1965), 425. The landed classes supported legislation for regulating the economy because of their fear of change and social unrest but above all because of their concern for defense. See also Penelope Corfield, "Economic Issues and Ideologies," in Conrad Russell, ed., *The Origins of the English Civil War* (London, 1973), 198, and Ramsey, *Tudor Economic Problems*, 165-167.

Table 2
Tonnage of Merchant Marines in Europe, 1300-1800

Year	England	Dutch Republic	Germany	France	Spain & Portugal	Italy	Venice	Genoa	Denmark-Norway & Sweden
1310[36]	70,000			40,000					
1424[37]								31,000	
ca. 1450[38]							80,500		
ca. 1470[39]		60,000	60,000						
1560[40]	45,000								
1562[41]	65,000								
1567[42]		175,000							
ca. 1567[43]							32,000		
ca. 1570[44]	42,000	232,000	110,000	80,000					
1582[45]	80,000								
1585[46]					225,000				
1600[47]		240,000							
1603[48]	60,000								
1629[49]	115,000								

got a high average tonnage early, not by concentrating on large ships but by having many bulk carriers. Other European states saw the size and character of their merchant marines directed and even determined in a similar way to the English, but in the late seventeenth and eighteenth centuries such efforts disappeared. The need for big ships was gone. The single exception was in trade to the East Indies. There large ships still had a place, and it was no coincidence that it was the last trade from which English shipbuilding bounties disappeared.[50]

The average size of ships increased in part for economic reasons. The larger the ship, the higher the ratio of tons served for each crewman. That was always the case. There were other factors which created productivity gains, but that one was consistent throughout the period.[51] It was especially obvious in bulky trades such as those in tobacco, sugar, coal, and grain. In 1582, 35.4 percent of English ships were from 100 to 499 tons, but in 1788 it was 67.3 percent.[52] By the latter date, purpose-built naval vessels could and did handle defense of merchant ships. Traveling in convoy, they did not have to worry about acting as fighting ships. The government policy to direct shipbuilding and shipping toward defensible ships was no longer needed. Rather, by creating navies for convoy duty,

36. Estimate by R. W. Unger, based on contemporary surveys and customs accounts.

37. Jacques Heers, *Gênes au XVe Siècle Activité économique et problèmes sociaux* (Paris, 1961), 280. Ships of five hundred tons or more only.

38. Lane, *Navires et Constructeurs*, 95.

39. Vogel, "Zur Grosse der europäischen Handelsflotten," 268-333.

40. Estimate by R. W. Unger, based on contemporary surveys and customs accounts.

41. Ibid.

42. H. P. H. Jansen, "Handelsvaart van de Noordnederlanders," *Maritieme Geschiedenis der Nederlanden*, vol. 1 (Bussum, 1976), 272. A 1564 report only gives ships of three hundred tons and more and gives a total tonnage of 90,000. The 1567 report includes only ships of two hundred tons or more.

43. Frederic C. Lane, *Venice, A Maritime Republic* (Baltimore, 1973), 480, includes ships of 240 tons or more only.

44. Vogel, "Zur Grosse der europäischen Handelsflotten," 268-333.

45. Estimate by R. W. Unger, based on contemporary surveys and customs accounts.

46. Hamilton, "The Decline of Spain," 215. The figure is for both Spain and Portugal and is an approximation.

47. Philippe Dollinger, *The German Hansa*, ed. and trans. by D. S. Ault and S. H. Steinberg (London, 1970), 344-345, 355, 358.

48. Vogel, "Zur Grosse der europäischen Handelsflotten," 268-333.

49. Jack, *Trade and Industry*, 98.

50. See Dietz, "Royal Bounty and English Merchant Shipping," 5.

51. Douglass C. North, "Sources of Productivity Change in Ocean Shipping, 1600-1850," *Journal of Political Economy* 76 (1968): 959.

52. Ibid., 958.

Year	England	Dutch Republic	Germany	France	Spain & Portugal	Italy	Venice	Genoa	Denmark-Norway & Sweden
1636[53]		500,000							
ca.1670[54]	94,000	600,000	104,000	80,000					
1676[55]	500,000	900,000		100,000					
1686[56]	340,000								
1700[57]		500,000							
1702[58]	320,000								
1750[59]		500,000							
1761[60]	460,000								
1775[61]	700,000								
1786[62]	751,626								
1786-7[63]	881,963	397,709	155,349	729,340	234,303	311,964	60,332	42,130	555,299
1788[64]	1,055,299								
1788[65]	1,200,000								
1790[66]	1,290,000								
1800[67]	1,856,000								

After 1707, all "England" figures are for the United Kingdom.

governments opened possibilities for shipbuilders to make choices based on the goods to be carried rather than the fighting to be done. The shipbuilding industry was left free to seek out the optimum size for cargo ships, a size which fell in the range of around four hundred tons. The Dutch merchant marine seems to have arrived at that solution first, in part because there was no central government to direct shipping, in part because shippers had great influence over the government that emerged in the Dutch Republic, and in part because of a well-established Dutch tradition of convoying in the fishery which dated from the fourteenth century. The long-term results were a more efficient shipping industry, changes and adjustments in ship design, and, above all, long-term growth in shipping and, therefore, in the total shipping tonnage of Europe.

53. Unger, *Dutch Shipbuilding before 1800*, 11, includes estimate for all trades. See also Bruijn, "De vaart in Europa," 200. His estimate of 310,00 tons does not include extra-European trades, fisheries, whaling, or inland vessels.

54. Vogel, "Zur Grosse der europäischen Handelsflotten," 268-333.

55. Petty, *Political Arithmetick*, 5.

56. Davis, *Rise of the English Shipping Industry*, 395-406.

57. Estimate based on the number of ships, extrapolating for size of ships and adding an approximate tonnage for extra-European trades, fisheries, and whaling.

58. Davis, *Rise of the English Shipping Industry*, 395-406.

59. Estimate based on the number of ships, extrapolating for size of ships and extra-European trades, fisheries, and whaling. J. V. Th. Knoppers, "De vaart in Europa," *Maritieme Geschiedenis der Nederlanden*, vol. 3 (Bussum, 1977), 226-227, gives an estimate of 365,000 for ships trading only within Europe.

60. Vogel, "Zur Grosse der europäischen Handelsflotten," 268-333.

61. Ibid.

62. Davis, *Rise of the English Shipping Industry*, 395-406.

63. Romano, "Per una valutazione della flotta mercantile europea ala fine del secolo XVIII," 573-591.

64. Davis, *Rise of the English Shipping Industry*, 395-406.

65. Waters, "'The English Pilot' and English Shipping," 332-334.

66. Vogel, "Zur Grosse der europäischen Handelsflotten," 268-333.

67. Ibid.

XVII

THE NETHERLANDS HERRING FISHERY IN THE LATE MIDDLE AGES: THE FALSE LEGEND OF WILLEM BEUKELS OF BIERVLIET

Since the fifteenth century, writers have been claiming that Willem Beukels of Biervliet invented a new way of processing herring so that the fish would keep for a long time. The story has taken various forms, but in most cases that citizen of a fishing village in northern Flanders, now in the province of Zeeland in the kingdom of the Netherlands, is credited with starting the great expansion in the Low Countries herring fishery. His invention around 1400 made it possible for herring to be stored for long periods and to be shipped over long distances. He, then, so the story goes, was responsible for the rise of the Dutch herring fishery in the sixteenth and seventeenth centuries and in turn in part responsible for the unprecedented economic prosperity of Holland in those years. Even the famous seventeenth-century Dutch poet, Jacob Cats, sang the praises of Beukels, naming him as the first to learn how to preserve herring.

The story is appealing for its simplicity. Problems arise almost immediately, however. It is difficult to establish when Beukels lived, what his occupation was and precisely what he did, if anything, that was revolutionary.[1] The pattern of change in the herring fishery in the fifteenth century is far from clear. Examination of custom, toll and shipping records along with reports on piracy show that technical innovation was far from instantaneous. But the greatest problem is the causal allegation which places the origin of Dutch success in the province of Flanders a century before the steady and impressive rise in total production of cured herring. Insisting on the

I am indebted to J. P. van de Voort of the Visserijmuseum, Vlaardingen, and Professor Lynn White, jr., for their comments on an earlier draft of this paper.

[1] See for example Robert Ergang, *Europe from the Renaissance to Waterloo*, ed. 2 (New York 1954) 289-290. G. Doorman, "Het haringkaken en Willem Beukels," *Tijdschrift voor Geschiedenis* 69 (1956) 371-372, gives a short bibliography of those who accepted the story in the sixteenth and seventeenth centuries and in the nineteenth and twentieth. Included in the list are the sixteenth-century Italian who lived in Antwerp, Ludovico Guicciardini, the seventeenth-century Dutch chroniclers, T. Velius, M. van Boxhorn, Joh. Potanus, M. Smallegange, among others, and the modern historian of the Flemish fishery, Roger Degryse. See also R. Degryse, "Oorsprong van het haringkaken in Vlaanderen," *Nederlandsche Historiebladen* 1 (1938) 204-205; H. Blink, "De Geschiedenis en Beteekenis der Nederlandsche Haringvisscherij," *Vragen van den Dag* 45 (1930) 985.

connection of Beukels to seventeenth-century Dutch prosperity unnecessarily obscures the technical change of the fourteenth and fifteenth centuries. Moreover it tends to minimize the importance of a complex process of innovation. It is not in an immediate contribution to the economy or to the nutrition of medieval Europeans that the significance of change in the herring fishery is to be found. Rather it is valuable as an illustration of technological change in a period when Europeans were finding innovation easier.

It will probably never be possible to isolate Willem Beukels's precise contribution. It is possible, however, to say a great deal about the development of the herring fishery especially in Flanders in the fourteenth and fifteenth centuries, its growth and the changes in technique used by Netherlands fishermen. The transformation was intimately connected to the commercial position of other suppliers in Europe and to the action and reaction of government to the pattern of supply. That process of transformation shows the incremental nature of technical innovation and the complex of forces which typically make technological change less than dramatic.

I

A chronicler writing in the sixteenth century claimed that Willem Beukels invented a new way of treating herring. The town of Ostend, in a declaration of 1483, also attributed the success with herring to the same man. The story, then, was obviously a popular one even in the fifteenth century, and the chronicler was probably just repeating an oral tradition. Other writers picked up the tale and repeated it in much the same form again and again over the centuries. There is little substantive difference in the accounts. In each instance Beukels is given credit for discovering *kaken* or curing herring, of finding the way to make *kaakharing* or *pekelharing*.

The method of curing is still in use. Immediately after the herring is caught it is slit open with a sharp cut from a short curved knife especially designed for the job. Then the gills, stomach, heart, liver, gall bladder and genitals are pulled out of the body cavity. That way the fish quickly loses its blood. It eliminates the parts of the fish most subject to rapid deterioration. The fish are then put directly into barrels where salt is placed above and below the layers of gutted fish. The remaining blood and the salt immediately form a brine. That cuts the fish off from the air which would cause it to rot. While in this brine the fish slowly change in color, odor and taste. The process of curing is aided by the enzyme, trypsin, of the pyloric caecae. Those appendices of the stomach are often missed by fishermen when they pull out the viscera. Leaving them in the fish speeds the process of curing and improves the quality of the final product. When the herring is brought to shore it is repacked; and a white brine, that is, one with no blood, is added. That improves the preservation of the fish and also makes it possible to fill the barrels and compensate for evaporation and shrinkage.[2] *Kaakharing*, then, was not just any herring in barrels, but herring specially treated and packed from the time it was caught. In that form it could last for some time, usually a year or more depending on the quality of the gutting and packing.

It is certain that Willem Beukels did not invent that method of curing herring. It was used in Europe well before 1300. Herring was treated in that way on shore in a number of places, most notably on the coast of southwestern Sweden in the then Danish province of Scania. Certainly by the mid-thirteenth century herring was caught there in the narrow waterway between the mainland and the Danish islands and landed on the small Skanör peninsula where they were prepared and packed. The fishermen were apparently Danish using open boats, 5000 to 7000 of them altogether, each with six to ten men. Merchants regulated the treating of the fish, bought them from the producers and then shipped them to markets throughout northern Europe. The herring were so numerous in the strait that ships, it was said, had trouble using their rudders. The fishery went on from August to November but the catch was concentrated in September and October. Since there was a serious problem with the garbage from the work on the herring, there can be little question that they were gutted before being packed. It was illegal to treat fish in the boats. Fishermen had to bring them to shore directly after catching. The merchants could only operate from *vitten*, small plots along the shore granted to them or more precisely to their town.

From 1200 the preparation and sale of Scania herring was largely under the control of merchants from German Baltic ports and especially from Lübeck. It was those men from towns belonging to the Hanseatic League that handled the shipping of the Scania *tonharing* to markets in Germany and in western Europe. By the second half of the fourteenth century production was probably above 8000 lasts per year.[3] The Low Countries formed an important market for Hansa merchants. In fact Scania herring was a major item in their trade with Flanders, a western terminus of their trading network. By the late fourteenth century Hansards delivered some 1700 lasts of herring annually in Flanders.[4] From there other merchants moved the

[2] A. van Dijk, in *The Technology of Herring Utilization*, Report of the FAO Meeting (Bergen 1950) 225; A. F. M. G. Luijpen, *De Invloed van het Kaken op de Rijping van Gezouten Matjesharing* (Utrecht 1959) 15-17, 37-39.

[3] The last usually refers to a wagon-load of grain equal to about two English tons in weight. For herring, however, at least by the fifteenth century, the last was made up of 12 casks or *tonnen*. Each of those barrels contained between 800 and 1000 fish but generally about 900. A herring last was thus about 12,000 fish. If it took some 9 herring to make a kilogram the weight of the herring last would have been about 1300 kilograms, or less than the weight of a last of grain. The barrels added some weight but the increase was marginal. In the Dutch herring industry one last was equal to 17 casks on board ship and 14 casks repacked on shore. The estimate for the size of the Scania catch is open to question. For lower limit estimates which, however, show confusion about the size of a last see H. Henking, *Die Ostseefischerei: Handbuch der Seefischerei Nord-Europas* 53 (Stuttgart 1929) 8-9.

[4] G. W. Coopland, "A Glimpse of Late Fourteenth-Century Ships and Seamen from 'Le Songe de Vieil Pélerin' of Philippe de Mézières (1327-1405)," *The Mariner's Mirror* 48 (1962) 190-19[illegible];

preserved fish on to markets in the Low Countries, in Germany and in France. Herring appeared as a trade item in many parts of northern Europe throughout the later Middle Ages. Netherlands shippers often carried herring to England. In 1348 Holland and Zeeland shipmasters landed at least 198½ lasts of herring at the port of Yarmouth. Herring is often mentioned in the lists of cargoes landed in England in the fourteenth century.[5] The great majority of the herring was imported either for immediate consumption or to be treated. It is doubtful that much if any of it had already been cured. That was less true of the herring shipped to Germany. Shippers carried the product up the Rhine to the important market of Cologne and as far as Basel. Cologne merchants visited towns in the Low Countries to buy fish and bring it back themselves. From 1439 to 1441 about 375 lasts of herring went from the province of Holland alone to Germany and that did not include smoked herring, a popular product in the German market.[6] The figure for the late fourteenth century was undoubtedly lower and the figure for cured herring even lower than that. The greater the distance that the herring had to travel the more likely that it had been given a full curing treatment. Low Countries exports to France in the fourteenth century, then, were more usually simply reexports of Scania herring. Ports like Saint-Malo, Dieppe and Rouen were common destinations for Netherlanders shippers carrying herring. From there the fish were taken to Paris and to many other markets in the interior reaching Franche-Comté, Toulouse, the Dauphiné, the papal table at Avignon and even Spain. Whether by sea or overland and by river, herring also reached the Mediterranean. Herring was one of the typical products of the North which formed the basis for the commercial connection with southern Europe. Herring was therefore naturally in the cargoes of the northern European ships which invaded the Mediterranean in the sixteenth century.[7]

Doorman (n. 1 above) 378-379; J. Travis Jenkins, *The Herring and the Herring Fisheries* (London 1927) 57-64; Aksel E. Christensen, "La foire de Scanie," *Société Jean Bodin . . . Recueils* 5 (1953) 244-255; R. Degryse, "Schoense en Vlaamse Kaakharing in de 14de eeuw," *Bijdragen voor de Geschiedenis der Nederlanden* 12 (1957) 102-105; *idem, Vlaanderens Haringbedrijf in de Middeleeuwen* (Antwerp 1944) 93-96.

[5] H. J. Smit, *Bronnen tot de Geschiedenis van den Handel met England, Schotland en Ierland* 1: *1150-1485* (The Hague 1928) no. 444; Nelly J. M. Kerling, *Commercial Relations of Holland and Zeeland with England from the Late 13th Century to the Close of the Middle Ages* (Leiden 1954) 89-95.

[6] J. A. Van Houtte, "Die Beziehungen zwischen Köln und den Niederlanden vom Hochmittelalter bis zum Beginn des Industrie-Zeitalters," *Kölner Vorträge sur Sozial- und Wirtschaftsgeschichte* 1 (1969) 12; F. Ketner, *Handel en Scheepvaart van Amsterdam in de Vifftiende Eeuw* (Leiden 1946) 55, 67; W. S. Unger, *De Levenmiddelenvoorziening der Hollandsche Steden in de Middeleeuwen* (Amsterdam 1916) 140-142.

[7] Z. W. Sneller and W. S. Unger, eds., *Bronnen tot de Geschiedenis van den Handel met Frankrijk* 1: *753-1585* (The Hague 1930) nos. 33, 51, 52, 61, 52, 122; Henri Touchard, *Le commerce maritime breton à la fin du Moyen Age* (Paris 1967) 83, Michel Mollat, *Le commerce maritime normand à la fin du Moyen Age* (Paris 1952) 125, 313-317; Philippe Wolff, *Commerces et marchands de Toulouse (vers 1350-vers 1450)* (Paris 1954) 214; Fernand Braudel and Ruggiero Romano, *Navires et marchandises à l'entrée du port de Livourne (1577-1611)* (Paris 1951) 34-35, 50-51.

The quantities of herring shipped in the fourteenth century were small relative to the total volume of commerce. The figures were even less impressive the greater the distance from the centers of production. But what made the trade important was precisely that the herring could be moved over great distances. The curing process made it biologically possible for at least a part of the herring catch. The use of barrels for shipping the product made it economically feasible. Romans used barrels in some cases to carry wine and other commodities including fish. They learned to build them from Celts, and it was principally in northern Europe that barrels were the containers for shipment. Compared to jars, barrels are fifty percent more efficient, that is, in terms of the proportion of cargo space taken up by the container.[8] Cured herring entered commerce already packed in its relatively efficient container. The result was a good which could form the basis of long-range commercial connections, no matter how tenuous. A fleet of about a dozen of the largest deep sea trading vessels could have handled the total quantity of cured herring going from Scania to Flanders. But of course the number of ships carrying the product was much greater. Ships of lesser tonnage made the voyage and few if any made the trip loaded only with herring. In that trade and to a lesser degree in others cured herring was a commodity which could be counted on to find a market. It traveled well and could be stored for some time. Shippers then could use cured herring as the basis of a shipment and include more speculative items. That in turn over time could form the basis for further commercial connections. The trade between the Baltic and the Netherlands is only the best example.

The comital government of Flanders controlled the trade in herring from Scania. The German merchants had to bring the fish to a specified staple port. The fish arrived starting in October. No herring could be shipped from Scania until 9 October. The cargoes may have included some herring from Norway where a coastal fishery had long been established. In 1164 the pope had granted the inhabitants of the coast the privilege of fishing on feast days and Sundays if and when the fish came close to shore. Hansards landed their fish at the port of Damme or, after 1372 when the staple was moved, at Sluis in Flanders. They then repacked it before sending it on to customers either in that county or inland. After serious difficulties with the Flemish authorities which led to a boycott of the port of Bruges the Hansards in 1360 got the right to repack and repickle their herring in Flanders. There can be no question that Netherlands fishermen were fully familiar with this curing of herring on shore. Some towns got the right to have their own district on the Skanör peninsula. Fishermen from Kampen and Harderwijk, two towns in the northeast Netherlands, were operating there by the end of the twelfth century. In the second half of the fourteenth century men from towns in the province of Holland also became involved in the Scania fishery. Amsterdam, Enkhuizen, Wieringen and Brielle in 1350 joined their

[8] F. C. Lane, "Progrès technologiques et productivité dans les transports maritime à la fin du Moyen Age au debut des temps modernes," *Revue Historique* 251 (1974) 278-279; R. J. Forbes, "Food and Drink," in *A History of Technology* 2, ed. C. Singer et al. (Oxford 1956) 136.

two Netherlands predecessors in getting the right to have *vitten* at Skanör. At one point eleven towns from the Low Countries had concessions in the Scania fishery.[9]

Netherlanders used the method very little if at all along their own coast. It was not because they did not have a herring fishery. Coastal fishing for the *clupea harengus* in the Low Countries dates from the Roman period and probably before. Crews of four to seven men caught the fish, working from open boats. They would go out in the evening and return the next morning with herring. The fishing was done at night because then the herring came toward the surface to feed and fishermen could see them more easily because of their luminous appearance. The fishermen sold the catch as fresh or lightly salted herring. In the latter form it was taken by water or wagon to nearby markets. The light salting could slow the deterioration of the fish but could not stop it. Salt came from local salt pans. A Flemish writer said in 1125 that the herring was a little fish which could be easily salted to make it last for a long time. He was apparently describing the coastal fishery. After 1150 that fishery expanded in Flanders with the establishment of a series of new fishing villages along the coast. Production could be large, a total catch of about 1000 lasts being reported in 1321 for Calais alone. But the herring did not get the same treatment as that given fish at Scania.[10] It was because of the time of the year. Herring was caught along the Netherlands coast in November and December and up to as late as Candlemas, 2 February. This fish found a ready market in the large towns of Flanders in those winter months as fresh herring, since supplies of fresh meat almost disappeared after the slaughtering in early November. The herring found along the coast in that period were full herring, that is, fully developed males and females. During December the quality of the herring deteriorates as the fish spawn and their fat content falls. In May through July the fish find their best food in the North Sea and in those months their fat rises most rapidly. It goes down slowly through the winter.[11] The full herring caught in the southern North Sea from the end of October to early December is best suited to consumption as fresh fish or as smoked herring. Smoking was the usual method of preserving the coastal catch if it was preserved at all. The *bokking* was packed in barrels and shipped to Cologne and the Rhineland. In the thirteenth and fourteenth centuries, given the kind of herring they caught and the nature of the market, Netherlands fishermen had no reason to use the curing method along their own shores.

Netherlanders did catch herring for curing, along the English coast off Norfolk near the port of Yarmouth. Fish taken there through November was sold as full herring. It received the same treatment as did the product of the Flemish coastal

[9] A. Beaujon, *The History of Dutch Sea Fisheries* (London 1884) 8-9; Nelly Gottschalk, *Fischereigewerbe und Fischhandel der niederlandschen Gebiete im Mittelalter* (Bad Wörishafen 1927) 12-13; Degryse, *Haringbedrijf* (n. 4 above) 96-97; Henking (n. 3 above) 7-8; Jenkins (n. 4 above) 8-9.

[10] Doorman (n. 1 above) 376; G. Doorman, "Nogmaals: de middeleeuwse haringvisserij," *Bijdragen voor de Geschiedenis der Nederlanden* 14 (1960) 104-105; Degryse, *Haringbedrijf* (n. 4 above) 10-27, 54-55. The 1321 figure was unusual. The normal catch was under 500 lasts.

[11] Luijpen (n. 2 above) 13-14.

fishery. By 1300, though, some of that herring was cured. That was especially done with fish taken earlier, in July or August. Yarmouth was the great market for all herring caught along the English coast by all fishermen. Fish was exported from there to other English towns and to the Continent. In 1295 King Edward I of England issued an order that fishermen from Holland, Zeeland and Friesland coming to fish off the coast near Yarmouth should not be molested or robbed by his subjects. It is likely that Netherlanders were visiting the "Deepwater" fishing bank off Norfolk in the eleventh century and that the 1295 order was just a repetition of earlier regulations. In any case the royal ruling apparently had little effect since it had to be reissued in 1327.[12] Fishing off England came a little earlier than the fishing off Flanders, so Netherlanders were able to extend their season. The equipment and personnel were the same, so there was no change in technique or increase in investment. Fishermen then could follow the herring as they moved south from late August until the beginning of the following February. The fish caught off the English coast were also of high quality, matching the best full herring of the Flemish coastal fishery. Englishmen handled the curing of that part of the catch which was to be preserved. All the work was done on shore. At the end of the fourteenth century some Netherlanders apparently were going ashore to do their own curing, since the English government complained about foreign fishermen clandestinely landing with casks and salt to treat their fish. But that was a late development and certainly exceptional. The Netherlanders typically caught the fish and sold them to English merchants who contracted to act as brokers. They arranged for supplies of food for the fishermen to be available on shore. They also handled the marketing of the catch. For this the brokers got a share of the revenue from sales, usually five percent. That practice and indeed the whole of the Norfolk fishery was well established by the thirteenth century.

II

In the fourteenth century there was a series of steps which moved the curing of herring from the shore to the fishing boats. In essence it was a problem of transportation. As Netherlanders fished further and further north along the English and then along the Scottish east coast they increased the distance to their markets in the Low Countries and France. Fishing further north had a number of advantages. It meant that the season could begin earlier, in June, so that fishermen could increase the returns from their investment and more than ever could specialize in their work.

[12] Eric Dardel, *La pêche harenguière en France* (Paris 1941) 23-29. R. Degryse and O. Mus, "De laatmiddeleeuwse Haringvisserij," *Bijdragen voor de Geschiedenis der Nederlanden* 21 (1966-1967), 82-87; R. Degryse, "De Vlaamse Haringvisserij in de XVe Eeuw," *Handelingen van de "Société d'Emulation" te Brugge* 88 (1951) 116 n. 3; Y. N. Ypma, *Geschiedenis van de Zuiderzeevisserij* (Amsterdam 1962) 14-15; G. Doorman, "De Haringvisserij in de Middeleeuwen," *Bijdragen voor de Geschiedenis der Nederlanden* 19 (1964-1965) 162-163; Smit (n. 5 above) nos. 238, 247, 268, 269, 274, 281, 300.

It meant catching large autumn spawning herring of that same high quality which was found off the coast of Norfolk, and with an even higher fat content. The advantages could only be exploited if a method could be found to preserve the catch during the long voyage home. The standard practice when fishing off Norfolk was to take the last catch home in the boat and sell it in the Low Countries. If the fishermen were operating on the Dogger Bank or off the coast of Scotland, as much as 400 kilometers from their home ports, and they wanted to do that, then they had to keep the fish from going bad for three or four days if the weather was good and longer if it was not.

The solution found and used by 1252 was a partial salting. The resulting product was called *korfharing*. The treatment may have been similar to that used on shore in the Flemish coastal fishery. The herring was lightly sprinkled with a layer of salt and packed loosely in hampers. There was no question of gutting the fish. Later, by the early fourteenth century, Flemish fishermen also sold a type of herring called *vivelo*. The term referred to Filey, a port on the Yorkshire coast, but it probably meant more than just that the fish were caught there. It indicated that the herring came from further north than was common before. So it may also suggest that some small improvement was made in the preservation method. Since the distances were greater and the fish could be preserved to some extent, it made sense to use a larger fishing vessel. The corver became the typical ship for that longer-range fishing. Corvers must have been seaworthy and capable of carrying a sizable crew because in wartime they found work as privateers. In fact the word came to be a generic term in Flemish for privateer. Moreover the vessels had to carry salt to the Yorkshire and Scottish coast for treating the fish, since local salt was too impure for the job. They landed their salt, then went out for herring, salting it lightly and then bringing it into ports like Scarborough. The fish was shipped from there to the Low Countries. Vessels late in the season would not bother with going into an English port but would go straight to the Continent. The brokers for those corvers were not English but Flemish, that is, from the home port of the fishermen. Clearly they expected to market some if not all of the catch. Small open boats with a capacity of two lasts or less were also used often in company with a larger corver.[13] No matter the type of vessel, the treating of herring on board, albeit in a simple manner, expanded the scope of the herring fishery.

It also increased the value of the catch. The salt content of the North Sea is greater than that of the Baltic because it is warmed by the Gulf Stream, warmer water having a higher degree of salinity. The salt content also rises toward the northern North Sea away from the outflow of rivers. The salt content of waters on the Dogger Bank is about twice that of the waters off Scania. The higher degree of salinity meant there was more plankton and thus more food for the herring. The

[13] Degryse, *Haringbedrijf* (n. 4 above) 90, 98, and "Schoense" (n. 4 above) 100-101; R. Degryse, "Van 'Buza' tot 'Buis': De Ontwikkeling van het Vlaamse Visservaartuig (12^e^-16^e^ Eeuw)," *Ostendiana* (1972) 139-145; Degryse and Mus (n. 12 above) 88-97; Doorman (n. 12 above) 163-165, 168-169, and (n. 10 above) 106-110.

herring in the North Sea were bigger, from one-third to half again as large as those of the Baltic or from the Netherlands coast. They had more vertebrae and incidentally more of the trypsin needed to get the best curing.[14] The move into the open North Sea and further north allowed fisherman to increase the amount of food per fish caught, with at most a marginal increase in effort.

By 1300 Netherlands fishermen were certainly capable of handling large catches and taking them in deep waters. They had already developed the *vleet,* a type of large net which was a combination of smaller existing nets. Open boats in the coastal fishery would carry as many as forty nets, while for fishing in English waters the number went up to 75. These long drag nets had to be combined into one great one so that fishermen could handle them. The fishing boats grew in size to accommodate the larger net. Moreover already in the fourteenth century there were regulations fixing the minimum size of the holes in the net to prevent taking fish which were too small. The goal may also have been to protect the supply of fish.[15]

Despite the diffusion of the method of treating herring to much of northern Europe through the thirteenth century, curing was still done exclusively on shore. The critical change, the change most important to the long-term development of the herring fishery in the Low Countries, was the transfer of the *kaken* method from land to fishing vessels. The change meant that the major part of the work of preserving the fish was done on board and the ship was saved from making daily or almost daily visits to any coast. It is in fact that innovation and not the development of the curing method which was the decisive step of the fourteenth century and the one which created the potential for the long-term growth of the herring fishery in northwestern Europe.

Netherlanders at some time in the fourteenth century developed the ability to bring cured herring onto the market. Such fish, without the head, are mentioned in toll tariffs of 1377. In the third decade of the fourteenth century Flemings were delivering *kaecharicx* to Paris. Indeed the French word for cured herring, *caqueharing,* was the Flemish term which suggests that fish treated in that way first came to France from Flanders and in the early fourteenth century. If Netherlanders were able to supply cured herring by the 1330s then the questions are whether or not they did the job of preserving and, if they did, whether or not that work was done on board ship.

In the 1320s and 1330s Netherlands fishermen had trouble gaining access to the English coast where they had traditionally landed their catches. Even from 1270, because of diplomatic troubles among England, France and Flanders and also because of English wars with Scotland and France, Netherlands fishermen were subject to piratical attacks in the North Sea. In fact Edward I's order of 1295 was to stop just those attacks. Though there was some improvement after 1320, the outbreak of the 100 Years' War in 1337 led to another deterioration in conditions. In that year the

[14] Ypma (n. 12 above) 14-15; Jenkins (n. 4 above) 37-54.
[15] Degryse, *Haringbedrijf* (n. 4 above) 62 and (n. 13 above) 139.

town of Ghent, fearful of a shortage of fish, required that none of the catch be salted or preserved and that it all had to be delivered fresh. So certainly some salting methods were being used. Deprived of their market for herring in England and, more important, of access to the shore where the fish could be treated, Netherlands fishermen in the years from 1315 to 1330 may have turned to working on the herring on board their ships.[16]

The evidence for the adoption of curing on board at that early date is not convincing. It is impossible to say whether the cured herring shipped by the Flemings to Paris had been prepared on board or not. The threat of attacks by other fishermen, of being closed out of ports and troubles like those were a commonplace of the medieval fishery. There was little that was unique about the years around 1300. Fishermen did not need the excuse of a war or letters of marque to attack other fishermen. They always found it easy to shift to privateering when the opportunity presented itself. At any point through the thirteenth and fourteenth centuries the problem of piratical attacks and reprisals existed. The *korf* and *vivelo* methods served before the second and third decades of the fourteenth century for fish caught as far north as Yorkshire, so distance was not a critical consideration. And there are a number of letters of complaint surviving from those years addressed to the government of Flanders from fishermen detailing what they had lost to pirates. The inventories mention nets, ropes, trade goods, cash, salt; but none of them mentions casks on board.

It is probable that in the course of the fourteenth century Netherlands fishermen experimented with a full curing of herring on their ships. If they did cure and pickle the herring at sea, the quantities involved were small. The figures were low not only for the total European catch and for the total Netherlands catch but also in terms of the catch for each fishing boat. If fishermen did gut and pack herring it was a small proportion of what they took, perhaps the first and best fish. A larger proportion was lightly salted and then a proportion, often far and away the largest, was brought to port as fresh herring.[17] The fishermen did that because of the presence of adequate supplies of cured herring on the Netherlands market, that is, herring imported from Scania. In 1400 German merchants still imported the great majority of cured herring into the Netherlands. The fact that local fishermen had the technical capability to prepare cured fish was of little consequence. A series of changes in the legal and political circumstances of the Netherlands herring fishery in the last decade of the fourteenth century and in the first third of the fifteenth indicates that it was in precisely those years that, at least in Flanders if not the rest of the Low Countries, the production of cured herring became important to the fishery.

[16] R. Degryse, "La pêche et la marine dunkerquoises vers 1300," *Marine Academie van België: Mededelingen* 13 (1961) 22-23; *idem*, *Haringbedrijf* (n. 4 above) 34-48, 88-89; Doorman (n. 1 above) 374-376.

[17] Degryse (n. 12 above) 125-126; R. Degryse, "Het Avontuur van Meeus de Crooc, Stuurman van Lombardsijde (1403)," *Marine Academie van België: Mededelingen* 13 (1961) 33-41; Degryse and Mus (n. 12 above) 94-99, 103.

III

In 1395 for the first time Flemish fishermen delivered their cured herring at Sluis, the staple port for herring imports. Already presumably Hollanders and Zeelanders were treating some of their herring in the same way. In fact French merchants at that time were travelling to Zeeland to buy *kaakharing*. The Flemish government outlawed the landing and sale of herring which had been treated in any way. The prohibition included all types of preserved herring, not only cured herring but also *vivelo*, dried *korfharing* and even fresh herring placed in casks. The edict presumably dates from 1396. The duke of Burgundy, who was the count of Flanders, may have been inspired to act in that way by petitions made to him by the town government of Bruges. There were two reasons for the legislation. First, the Hanseatic League was never reluctant to use political and economic pressure to protect its markets. The Hansa merchants living at Bruges undoubtedly did not like the idea of having to face competition from locally produced cured herring which would damage their virtual monopoly in the county and through much of the region. Second and probably more important was the desire to protect the supplies of fresh herring for Flemish cities. If Netherlands fishermen had started to cure some of their catch, urban governments may have seen the potential for increases in that proportion. Cured herring would be for distant markets and that in the short run could only mean supply problems for local markets.

The government prohibition did not work. Protests from fishing villages along the coast pointed to the loss of sales of cured herring. Foreign and native merchants, they said, went to Zeeland outside the lands of the duke of Burgundy to buy the fish from Hollanders, Zeelanders and Englishmen. The duke in 1399 rescinded the earlier prohibition but clearly did so grudgingly. He allowed Biervliet fishermen to place herring in casks from 24 August to 21 September. The fish had to come from north of Harwich on the English coast. After the closing date all herring caught had to be either smoked or sold as fresh fish. The cured herring could only be landed at the town of Biervliet and had to pay a tax of one noble for each last. This was consistent with the government policy of establishing fixed ports for specific trades. It was also consistent with the policy of taxing any trade or industry that the government could. The grant was only for one year but it was renewed. It was still against the law to sell the cured herring in the interior of Flanders, so the product was for the export market.[18] That ruling was presumably to placate the Hansa merchants. So was the requirement that the casks of Flemish herring had to have a special mark or brand which would distinguish the fish from Scania herring. The grant was specifically for Biervliet and may have been first given by the lady of Biervliet, the duke's wife.

[18] Degryse (n. 1 above) 203-204, 207-208, 216-218. He gives a full transcription of the order of 5 August 1399. R. Degryse, "Het Begin van het Haringkaken te Biervliet (±1400)," *Handelingen van de "Société d'Emulation" te Brugge* 95 (1958) 72-74; Degryse and Mus (n. 12 above) 101-102.

Biervliet was at least for a short time the center of the production and the trade in Flemish cured herring. And that may be the origin of the notoriety given to Willem Beukels. The name is not uncommon in coastal Flanders and certainly there was more than one Willem Beukels who lived there in the fourteenth century. There is a difference in the reports, some writers giving the date of his death as 1347 and not the usual 1397. It may well be that the important man was a certain Willem Beukels who lived at Biervliet in the early fourteenth century and who at that time experimented with the curing method used at Scania. Conversely, the story may have originated with a Willem Beukel who is mentioned at Biervliet in 1392, or even with a Willem Beukel of Hughevliet, a nearby village closely associated with Biervliet, mentioned in 1388 and again in 1396. The last was a captain and probably an owner of fishing vessels who at times turned his hand to privateering. A Pieter Beukels was involved in salt boiling in the first half of the fifteenth century at Biervliet and he may well have been the son of Willem Beukel of Hughevliet. The two were associated with men later involved in landing cured herring at Biervliet. So it is possible that the late fourteenth-century Willem Beukel of Hughevliet was a promoter of the curing of herring at sea and that he was a man of enough stature to be able actively and effectively to lobby the government to change the law in favor of Biervliet. It is also possible that he made some change in the method of packing the fish on board, perhaps in the way the herring were placed in the cask.[19] But that, like so much surrounding the story, is conjecture. The most likely explanation is that the family assumed a coat of arms which included two of the small knives which were used in gutting herring. The legend may have grown out of that mark on the grave of one of the Beukels who lived in Biervliet and who was buried in the church there. As memories of events around 1400 faded it was not illogical to associate that family and then one Willem Beukels with the increase in curing of herring in the first half of the fifteenth century. If there was any change in technique around 1400 it certainly was not a major one. The transfer of curing to shipboard may well have been practiced and promoted by a Willem Beukels. His influence on that change, however, was certainly much less than that attributed to him by legend.

The legal and political battle over the sale of cured herring by Flemings continued through the first two decades of the fifteenth century. Once the grant had been made to Biervliet other ports wanted the right to receive cured herring. After being petitioned the government responded by granting such privileges but the fish still had to be sold at Biervliet. By 1412 four more ports were allowed to receive cured herring but in all cases it had to be exported and could not be sold for consumption in Flanders. In 1413 even that restriction was dropped, at least for the fishermen of the village of Nieuwpoort. The required landing port was moved in 1413 from

[19] Dardel (n. 12 above) 46; Degryse, "Schoense" (n. 4 above) 105-107; R. Degryse, "Willem Beukel en het Begin vat het Kaken: Een Antwoord," *Bijdragen voor de Geschiedenis der Nederlanden* 15 (1960) 215-220; Doorman (n. 1 above) 377, 383-385. The town of Ostend claims a citizen, one Jacob Kien, as the inventor of the process. Ouddorp on the island of Goeree in South Holland has its own candidate, one Jan Machiel Duffel.

Biervliet to Damme. In the last years of the 1410s the pressure mounted again to outlaw the landing of cured herring. In 1419 the town of Ghent prohibited the sale of any cured herring from Flanders, Holland and Zeeland in its markets. The reason given was the shortage of fresh fish. Ghent was joined by other large towns in Flanders, apparently convinced that they could get adequate supplies of cured herring from Scania or convinced that unless they acted there would be nothing but *kaakharing* available. The reaction from the fishing ports along the coast was strong and immediate. After delays a meeting was finally held in 1420 at which the towns withdrew their prohibition and Flemish fishermen immediately went back to curing herring and selling it in Flanders.

That finally swept away the remnants of restriction placed on the Flemish herring fishery in about 1396. It left the industry almost free to pursue the opportunities created by curing fish on board ship. And the new duke of Burgundy, Philip the Good, proved more interested in aiding the herring fishery than had his predecessor, John the Fearless. In 1424 Philip sharply reduced the duty charged on cured herring. A treaty negotiated with England in 1407 gave Flemish fishermen a guarantee of safety while working in English waters. In 1433 Holland and Zeeland were officially added to the possessions of the duke of Burgundy. That gave fishermen from those two northern provinces the same protection as Flemings, protection reaffirmed in a treaty with England of 1439. In Zeeland and Holland the herring fishery never had to face the political pressure of the Hansa. By the mid-fourteenth century governments there were trying to develop markets in the two provinces for cured fish. The presence of potential outlets for the product acted to increase the curing of herring on board ship. The port of Brielle by the early years of the fifteenth century, perhaps drawing on the experience of fishermen at Scania, had established itself as a center for the Dutch herring industry.[20] By the 1430s all Netherlands fishing ports were under the same government, one which enjoyed some power in international affairs. The dukes of Burgundy were less dependent on the Hanseatic League for the supply of all kinds of goods and of shipping services for their lands, once they had added Holland and Zeeland to their domains. The Dutch were already competing effectively with German traders. So the government, if not positively committed to the promotion of the curing of herring by Netherlands fishermen, at least was not going to impede those men from finding and using the most profitable techniques.

IV

Even in the first two decades of the fifteenth century the quantity of cured herring landed in Flanders could not compare with the imports of Scania *tonharing* of the

[20] Degryse, *Haringbedrijf* (n. 4 above) 49-51, 102-105, and (n. 1 above) 208-210, 218; Gottschalk (n. 9 above) 10-11; Doorman (n. 12 above) 171; S. Haak, "Brielle als vrije en bloeiende Handelsstad in de 15de eeuw," *Bijdragen voor Vanderlandsche Geschiedenis en Oudheidkunde*, ser. 4, 6 (1907) 35-36.

last years of the preceding century. Biervliet, the staple for all Flemish cured herring, received an annual average of less than forty-six lasts of herring from 1398 to 1408. The years were somewhat unusual since the area was devastated by a flood in 1404. The total value of the herring catch at Ostend rose some eight times from 1418 to 1467 and by the latter date volume was up to some 1600 lasts for that single port. The same kind of turnaround in the Netherlands fishery dating from the 1420s occurred in other ports. At Biervliet in 1426-1427 the annual average production of cured herring was 324 lasts. At Damme, then the staple port, the average annual landing of cured herring was 754 lasts in 1437 through 1441. About twenty-two percent of the total was brought there by fishermen from Holland and Zeeland. And those years were not especially good ones. There was a breakthrough some time in the second third of the fifteenth century when Netherlands production of cured herring replaced imports from Scania in the large Flanders market. The records of institutions confirm the shift away from traditional supplies. Netherlands cured herring was also exported, not just to inland markets in Brabant but also to Germany. Still the quantities were a far cry from the production totals of Scania in the fourteenth century. The towns of Flanders in 1423 after an investigation called for the promotion of putting herring in casks at sea because they felt production was not rising fast enough.[21] That was a sharp change from the attitude of just four years before.

The change in production methods and in government policy depended on related developments in technology and business methods. More important were certain specific and unique circumstances which made it commercially advantageous for fishermen to raise the proportion of their catch which they treated on board. Taking more herring further north and curing it, although technically feasible, was advisable only if ships and working capital could be found to make the voyages. It took time for Low Countries shipbuilders and fishermen and merchants to implement the necessary changes. The spread of the use of the large *vleet* net throughout the Netherlands did not take place until the second decade of the fifteenth century. It was in that same ten year period that shipbuilders produced the first herring buss. That ship type was probably a modification of the standard Scandinavian cargo ship of the thirteenth century. It was built to handle the heavy seas that herring fishermen met off the coast of Scotland and built to deal specifically with problems of using the big net. Men with more capital and with more extensive trading connections were drawn into the fishery and into the shore-based industry devoted to repacking and then distributing the fish. More money then could be advanced to crews for their longer voyages. The wealthier brokers were in a position to sell the product over a

[21] Degryse and Mus (n. 12 above) 105-106; Degryse (n. 18 above) 74-81; R. Degryse, "De Crisis in het Haringbedrijf te Oostende en te Damme van 1437 tot 1441," *Handelingen van de "Société d'Emulation" te Brugge* 102 (1965) 53-64; *idem*, "De Biervlietse Kaakharingproductie in de Jaren 1426 en 1427," *Marine Academie van België: Mededelingen* 13 (1961) 43-58; E. Vlietinck, *Het oude Oostende en zijne driejarige Belegering (1601-1604)* (Ostend 1897) 73; Ypma (n. 12 above) 37.

wider market.[22] The first signs of such changes in the organization of the industry date from the first half of the fifteenth century. The expansion of Low Countries production could only proceed as rapidly as constraints were removed, as rapidly as herring busses came into use and as investment increased.

While those developments in related technologies acted to remove restrictions, the commercial situation in the years around 1400 was a positive impetus to Netherlands fishermen to make more *kaakharing.* Their competitors, the merchants of the German Hansa, were periodically handicapped. There were short run problems of supply of cured herring from Scania. Production in Scania was always subject to variation, but in the late fourteenth and early fifteenth centuries the problems became more acute. The fishery there was always highly vulnerable both strategically and biologically. Supplies could be and were intermittently interrupted and each time that happened consumers were forced to seek a substitute. That meant using other types of fish or, increasingly, as knowledge of curing herring spread, looking for alternate sources of *kaakharing.*

Netherlands fishermen consistently exploited the opportunities created by the difficulties of the Scania fishery. Moreover, the Hansa merchants inadvertently helped them to do just that. The war between the Hanseatic League and Denmark which ended with the Treaty of Stralsund in 1370 forced the suspension of fishing in Scania in 1367 and 1368. The victory of the League was so complete that it then tried to eliminate all non-Hansa merchants from Scania. The League was not entirely successful. From 1388 to 1392 Hansards boycotted the Bruges market, so that none of their goods reached the Flemish staple of cured herring. In both cases a restrictive policy suddenly cut off supplies and drove more Netherlands merchants to take an interest in buying and marketing herring caught and cured on the North Sea. In 1393 fishing in Scania had to be suspended because of the activity of pirates. Presumably no cured herring was shipped to the Netherlands in that year and local fishermen were again put under pressure to supply the product. Since the herring was probably not edible for much more than a year it was impossible to stockpile it against such interruptions in supply. In any case it is doubtful that any merchant or government was in a position to finance such stockpiling. In 1410 the herring simply did not appear at Scania. There were other instances of such a sudden change in the spawning habits of the fish in the fifteenth century. There were poor catches reported in 1411, 1412, 1416 and 1425, for example.[23] By no means did the herring permanently disappear from the coast of Scania. It was not some shift in the Gulf Stream which led to the disappearance of the herring. Rather, coastal herring along the eastern side

[22] J. Ploeg, "Speurtoch naar Haringbuizen," *Mededelingen van de Nederlandse Vereniging voor Zeegeschiedenis* 25 (1972) 25-31; Theodorus Velius, *Chroniick van Hoorn* . . . (Hoorn 1648) 17; Degryse (n. 12 above) 121-127, and (n. 13 above) 146-150. He is correctly sceptical about giving an early fifteenth-century date for the full development of the herring buss.

[23] Gottschalk (n. 9 above) 11; Degryse, *Haringbedrijf* (n. 4 above) 99-100; Doorman (n. 1 above) 381-382; H. van der Wee, *The Growth of the Antwerp Market and the European Economy (Fourteenth-Sixteenth Centuries)* (The Hague 1963) 2.42-43, n. 76.

of the North Sea were subject to such sudden but short term changes in habits. They reappeared with the same irregularity.

The troubles with supply of Scania *tonharing* in the 1390s, the 1400s and the 1410s certainly affected government policy in Flanders. Indeed the laws seemed to change with the fortunes of imports from the North. In years when the Hansa merchants could not supply fish the government was defensive. It tried to prevent great increases in the output of domestic *kaakharing* for fear that it would add to their problems by generating a shortage of fresh herring. That had the political advantage of gaining the friendship of the Hansa merchants and increased the likelihood that, when supplies were available, those men would send them through Flemish ports. When there was enough cured herring for local markets and for those along the Rhine and in France the Flemish government lessened the restrictions on local producers but not enough to disturb their relations with the German traders. After about 1420 the interruptions led to a permanent change in government policy. The war between the towns of the Wendish coast, members of the Hanseatic League led by Lübeck, and Holland from 1437 to 1441 cut off supplies from Scania again. It was exactly in those years despite the activity of Dutch privateers and the generally high level of violence at sea that Netherlands fishermen proved themselves able to fill the vacuum left by a suspension of Scania supplies. It would be wrong to point to any single interruption or event which destroyed competition for Netherlands fishermen in preparing cured herring. A few years of peace in the first half of the fifteenth century which their principal competitors did not enjoy certainly helped. Earlier the change in the course of the river Yare and the silting of the harbor at Yarmouth from 1347 worked to the disadvantage of English fisherman and forced Netherlanders to look for alternate ports or simply to cure their herring at sea.[24] The fact that no other producer could consistently and reliably supply the Netherlands market graphically demonstrated again and again the advantages of curing herring on board. Each of the sudden collapses of supply gave Netherlands producers of cured herring windfall profits. Each time the gains led to increased production and to increased investment in equipment to raise production. Netherlands fishermen ranged over a wide area of the North Sea. Though there were good and bad years in the deep sea fishery a complete collapse of that catch is unheard of, unlike coastal fisheries. Netherlanders therefore always had a greater security of supply. Improvements in the safety of fishing vessels at sea through treaties or through arrangements for their protection along with the construction of ships of better quality all served to increase that advantage of the deep sea fishery. Each time that competitors had problems, Netherlands merchants were able to make inroads into new markets with herring produced by local fishermen, cured on board ship. And they could keep those markets as output rose. By the late fifteenth century even Hansa towns found themselves buying cured herring from Dutch fishermen.[25]

[24] Christensen (n. 4 above) 264-265; Degryse, "De Crisis" (n. 21 above) 62-63; Doorman (n. 12 above) 172; Jenkins (n. 4 above) 9-11.

[25] R. Degryse, "De Omvang van Vlaanderens Haring- en Zoutevisbedrijf op het Einde van het Frans-Bourgondisch Conflict (1482)," *Marine Academie van België: Mededelingen* 15 (1963) 45-47; Vlietinck (n. 21 above) 98; Ketner (n. 6 above) 18-19.

The prices of cured herring in the Netherlands reflected the sudden interruptions in the supply of fish from Scania. They also show the mechanism which existed for keeping the Flemish government informed of the situation in the herring market. The price data for the Low Countries around 1400 are scanty at best. The graph shows the pattern of price changes at Antwerp from 1387 to 1450. The figures reported are the three-year moving average of prices. While using such averages obscures the most violent fluctuations, at the same time it more clearly shows the trends over a long term, the kinds of changes which influenced investment decisions. Antwerp in those years was an increasingly important market for all goods, acting both as a center of consumption and a transshipment point for taking goods inland.[26] The Antwerp prices then are probably a good reflection of the general situation in the Netherlands.

In the 1390s the prices of cured or wet herring as they were called rose sharply at Antwerp. In 1395 the average price was at almost two and one-half times the level of 1386. The sudden jump can only be explained by a failure of imports of *tonharing* from Scania. With amounts like that being paid for a cask of cured herring it is not surprising that Flemish fishermen brought their own cured herring to the staple port for the first time in 1395. The high prices, and the greater concentration on curing the herring catch which followed from them, were probably the reasons for the government's acting so sharply to prevent the production and sale of preserved herring by Flemings. The grant by the duke of Burgundy in 1399 to allow limited production was only a small concession, since there had been a dramatic collapse in prices. The level was even below that of 1388 when the troubles with the Hanseatic League began. With the economic incentive for curing herring largely removed the destruction of the legal barrier was little more than symbolic. Prices began to rise again after a period of decline from 1404. That was the year of the flood which destroyed Hughevliet, badly damaged Biervliet and also disrupted salt production in the province of Zeeland. All that certainly limited the output of cured herring in the Low Countries. The flood may explain the small and slow rise in prices after 1404. But it was still what happened in the Sound that had the greatest and indeed the overwhelming influence on prices in Antwerp. The failure of the catch in Scania in 1410 sent the average price of *tonharing* up more than eighty-five percent from 1409 to 1413. After abating in 1414 and 1415 the rise became even more marked, reaching a maximum in 1419. The price then was more than three times that in the year before the flood, 1403. It was in 1419 that the Ghent city government took strenuous action to stop Flemish fishermen from preserving their herring. Undoubtedly those men made every effort to take advantage of the higher prices and cured higher proportions of their catch. It was a repetition of the situation in the mid-1390s. The sharp fall in prices from 1420 through 1422 explains the decision of Ghent and other towns to withdraw from their policy of restriction and indeed explains why they gave in relatively easily to petitions from the fishing villages in

[26] Van der Wee (n. 23 above) 1.227-280. This is the source for the data presented in the graph. Another price series from Antwerp appears in C. Verlinden et al., *Dokumenten voor de Geschiedenis van Prijzen en Lonen in Vlaanderen en Brabant* (Bruges 1959-1972) 2.2.731-737. The series begins in 1426 and is not such as to contradict conclusions drawn from the fuller series of van der Wee.

GRAPH: Herring prices at Antwerp, 1387-1450 3-year moving average in Brabant money of account

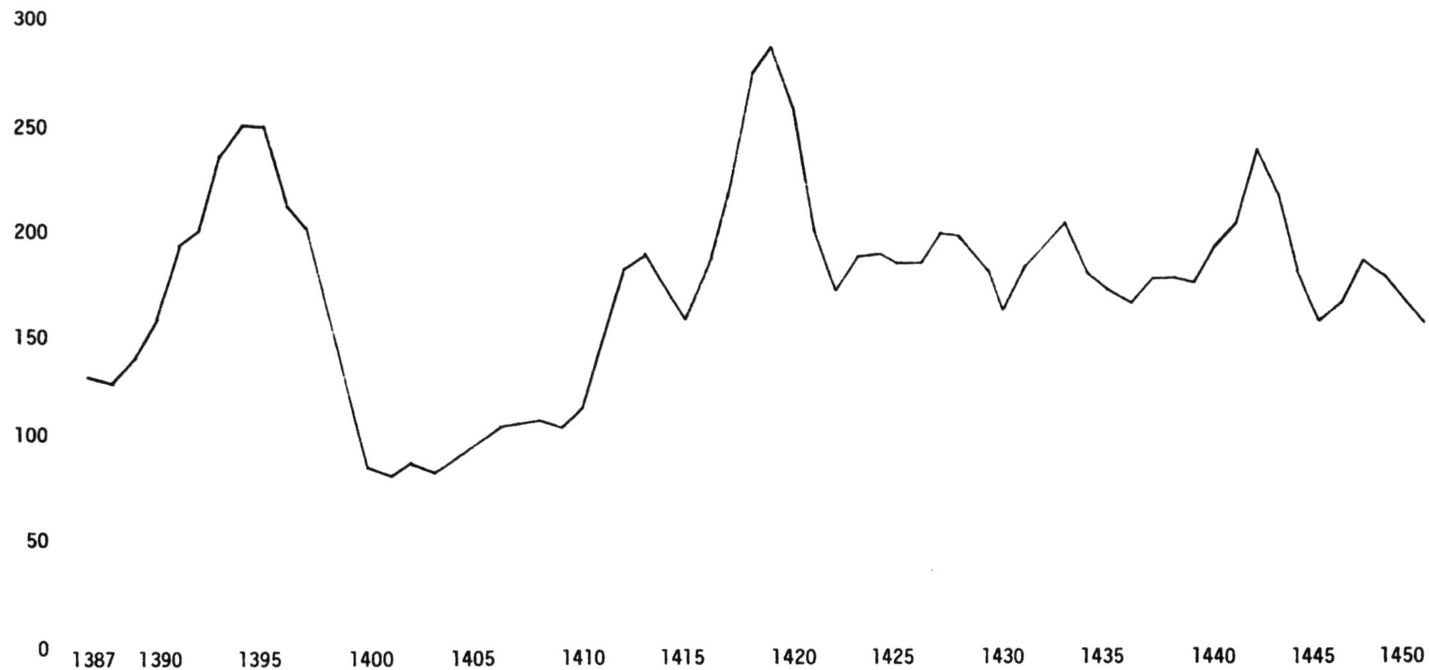

1420. Prices were already falling. The relative stability of prices but at a high level over the following decade undoubtedly contributed to the newfound government interest in promoting domestic production of cured herring and also to the interest of Netherlands fishermen in increasing their output of the product. The Wendish War in 1438-1441 led to another increase in the delivered price of herring at Antwerp. But the average annual increase in the worst years, 1440 through 1442, was only eleven percent. The rise was not anywhere near the scale of the increase during the Hansa boycott of Bruges from 1388 to 1392, though admittedly the latter was from a lower base. Indeed the fluctuations in prices were less and less extreme over the whole period. It seems that by the 1440s production in the coastal provinces of the Low Countries, in Friesland, Holland, Zeeland and belatedly in Flanders, was at such a level that it could offer an effective substitute for at least a significant part of the imported Scania *tonharing.*

V

The troubles in the last two decades of the fourteenth century, reflected in the high prices of Scania herring, created the necessary impetus for Low Countries fishermen to increase sharply their curing of fish on board ship. Moreover the problems gave governments the incentive to consider ways to promote that production. Once the sudden price increases had created the opportunities for profits from curing herring, then investment in the technique in the form of ships and casks and nets was created either from the profits themselves or from the wealth of merchants interested in getting part of those profits. The increase in curing also gave more fishermen a chance to learn how to use the method and become accustomed to gutting and packing fish on board. This learning-by-doing combined with the windfall profits created a compulsive sequence in which advance on one front made for opportunities for advance on another. Once started the process of adoption of the technique gathered momentum. By the middle of the fifteenth century that sequence along with the changes in prices had generated a new and sizable industry in the Low Countries producing a commodity which had formerly been imported.

The change to domestically produced cured herring did not imply any change in the volume shipped. With the general increase in population in Europe through the fifteenth century the demand for herring rose. But there is no indication that the transfer of curing from the shores of Scania to Netherlands fishing boats in the North Sea created any major shift in consumption. The trading patterns of the fourteenth century, then, remained the same. The Hansa ships which carried herring to the Low Countries did disappear. But they were partially replaced by Dutch ships carrying the same product in the opposite direction. That gave some small aid to the expanding Netherlands merchant marine. More important was the increase in the tonnage of the fishing fleet, a direct result of the change to deep sea fishing. Given the level of production there were probably at least 50 and as many as 200 herring busses operating from the Netherlands coast in the fifteenth century. In the off-season owners used those vessels to carry other cargo, thus increasing the available tonnage at Low Countries ports. Beyond the fishermen and sailors the domestic herring fishery also meant work for packers on shore and suppliers such as coopers and shipcarpenters among others. Still the quantity of herring produced was small. The trade in wine from Bordeaux to England suffered badly in the first half of the fifteenth century because of wars but still the annual average export of wine from Bordeaux from 1400 to 1440 was about 11,000 tons. That was eight to ten times the volume of herring cured on board ship by Netherlanders at that time.[27]

Netherlands production of cured herring in the fifteenth century rose to a point where it could replace imports of Scania *tonharing.* But it was not until the mid-sixteenth century that Netherlands output reached that of the Baltic fishery at its height. By 1580 the Netherlands herring fishery produced about 20,000 lasts annually, more than twice the figure for the fourteenth century Scania fishery. The annual average export of herring to the Baltic in the 1560s was about 2800 lasts, almost two-thirds more than the quantity traveling in the opposite direction in the late fourteenth century.[28] Netherlands fishermen and merchants reached and surpassed the record of the Scania fishery by tapping the growing market for food. They were able to take advantage of that potentially larger market for herring because of developments in their own fishery, the development of complementary techniques and the evolution of a legal and commercial framework for the industry at home and elsewhere in Europe.

No sudden breakthrough in herring production or trade was possible in the later Middle Ages because it was not a product for mass consumption. Even if all the cured herring brought from the Baltic in the late fourteenth century had been consumed in the three great Flemish towns of Ghent, Bruges and Ypres, at most each resident would have had one herring per day. The population of the Low Countries around 1400 was probably under 2,000,000; but even so, imports from Scania could only supply each person with about 12 fish each year.[29] Since there was extensive reexport of cured herring, that figure should be at least halved. There was good reason for the limited market for herring, despite the method of preservation and the relative ease of transportation. Herring unquestionably was and is a good source of protein. It compares favorably with other fish as a source of fat and vitamin D. In

[27] Margery K. James, "The Fluctuations of the Anglo-Gascon Wine Trade during the Fourteenth Century," in *Essays in Economic History* 2, ed. E. M. Carus-Wilson (New York 1962) 143-144; Yves Renouard, "Les transformations économiques et sociales," in *Histoire de Bordeaux* 3, ed. Charles Higounet (Bordeaux 1965) 423-424.

[28] H. A. H. Kranenburg, *De Zeevisscherij van Holland in den Tijd der Republiek* (Amsterdam 1946) 39-43; W. S. Unger, "De Sonttabellen," *Tijdschrift voor Geschiedenis* 41 (1926) 143-144.

[29] Population data are notoriously difficult to find and inaccurate at best. See David M. Nicholas, *Town and Countryside: Social, Economic and Political Tensions in Fourteenth-Century Flanders* (Bruges 1971) 234. The estimate for the three towns assumes that they all had a population at least as great as that of Ypres in 1412. That much is certainly true. See B. H. Slicher van Bath, "Historical Demography and the Social and Economic Development of the Netherlands," *Daedalus* 43 (1968) 609-610.

general the mineral content is high as is the calorific value per unit weight.[30] The problem at least in the first half of the fifteenth century was cost. In general consumers in Brabant paid less for fat, calories and even protein when it came in cheese, rye or wheat than when it came embodied in herring. Data for meat are not as complete but they suggest that in the third quarter of the fifteenth century herring was a cheaper source of calories and about equally as expensive a source of protein as meat. In famine years when grain prices skyrocketed to record levels then herring was a less expensive source of protein than rye. That was true in 1437 and 1438. Wheat was about as good a source of protein for the same expenditure as herring. But even in years of record grain prices herring was still a much more expensive source of calories. And in any case cheese was always a better buy, except in one year, 1438, when protein from cheese cost slightly more than protein from herring.

Ratios of the Cost of Calories, Protein and Fat from Cheese, Rye, Wheat and Meat Relative to the Cost of Calories, Protein and Fat from Herring in Brabant[31]

Food	*Period*	*Calories*	*Protein*	*Fat*
Cheese	1386-1450	.41	.57	.56
Rye	1392-1450	.16	.53	.04
Wheat	1427-1450	.29	.91	.03
Meat	1453-1474	1.84	.95	.04

The estimates are subject to sizable but unspecified error. Still it would be surprising if all errors were consistently one-sided. Even if the ratios are off by as much as fifty percent, the conclusion that herring was a relatively expensive source of nutrients is not in jeopardy. Cured herring was valuable in diversifying the diet of medieval Europeans. It was valuable as a source of protein at times when other sources were not available. It was valuable in satisfying religious prohibitions against the eating of meat. But, economically, consumers were wise to choose other foods. Grains were a significantly better source of nutrients for equal expenditures and in any case that was all the less well-off could afford.

The adoption of curing herring on board ship in the early fifteenth century did not mean a major change in nutrition for Europeans. Nor did it mean an immediate or massive change in employment or industrial practice in the Low Countries. Rather what the technical innovation did do was lay the basis for the long-term development of a large industry which depended on international trading connections both for

[30] Olav Notevarp, "Herring, the Raw Material," in *The Technology of Herring Utilization* (n. 2 above) 38-39; Charles Louis Anderson, "Preserved Pickled Herring," *University of Washington Publications in Fisheries* 1 (1925) 24.

[31] Prices are taken from van der Wee (n. 23 above) 1.175-176 (rye in Antwerp), 184 (wheat in Lier), 218-220 (cheese in Antwerp), 225 (meat in Malines), 279-281 (herring in Brabant). Relative food values are from B. K. Watt et al., *Composition of Foods*, rev. ed., Agriculture Handbook no. 8 (Washington 1963) table 1. Adjustments are made for the historic change in the specific gravity of grain following Lord Beveridge, "The Yield and Price of Corn in the Middle Ages," in *Essays in Economic History* 1, ed. E. M. Carus-Wilson (London 1954) 16-17.

supplies of raw materials and for sale of the finished product. The trading complex which emerged over time stretched throughout Europe. The innovation also made possible the more extensive use of a previously untapped resource, that is, fish in deep water. It also involved a shift in the source of supply of cured herring to western Europe from the Baltic. But most important the development in the Netherlands herring fishery offers a good example, despite some confusion, of the process of technical innovation in the later Middle Ages. In the face of the economic troubles of the fourteenth and fifteenth centuries Europeans produced a number of important changes in technology. The absolute and relative figures of output in this case as in many others tend to mask the importance of the qualitative change as well as to diminish the significance of the process of innovation itself.

The *kaken* method of preserving herring was well known and widely used in Europe before 1300. Its use on board ship probably began early in the fourteenth century. But it took the sudden and sharp falls in supply created both by political decisions and by the variations in the catch of the coastal fisheries to lead Flemish and all Low Countries fishermen to adopt extensively the curing method on board their herring boats. Apparently there was a Willem Beukels living at the time of that change in practice and apparently he was to some extent involved in promoting that change. His presence and the association of his family with the herring fishery at Biervliet explains the traditional association of his name with the adoption of the technique. It does not explain the concentration of so many writers on this one man or their insistence on his action as the reason for the success of the Dutch herring fishery in the first half of the seventeenth century. The adoption of any innovation depends on a great deal more than simply the presence of the invention. The survival of the legend of Willem Beukels of Biervliet serves only to obscure the complexity of that process. It also serves to misrepresent the intimate relationship in this case of the economy and government policy to technical change. The history of medieval technology is undoubtedly filled with many similar examples.

XVIII

Dutch Herring, Technology, and International Trade in the Seventeenth Century

Herring exports to the Baltic from the Netherlands in the seventeenth and eighteenth centuries were closely related to exports of the previous year rather than to aggregate levels of trade. Dutch domination of the European market for salted herring in the seventeenth century thus cannot be explained by some external factor but rather by the internal nature of the Dutch fishery: by technology, organization, and the institutions which administered it. Regulation was designed to maximize rents but, as other fishermen gained the skills of their Dutch competitors, that strategy turned into one which at first limited sales and then returns to the Dutch industry.

. . . O, wat een gulden Neeringh
en voedsel brengt ons toe de Coninckltjke Heringh;
hoe menig duysend ziel bij dezen handel leeft en
winnende sijn brood God dank en eere gheeft.[1]

THOSE were the words of Joost van den Vondel, the greatest Dutch poet of the seventeenth century, in adulation of the "royal herring." As he suggested, the herring was an important commodity in the international trading network of the Dutch Republic. The herring fishery was a transforming industry, a *trafiek*. Netherlanders caught the fish at sea, treated them using imported salt, and packed them in casks of imported wood. They exported the final product. Herring played an integral part in the "mother trade," the shipping of corn and forest products from Baltic ports to the west coast of France and Iberia to be exchanged for salt, wine, and other goods which in turn were brought back to the Netherlands. Those goods were shipped on to the Baltic in their original form or in some processed form or, in the case of some of the salt, transformed by combination with herring. It was that and related exchanges that made the Dutch Republic unquestionably the leading trading state per person in seventeenth-century Europe. Though it is true that Dutch herring exports were only possible because of the existence of the trading network, the quantity of fish sent overseas was not a function of the quantity of any

The Journal of Economic History. Vol. XL, No. 2 (June 1980). ISSN 0022-0507.

The author is Associate Professor of History at the University of British Columbia. The analysis and preparation of this paper depended on the assistance of Virginia Green. The University of British Columbia supplied computer time. The author is indebted to Piet van der Veen for his personal help and to Robert Allen, Don Paterson, Jan de Vries, and especially John Norris for reading and commenting on an earlier draft.

[1] Joost van den Vondel, "Lofsangh op den Scheepsvaart," *De vernieuwde Gulden Winckel* (Amsterdam, 1622), lines 197–200. "O what a golden industry is created for us by that food, the royal herring. How many thousand souls, thank God, live by this trade and earn their living from it."

or all of the other goods exchanged in the "mother trade." Rather, herring exports depended on factors internal to the Dutch herring fishery and the herring fisheries of other northern European states.

An examination of the short-run relationship between Polish export earnings and Polish expenditure on herring imports shows little causal connection. Grain exports fluctuated widely, depending on the weather, levels of violence, and other exogenous factors. Moreover, Polish landowners had many things to spend their earnings on other than herring. While over the long term Dutch herring sales in the Baltic showed some connection with Polish exports, year on year the relation was very weak. The principal reason for Dutch success in exporting herring to the Baltic has to be found elsewhere.

Price differentials and profits offer a more complete explanation. Above all, however, it was certain specific technical changes and the development of certain political institutions in the course of the fifteenth and sixteenth centuries that allowed the Dutch herring fishery to gain a dominant position in European markets. Over time, Dutch technical superiority was eroded as competitors developed the same skills. As alternate sources of supply emerged, the Dutch chose to limit production in order to maintain the premium prices their herring commanded. This led first to a decline in the volume of fish exported and then to a decline in value. In these new circumstances, the strategy that had previously led to market dominance and high rents became a contributor to falling total output and falling returns. The contraction of the Dutch herring fishery developed into just another part of the relative stagnation of the Dutch economy in the eighteenth century.

Vondel was not the only writer who was impressed with the value of the herring as a source of food, as a popular medicine, and as the product of a major industry. Commentators both in the Netherlands and elsewhere in Europe remarked on the size of the Dutch herring catch and its contribution to the economic growth of Holland in the years after 1600.[2] By the eighteenth century the Dutch herring fishery had taken on something of a

[2] H. Blink, "De Geschiedenis en Beteekenis der Nederlandsche Haringvisscherij," *Vragen van den Dag*, 45 (1930), 985–86. Adriaen Coenen Zn., *Visboeck*, Handschriftkamer, Koninklijke Bibliotheek, begun 1577, fol. 15r–16v. In this lavishly illustrated short manuscript on the fishery the author twice pictures the herring with a crown on its head and calls the fish, "our noble herring, the king above all other fish." John R. McCulloch, ed., *A Select Collection of Scarce and Valuable Tracts on Commerce* (London, 1859), pp. 21–22. Sir Walter Raleigh estimated for his king, James I, the employment which grew directly and indirectly out of the Dutch herring fishery. Pieter de la Court, *The True Interest and Political Maxims of the Republick of Holland and West-Friesland... Written by John DeWitt and other Great Men in Holland* (originally published in Dutch in 1662; London, 1702), pp. 37–42, added recognition of the secondary jobs created in shipping and manufacturing, the value of the fishery as a school for seamen, and the value of herring as an exportable good. His estimate of 19 percent of the population earning their living from the fisheries is too high. Raleigh was also much too extravagant: his claim that the net gain to the Dutch Republic from the herring fishery was 21,500,000 guilders was well above the actual figure of about 2,500,000 guilders. See H. A. H. Kranenburg, *De Zeevisscherij van Holland in den Tijd der Republiek* (Amsterdam, 1946), pp. 39, 212. The contribution of the herring fishery to total Dutch output had been stated officially as early as 1476.

mythical quality for writers—Voltaire, for example—and it is through that myth that historians in later years have come to write about the industry. The claims in some cases go to the extreme of explaining the Dutch navy, the trade of the Netherlands, and the overseas colonies all as children of the North Sea fisheries.[3] Even less extreme writers point to the herring fishery as one of the bases of seventeenth-century Dutch prosperity, noting the fishery's chief contribution as a commodity-return in multilateral trade, as well as its being a direct source of income. Certainly, it was already an important contributor to gross output in the sixteenth century, when Charles V's personal physician said that the Dutch got more gold and silver by catching and selling fish than other countries did by digging the metal out of the ground. The Dutch government in 1624 called the fishery the gold mine of the republic. The estimates perhaps better embody the moral the authors wanted to draw than they do actual output figures, and so they should not be taken seriously.[4] The history of the herring fishery—especially the internal history of the whole range of activities associated with it—has then been typically obscured, the contribution of the industry being seen in gross terms and never examined as a result of what went on in the fishery itself.

The method for curing or pickling herring was well known during the Middle Ages. Soon after the herring were caught, the packer eviscerated the fish, mixed them with salt to form a brine, and then packed them into casks with more salt. The contribution of Low Countries fishermen was to adapt this method for use on board ship, which meant that the herring had to be repacked when it was brought to port. By doing the work of preserving at sea, Dutch fishermen could stay away from shore longer. That in turn enabled them to seek out and exploit new deepwater fishing grounds off the coast of Scotland, off the Shetland Islands, and off Iceland. Netherlanders cured herring on board ship before 1400, and in the second third of the fifteenth century market phenomena and government policy combined to allow a sharp rise in the production of salted herring in the Low Countries.[5] Salt importing began in the fifteenth century. The

[3] For Voltaire see Gerard Doorman, "Nogmaals: de middeleeuwse haringvisserij," *Bijdragen voor de Geschiedenis der Nederlanden,* 14 (1960), 104. Nels A. Bengston and William Van Royen, *Fundamentals of Economic Georgraphy* (Englewood Cliffs, N.J., 1956), pp. 314–15, made the most lavish claims for the importance of the herring fishery. The extreme statement appeared in the first (1935) through the fourth (1956) editions, but was dropped in the fifth (1964) and subsequent editions.

[4] Robert Fruin, *Tien Jaren uit den Tachtigjarigen Oorlog, 1588-1598,* 5th ed. (The Hague, 1899), p. 185. McCulloch, *Tracts on Commerce,* p. 97. The implied comparison was presumably with mines in the New World. In the first half of the seventeenth century, even in the best years for the fishery, specie of a value almost four times that of the Dutch herring catch arrived annually in Spain from America. Compare Earl J. Hamilton, *American Treasure and the Price Revolution in Spain, 1501-1650* (Cambridge, Mass., 1934), pp. 32–35, and Kranenburg, *De Zeevisscherij,* pp. 133, 212. Nicolaas W. Posthumus, *Inquiry into the History of Prices in Holland* (Leiden, 1946–1964), vol. I, pp. cxv–xvi. The value of the herring catch in the 1630s, one of the best decades for the fishery, was annually about 30 metric tons of silver. Incidentally, in the same decade Spain received an annual average of 140.5 metric tons of silver.

[5] Richard W. Unger, "The Netherlands Herring Fishery in the Late Middle Ages: The False Legend of Willem Beukels of Biervliet," *Viator,* 9 (1978), 335–56.

herring fishery was the chief consumer of that salt brought from France, Spain, and Portugal. The sea salt, becasue of its relatively high magnesium sulphate and magnesium chloride content, was well-suited for preserving the herring. It was also cheaper than domestic salt which was supplied by burning peat from coastal bogs, impregnated over the centuries with sea salt.[6] Despite the fact that the transfer to curing on board had been made by 1400, and that supplies of sea salt from the Atlantic coast were available well before 1500, it was not until the seventeenth century that Dutch herring production reached its peak. The explanation for the long delay lies in the history of the fishery itself, in developments in both the economics and the technology of the fishery. Those two factors also help to explain the decline in output after about 1650 and then the collapse in the eighteenth century.

I

The development of technology in the herring fishery extended from the fourteenth to the mid-sixteenth century and took many forms. The wide range of new techniques and new equipment laid the basis for the long-term growth in output. By the time of the Dutch Revolt against Spanish rule beginning in 1568, the Netherlands fishery enjoyed a marked superiority in Europe. There was little improvement in techniques during the period of the Republic down to 1795. The technical changes in the fifteenth and sixteenth centuries included, first, improvements in the techniques of curing on board ship; second, changes in the organization of the herring fishery; third, improvements in the equipment, in the capital goods; and fourth, the development of political institutions which protected fishing boats and regulated production to maintain quality. Changes in method often set up compulsive sequences whereby one technical development leads to the use of others. In the herring fishery, such a sequence occurred, for example, with the design of ships. Moreover, the long-term process of learning-by-doing gave the Netherlands a large pool of experienced and knowledgeable personnel at all steps in the preparation of herring. The greatest impetus to the use of all the superior methods

[6] W. Brulez, "De Zoutinvoer in de Nederlanden in de 16e eeuw," *Tijdschrift voor Geschiedenis*, 68 (1955), 181–84. Johannes van Dijk, "The Technology of Herring Utilization," *Report of the FAO Meeting* (Bergen, 1950), pp. 224–25. H. de Jager, *De Middeleeuwse Keuren der Stad Brielle* (The Hague, 1901), pp. 161–62, 190–91. Herman van der Wee, "De groei van de Nederlandse haringindustrie en het raadsel van het Zeeuwse Zout, 14e–16e eeuw," *De Vier Ambachten* (1964–1965), pp. 18–23.

Production in the Zeeland coastal salines seems to have fallen off in the fifteenth century, making the importation of salt from the Atlantic coast of Europe even more advisable. The cause was probably the frequent and disastrous floods. Herman van der Wee, *The Growth of the Antwerp Market and the European Economy* (The Hague, 1963), vol. I, pp. 287–91. The advantages of imported sea salt were partly offset by its higher level of impurities, which meant that it had to be extensively refined. Moreover, it took only four casks of Zeeland salt to treat fourteen lasts of herring whereas it took five and one-half casks of refined sea salt.

was the presence of a market for the preserved herring and a market that had potential for growth.[7]

When Dutch fishermen first began to cure herring on board ship in the fourteenth century, the product was of lower quality than fish treated on shore. By the end of the sixteenth century, however, that was no longer the case.[8] In fact, in the seventeenth century Dutch herring sold at a premium over herring pickled in France or England. The experience gained over time in gutting and treating the herring at sea may help to explain the improvement in quality. The same may be true for the job of repacking the fish in port. Dutch fishermen may have accidentally stumbled on the advantages of leaving part of the stomach, the pyloric caecae, in the fish to promote curing. Those appendices of the stomach contain trypsin, which speeds the curing process and also improves the aroma of the final product. There is some indication that seventeenth-century Dutch fishermen did not remove all of the stomach and pancreas simply because the work was done so rapidly. Typical Dutch practice was to gut the fish the morning after they were caught, which minimized deterioration. This made the gutters work quickly, handling up to 2000 fish per hour, and so they may have often failed to remove all of the stomach. An illustration dated 1652 shows gutted herring with parts of the viscera left behind. A modern survey shows that from 10 to 50 percent of herring gutted using the same process still had the entire stomach; therefore, an even higher proportion had at least the pyloric caecae.[9] While Dutch producers may have taken advantage of higher concentrations of trypsin without understanding their value, it is probable that they did learn by experiment the optimal salt concentrations both for packing on board ship and for the repacking done on shore.

The shift of the Dutch from coastal to deep-sea fishing for herring also increased the complexity of investment and marketing in the fishery. The increase in the duration of voyages—from overnight to from five to eight weeks—increased the turnover capital requirements of fishing ventures. They required larger and more expensive boats and crews. Under local sea law, the men on board had to be fed at the expense of the investors for the entire trip. More casks and salt were needed for curing. All this was very different from the modest capital demands in the early fifteenth century when the herring fishery was pursued by small boat owners who re-

[7] The pattern is similar to that noticed in general for the adoption and widespread use of any technical change. Nathan Rosenberg, "The Direction of Technological Change: Inducement Mechanisms and Focusing Devices," *Economic Development and Cultural Change*, 17, no. 1 (1969), 1–24; idem, "Factors Affecting the Diffusion of Technology," *Explorations in Economic History*, 10 (Fall 1972), 7–28.

[8] Eric Dardel, *La Pêche Harenguière en France: Etude d'historie économique et sociale* (Paris, 1941), p. 153. Ysbrand N. Ypma, *Geschiedenis van de Zuiderzeevisserij* (Amsterdam, 1962), p. 40. Van der Wee, *Growth of the Antwerp Market*, vol. I, p. 278.

[9] Gerard Doorman, "Het Haringkaken en Willem Beukels," *Tijdschrift voor Geschiedenis*, 69 (1956), 373. Luijpen, *De Invloed*, pp. 37–39, 61–73.

lied on brokers for financing and marketing, all for about 5 percent of gross income. By the mid-fifteenth century the brokers were becoming owners and operators of ships as well. They were merchants with an interest in more assured supplies of preserved fish. They usually divided the functions in a partnership, one partner acting as broker-merchant and another as skipper. Other merchants, ship chandlers, and even individuals with no direct connection with fishing could and did invest in the boats and their supplies. The status of the fishermen changed, too, from being owner-operators of boats to being wage laborers. The trend toward concentration of capital and of marketing in the hands of a smaller number of men continued in the sixteenth and seventeenth centuries. Ownership was vested increasingly in the hands of greater merchants in the large ports on rivers and inland seas with international trading connections.[10] After about 1600, financing was subjected to even greater specialization. Increasingly, single fish merchants replaced partnerships supplying all of the capital as impersonal investors lost interest in the herring fishery. At the same time the international herring traders became more interested in gaining control over supplies.[11] The seventeenth-century Dutch fish merchant pressed vertical integration to the point where he supplied all the capital and owned the product from the time it was caught, through processing and shipment, until it was sold to the final consumer.

Such concentration was not common in the sixteenth and seventeenth centuries. It occurred in the Duch herring fishery for a number of reasons. Falling capital costs—the average herring boat cost less over time—and rising merchant incomes combined to put ownership of the vessels within reach. By owning the boats and paying a wage to fishermen, merchants took the risk of failure into their own hands. But with the catch rising, risks were falling. The merchants effectively appropriated any rent which the fishermen might have earned. There were advantages to extending investment into production and also good reasons for merchants to extend their interest in the other direction, into marketing. As the final consumer became more distant from the producer, access to knowledge of markets and prices became more critical. A well-informed merchant was in the best position to sell the catch and to get the highest possible price. The

[10] Coenen Zn., *Visboeck*, fol. 20v. Renée Doehaerd, "La Genèse d'une entreprise maritime: les pêcheurs de Wenduine au XVe siècle," *Contributions à l'Histoire Economique et Sociale*, 1 (1962), 9–25. Dardel, *La Pêche Harenguière*, pp. 55–56, 86–92. H. A. H. Kranenburg, "Het Visserijbedrijf van de Zijdenaars in de 15e en 16e Eeuw," *Tijdschrift voor Geschiedenis*, 62 (1949), 328–32. Towns established rules to protect investors from unscrupulous skippers who might not pay them what they deserved. For example, Klaas Heeringa, *Rechtsbronnen der stad Schiedam* (The Hague, 1904), p. 245. Also, H. de Jager, *De Middeleeuwse Keuren der Stad Brielle*, p. 162, paragraphs 6, 7.

[11] The van Adrichems, a prominent Delft business family of the late sixteenth century, is a good example of these structural changes. Algemeen Rijksarchief, The Hague, Archief van Adrichem, 12, 13, 126, 127. H. Enno van Gelder, "Gegevens Betreffende de Haringvisscherij op het einde der 16de Eeuw," *Bijdragen en Mededeelingen van het Historisch Genootschap*, 32 (1911), 1–62, publishes 3 of the 29 surviving accounts of the van Adrichems' herring fishery ventures. Kranenburg, *De Zeevisscherij*, pp. 61–71, 117–25.

work on land, the repacking of the herring, was important to the quality of the final product. Having a resident merchant who was in a position to organize and oversee that work was necessary for the success of the entire operation, from catching to selling the fish. Above all, though, the herring industry was subject to integration because it was a transforming industry relying on imported raw materials and on overseas markets. Greater merchants dominated the industry because they had access to information about and control over the prices and supply of inputs and of output.

Improvements in equipment for the herring fishery were made mainly in the principal capital good, the boat. Low Countries shipbuilders around 1400 developed the herring buss, a vessel specifically suited for use in the deep-sea fishery. Herring busses were much more efficient than the small, flat-bottomed, keelless boats of the coastal fishery. Busses, purpose-built for the herring fishery, were certainly in widespread use in Holland in the 1440s. They were large enough to survive North Sea storms and to carry all the necessary gear including the big nets and the casks. There was space on board for men to work at gutting and packing the fish. Over time, builders modifed the buss so that by the early sixteenth century it was a three-masted vessel with sharply curved bows. There was a full deck with cover for the crew and for the empty and full casks. A ship with a relatively high ratio of length to breadth is better able to keep pressure on a long drag net when fishing, so busses were designed with higher ratios—usually about 4.5:1—than other seagoing ships.

In the seventeenth and eighteenth centuries the buss underwent significant changes, making it even more efficient. The flat stern was replaced with a rounded one which increased the ships' manageability. The three sails, one on each mast, were orignally square and remained so until early in the eighteenth century when rigging changed completely. The three masts were reduced to two, and one of those carried a fore-and-aft sail which needed fewer men to handle it. In general, herring busses were highly durable, lasting on average more than twice as long as cargo ships of similar size. A cross-section of the hull near the center would give the impression of an oblong rectangle with the corners not quite square. That shape and the high ratio of length to width gave the buss sizable carrying capacity for its length compared to similar boats. Carrying capacity grew over time as well. In the early fifteenth century busses were probably about the same size as coastal craft, but by the sixteenth century busses of 60 tons were not uncommon. The maximum feasible size was about 200 tons, and in the late sixteenth century busses of about 140 tons were typical. In the seventeenth century, however, builders and fishermen found that 60 tons and lengths of less than 20 meters overall were optimal. The smaller vessel cost less to build and much less to operate since the crew was only about 13 men instead of between 18 and 30. The change to smaller busses may have also been a result of increased specialization in shipping, with busses used exclusively for fishing and not carrying cargo in the off-season. While the ability to earn in alternative employment may have eased the adoption of the buss at the outset, by about 1600 the type was fully job-specific. The increasing efficiency of the buss contributed to the greater effectiveness of Dutch fishermen going after North Sea herring. The Dutch government demonstrated its recognition of the contribution of the buss design by consistently prohibiting the export of busses.[12]

Political institutions emerged to provide protection for herring fishermen because the busses, being equipped solely as fishing boats, were highly vulnerable to attack. In the fifteenth century herring fishermen organized convoys for mutual protection, and they fitted out vessels to defend the convoys. By the 1440s town governments were cooperating in the convoying of fishing vessels from the coastal provinces of the Low Countries. By the mid-sixteenth century the government of the Low Countries had assumed responsibility for supplying protection for the herring fleet, assessing taxes, and administering and paying for warships doing convoy duty.[13] Convoying continued under the Dutch Republic and became much better organized. The attacks of Dunkirk privateers and the increasing capabilities of defending warships broke down residual opposition to convoys and convoy charges. In the seventeenth century Dutch convoys were effective against most privateers and enemy warships, except in certain wars and at certain times. Convoys served a valuable purpose: they allowed Dutch fishermen to range widely without as much fear of attack and they allowed shipbuilders to construct even more job-specific fishing vessels.

Government in the Low Countries also developed an elaborate set of regulations governing all phases of the production of herring. The legislation was directed largely at maintaining the quality of the domestic product. The body of rules first began to develop in certain port towns, and in

[12] Jan van Beylen, *Schepen van de Nederlanden Van de late middeleeuwen tot het einde van de 17e eeuw* (Amsterdam, 1970), pp. 135–41. The earliest trustworthy illustration of a herring buss dates from 1504 or 1540. The change from a flat to a rounded stern on larger busses has been dated to between 1600 and 1650. Nicholaes Witsen, *Architectura Navalis et Regimen nauticum* . . . 2nd ed. (Amsterdam, 1690), pp. 186–87. Johannes E. Tillema, "Ontwikkeling van de Nederlandsche Haringvisscherij in den Loop der Eeuwen," *Het Nederlandsche Zeewezen*, 16 (1917), 66–67. Kranenburg, *De Zeevisscherij*, pp. 15–18, 56–58, 200–01. J. Ploeg, "Speurtocht naar Haringbuizen," *Mededeelingen van de Nederlandse Vereniging voor Zeegeschiedenis*, 25 (1972), 25–31. Two-masted busses apparently existed as early as the sixteenth century but did not dominate the three-masted type until after 1700. Coenen Zn., *Visboeck*, said that busses of his day could land 30-36 lasts of herring, a last being made up of fourteen casks each containing about 900 fish.

[13] Roger Degryse, "De Omvang van Vlaanderens haring- en zoutevisbedrijf op het einde van het Frans-Bourgondisch conflict (1482)," *Académie de Marine de Belgique, Communications*, 15 (1963), 37–38. Rudolf Häpke, *Niederländische Akten und Urkunden zur Geschichte der Hanse und zur Deutschen Seegeschichte* (Munich, 1913–1923), vol. I, #14, #115, #628. Algemeen Rijksarchief, The Hague, Archief van de Rekenkamer der Domeinen van Holland, 4990, is an account, dated 1523, for the fitting out of 11 warships for protection of herring boats. Roger Degryse, "De Konvooieering van de Vlaamsche visschersvloot in de 15de en de 16de eeuw," *Bijdragen voor de Geschiedenis der Nederlanden*, 2 (1948), 1–24. Roger Degryse, "Het tucht- en politiereglement voor de Hollands-Vlaamse krijgsvloot van buiskonvooiers van 1547," *Académie de Marine de Belgique, Communications*, 15 (1963), 17–30.

1424 the province of Holland started its regulation of the herring catch, salting, packing, and the size of casks. Apparently, governments were often inspired to greater regulation by complaints from overseas;[14] thus the rise in regulation after 1424 was partly attributable to the growth of herring exports. In 1519 Charles V issued the first general law dealing with the entire Low Countries herring fishery. The law, which continued in force with minor changes into the nineteenth century, subjected the fishery for the first time to one undivided authority.

After the Revolt the States of Holland carried on the policy, leaving intact a standing committee, first set up in 1567, of representatives from the major producing towns. The committee, the *College Van Commissarissen van de Groote Visscherij*, was originally intended by the States to advise lawmakers on the best legislation for the herring fishery. By 1600, though, the committee had acquired the power to lay down laws limiting the operation of the deep-sea fishery, and it used that power to systematize the variety of existing rules. The frequently expanded legislation dealt largely with fixing precise dates for the fishing season and preventing the use of inferior materials in packing. The committee was also responsible for organizing convoys, paid for by a tax on salt imports. Although producers were independent, each of the many individual firms was subject to the precise rules of the *College*. Moreover, each producing town took on the job of enforcing those regulations, and so surveillance was close. Size of casks and the minimum weight of fish per cask were fixed, as was the volume of salt used in packing. Casks had to be branded by inspectors, the brand serving to differentiate Dutch from other herring. The *College* met annually at Delft at the start of the herring season and issued licenses to busses. A boat could not go out for herring without this license; thus, regulation effectively controlled production. The *College* combined rules to dominate European markets and manipulate production and price, as best it could, to the advantage of all Dutch producers. To do that it forced the producers to act in consort, like one producer.[15] Regulation certainly limited the scope of activity for Dutch fishermen but it enabled them to command a higher price for their herring than could competitors.

Essentially a federation of producers' representatives, the *College* tried

[14] Rijksarchief in Noord-Holland, Verzamling aanwinsten, L. 504, fols. 99r–100r, is a set of rules established by Duke Philip for the herring fishery, both deep-sea and in inland waters. J. A. Fruin, *De Oudste Rechten der Stad Dordrecht en van het Baljuwschap van Zuidholland* (The Hague, 1882), vol. II, #229, is a town ordinance on herring selling and packing dating from 1494. Heeringa, *Rechtsbronnen*, pp. 232–50, is a town ordinance on the proper practice of commanders of herring boats and on packing and salting the herring dating from 1434. S. Haak, "Brielle als vrije en bloeinde Handelsstad in de 15de eeuw," *Bijdragen voor Vaderlandsche Geschiedenis en Oudheidkunde*, 4th ser., 6 (1907), 36–37.

[15] The government of the Netherlands began its first tentative regulation of the herring fishery in 1509. Nelly Gottschalk, *Fischereigewerbe und Fischhandel der niederländischen Gebiete im mittelalter* (Bad Wörishofen, 1927), pp. 16–19. J. Travis Jenkins, *The Herring and the Herring Fisheries* (London, 1927), pp. 68–75. Kranenburg, *De Zeevisscherij*, pp. 73–79, 151–57. Tillema, "Ontwikkeling," 15 (1916), pp. 348–49, 360–63, 371–72; and 16 (1917), 19–20.

to keep poor herring or poorly cured herring off the market. Its legislation prevented Dutch producers from doing damage to their markets through either overproduction or gaining a poor reputation.[16] Restricting supplies meant indirectly raising prices, but the market for herring was less sensitive to increases in price than it was to decreases in quality. The *College* on many occasions made rulings about *ventjagers*, fast ships sent out with the fleet to rush back the first catch which was loaded directly on board from herring busses. Such regulation affected only a very small percentage of the total herring catch; however, the concern over the dates when herring for the *ventjagers* was taken is another illustration of the regulators' consuming interest in quality control.[17]

The technical changes in equipment, methods, and institutions over the fifteenth and sixteenth centures were the basis for the strong commercial position and the relatively sizable output of the Dutch herring fishery at the beginning of the Republican period. The change in technology contributed to and in part induced the long-term rise in output and the long-term rise in exports, which culminated in the record catches and sales of the first half of the seventeenth century.

II

The peak of Dutch herring output and of exports coincided with the peak in Dutch trade to the Baltic. It is difficult to estimate the maximum level of the herring catch. Various contemporary guesses, usually made in the context of polemics, are highly suspect, and official returns are few. All data, however, point to a rise in the number of herring busses and a rise in total production of cured herring from the sixteenth through the early seventeenth century.[18] In the 1630s and 1640s, the fleet of some 500 busses annually caught an average of about 50 lasts of herring each, that is, in their two or, more commonly, three voyages. Total output was then

[16] The same goal of quality control to satisfy export markets was recognized in Lübeck regulations of the Scania and other fisheries of 1576. Dietrich Schäfer, *Das Buch des Lübeckischen Vogts auf Schonen*, 2nd ed. (Lübeck, 1927), pp. 132–34.

[17] Regulation of *ventjagers* has certainly been too much emphasized by historians. A. R. Michell, "The European Fisheries in Early Modern History," *The Cambridge Economic History of Europe*, vol. V (Cambridge, 1977), pp. 152–53. Van Beylen, *Schepen*, p. 139. Tillema, "Ontwikkeling," 16 (1917), 3–4. Häpke, *Niederländische Akten*, vol. II, #1032.

[18] The long-term trend was interrupted by the Revolt and the attacks of the Sea Beggars. For a list of overestimates see Hermann Wätjen, "Zur Statistik der holländischen Heringsfischerei im 17. und 18. Jahrhundert," *Hansisches Geschichtsblätter*, 28 (1910), 131–32. A 1476 government document includes a figure of 250 for the number of busses in Holland, Zeeland, and Friesland. Kranenburg, "Het Visserijbedrijf," p. 327. An official estimate of 1552 put the buss fleet of Holland at 300 vessels. Häpke, *Niederländische Akten*, vol. I, #628. The number was larger when the provinces of Zeeland and Flanders were added, and so the estimate of a fleet of 700 busses for the entire Netherlands in the late sixteenth century seems just possible. Brulez, "De zoutinvoer," pp. 183–84. Also Degryse, "Het tucht- en politiereglement," p. 18. Estimates by contemporaries in the early seventeenth century of a buss fleet of 2000 or more were clearly inflated. The figure of 500 for the province of Holland alone was much nearer the mark. The best modern calculations on herring output are those of Kranenburg, *De Zeevisscherij*, pp. 25–43. Unfortunately, he limited his estimates to the province of Holland.

at least 25,000 lasts, or more than 32,500 metric tons. Holland fishermen in good years caught and cured better than 200,000,000 herring, which may well have been about one-half the catch of all European herring fisheries.[19] Output, the number of herring busses, and the average catch per buss all fell after about 1650. In the eighteenth century in general the catch was about 6000 lasts annually.

The ports of the Baltic littoral constituted the largest single market for Dutch cured herring. Exports to Eastern Europe began in the fifteenth century, and about 1500 half the herring imported into the port of Gdansk came from Holland.[20] The volume increased over time. Betweeen 1562 and 1657 the average annual import of Dutch herring into the Baltic was 5415 lasts. The best years for exporters were from 1600 to 1629, when an average of 8245 lasts of herring entered the Sound in Dutch ships. The Netherlands produced about 82 percent of all herring shipped into the Baltic from Western Europe in the century ending in 1660. About 75 percent of the total went in Dutch ships. Dutch producers reached the maximum in 1602, when they sent more than 12,000 lasts of herring into the Baltic. The herring was consumed in Baltic ports or shipped inland along the rivers of Poland and Russia. On average, some 11 percent of the herring that entered the Baltic in the first half of the seventeenth century found its way up the Vistula as far as Warsaw.[21]

While Baltic markets took about 40 percent of Dutch herring production, other export markets absorbed about the same amount. Cologne and the Upper Rhine Valley was long an important market for cured fish. So, too, was the North Sea coast of Germany, which took about 20 percent of the total catch. Despite the competition from local fishermen, Dutch shippers were even successful in exporting cured herring to ports in Normandy, whence it presumably went on to Paris. In the eighteenth century, however, production and exports fell. The Baltic declined as a destination for Dutch herring as early as the 1660s.[22] The domestic market never took

[19] Michell, "The European Fisheries," pp. 148–49, 155. Employment in the herring fishery in North Holland alone at the peak was some 3500. Adriaen M. van der Woude, *Het Noordkwartier, een regionaal historisch onderzoek in de demografisch en economische geschiedenis van westelijk Nederland van de late middeleeuwen tot het begin van de negentiende eeuw* (Wageningen, 1972), p. 408. A figure of 10,000 employees for the entire Republic is probably low. Still the total was less than 2 percent of the labor force.

[20] Werner Böhnke, "Der Binnenhandel des Deutschen Ordens in Preussen und seine Beziehung zum Aussenhandel um 1400," *Hansische Geschichtsblätter*, 80 (1962), 27. Victor Lauffer, "Danzigs Schiffs- und Waarenverkehr am Ende des XV. Jahrhunderts," *Zeitschrift des Westpreussischen Geschichtsvereins*, 38 (1894), p. 44. Klaus Spading, *Holland und die Hanse im 15. Jahrhundert. Zur Problematik des überganges vom Feudalismus zum Kapitalismus* (Weimar, 1973), pp. 50–51. Haak, "Brielle," pp. 40–43. Gottschalk, *Fischereigewerbe und Fischhandel*, pp. 39–42.

[21] Willem S. Unger, "De Sonttabellen," *Tijdschrift voor Geschiedenis*, 41 (1926), 144. Nina E. Bang and Knud Korst, eds., *Tabeller over Skibsfart og Varetransport gennem Øresund, 1498–1783* (Copenhagen, 1906–1953). Compare Bang with Honorata Obuchowska-Pysiowa, *Handel Wiślany W Pierwszej Połowie XVII Wieku* (Warsaw, 1964), pp. 16–17, tables 44–45. There are a number of years for which data are missing.

[22] Dardel, *La Pêche Harenguière*, pp. 41, 154–55. While the Dutch may have sent as much as 10,000 lasts of herring to Rouen each year around 1600—and the estimate appears high—that was still only

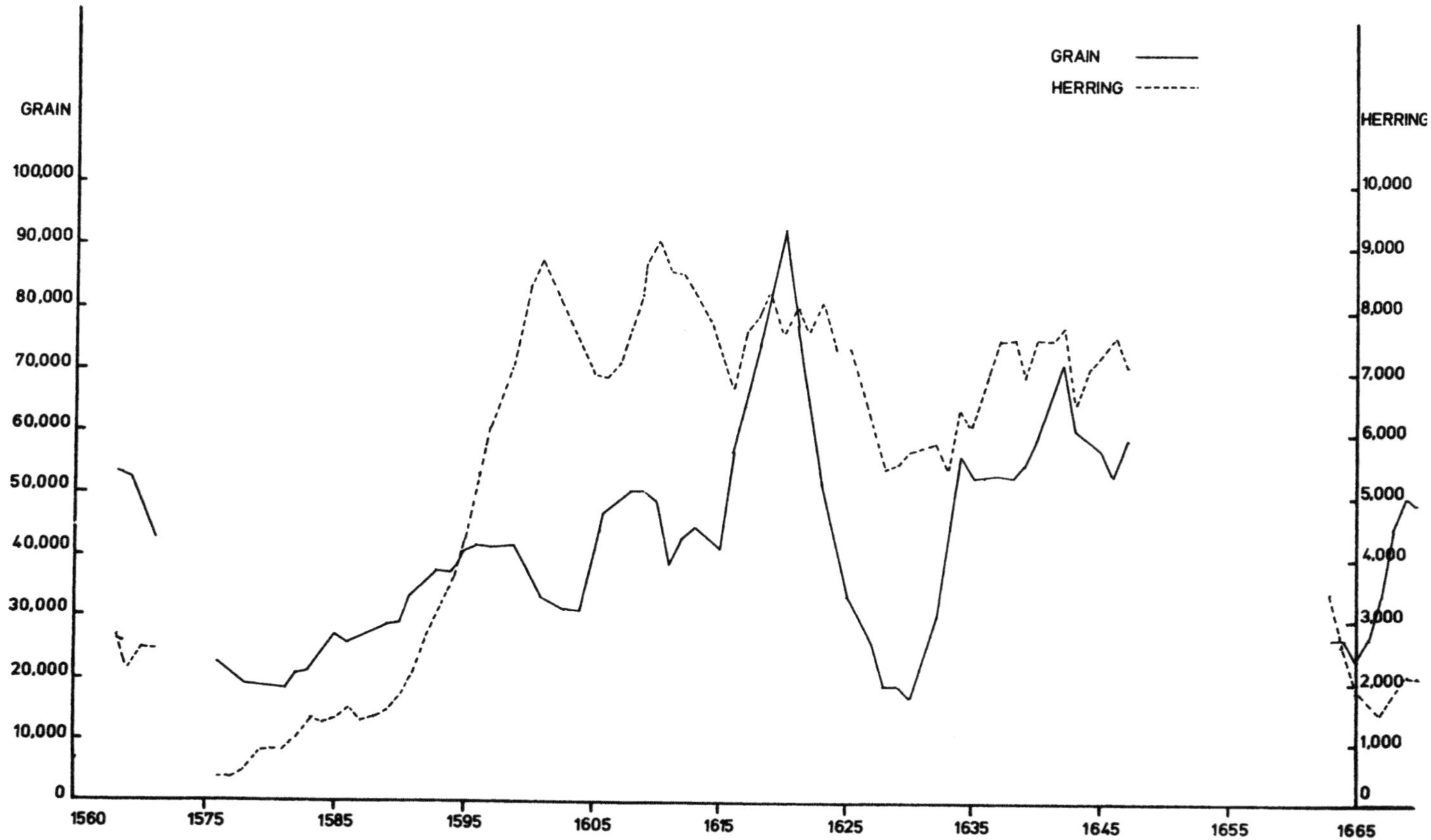

FIGURE 1
HERRING AND GRAIN: VOLUME IN LASTS THROUGH THE SOUND CARRIED IN DUTCH SHIPS, 1564–1779
(five-year moving average)

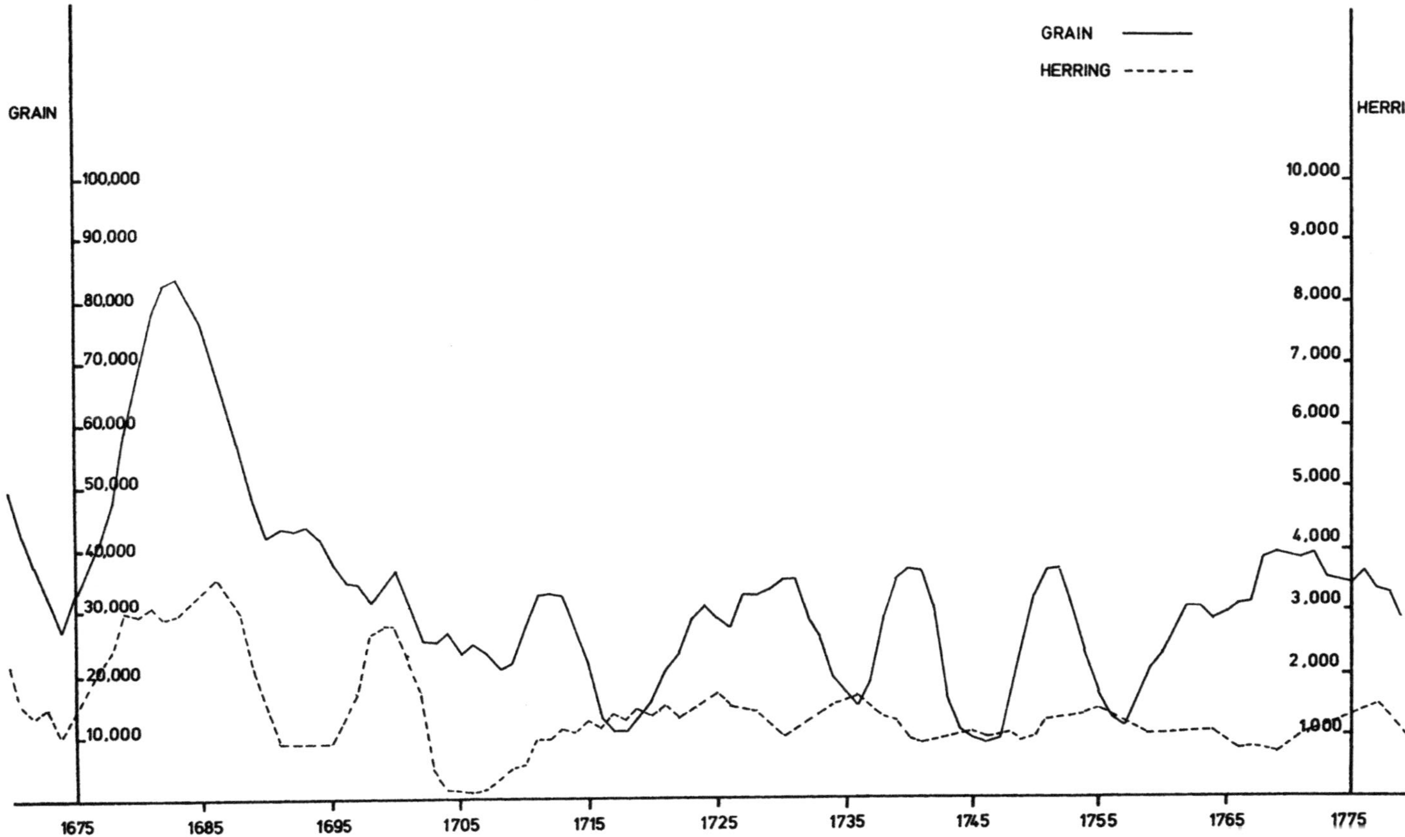

FIGURE 1 (CONT'D.)

a large proportion of the herring: even at the height of the fishery, only about 20 percent of total annual output, or at most 5000 lasts, stayed in Holland. The principal market for Dutch pickled herring both at home and overseas was always individuals of relatively high income.

The records of the tolls levied on ships passing through the Sound, the strait between modern Sweden and Denmark, indicate an increase in the already high level of Dutch exports of herring to the Baltic in the last years of the sixteenth century. Figure 1, giving five-year moving averages, eliminates the short-term fluctuations and shows the tendency toward a high level of herring shipment into the Baltic in Dutch vessels in the first half of the seventeenth century. Equally, it shows the decline after 1650 and then the relative decline and stability at a low level in the eighteenth century. That rise in exports in the late sixteenth century is reflected not only in the Sound dues but in the toll records of ports such as Kaliningrad, Elbląg, and Gdansk, and in the records of river tolls charged at Warsaw.[23]

As much as two-thirds of the Dutch tonnage going into the Baltic was unused, for ships went in ballast to Polish ports to fetch grain. Salt and herring were the two principal goods used to fill entering ships' holds, even if only partly. Because the shipment of grain out of the Baltic could pay for the cost of the entire voyage, as it did in the majority of the cases, the marginal cost of shipping herring to Baltic ports could approach zero on any single voyage. The only costs that had to be covered were loading, unloading, and administration.[24]

III

The boom in Dutch herring shipments coincided with the entry of Baltic grain into Mediterranean markets. The harvest collapses in southern

about one-fifth of total imports by all shippers into that Norman port. Kranenburg, *De Zeevisscherij*, pp. 35, 53, 133–35. Hamburg in the third and fourth decades of the seventeenth century took 3500 to 5000 lasts of herring each year from Holland. Already by the second half of the seventeenth century (1661–1700) the Baltic took only 12-20 percent of Dutch output. H. A. H. Kranenburg, "De Haringexport naar het Oostzeegebied in de 18e eeuw," *Tijdschrift voor Geschiedenis*, 72 (1959), 257–58.

[23] P. H. Winkelman, *Nederlandse Rekeningen in de Tolregisters van Koningsbergen 1588–1602* (The Hague, 1971). Horst Kempas, *Seeverkehr und Pfundzoll im Herzogtum Preussen. Ein Beitrag zur Geschichte des Seehandels in 16. and 17. Jahrhundert* (Bonn, 1964), p. 369. Franciscus B.M. Tangelder, *Nederlands Rekeningen in de Pondtolregisters van Elbing, 1585–1602* (The Hague, 1972). In 1594, 16 of 85 Dutch ships entered carrying herring and in 1602 it was 10 of 55, but the average cargo had almost doubled in those years to 21.4 lasts per ship. From 1605 to 1651 the annual average of herring passing Warsaw was 965.3 lasts, or almost twice the annual average of 504.5 lasts which passed the nearby Wloclowek customs house in the previous century. Obuchowska-Pysiowa, *Handel*, p. 210.

[24] Władysław Rusiński, "The Role of Polish Territories in the European Trade in the Seventeenth and Eighteenth Centuries," *Studia Historiae Oeconomicae*, 3 (1968), 123–25. Aksel E. Christensen, *Dutch Trade to the Baltic about 1600* (The Hague, 1941), pp. 369–79. The records of the van Adrichems of Delft confirm the expectation of low shipping costs for herring going to the Baltic. Dutch shippers even carried bricks and roofing tiles into the Baltic to fill their holds. The decrease in salt shipments from the 1590s to the 1630s is in part explained by the increase in local production near Gdansk.

Europe in the 1590s brought Dutch shippers and merchants to Iberia and to Italy carrying cargoes of rye and also of wheat, grain which originally came from the farms of Poland. Landowners there must have noticed an increase in their incomes. And it certainly would appear to be more than a coincidence that Polish merchants and landowners should increase their purchases of herring just at the time when their incomes were rising from growth in their sales of grain.

To test the hypothesis that the level of Baltic herring imports depended on the income from the sale of grain, calculations were made of the grain sales revenue earned by Polish exporters and also of the payments for herring made by Polish buyers. The revenue from grain sales was estimated by multiplying the volume of rye shipped out of the Baltic by the price of rye at Gdansk. Rye was by far the biggest earner for Baltic exporters, with wheat in second place. A similar calculation was made to estimate the revenue earned from wheat exports. For expenditures on herring, the volume of the fish passing eastward through the Sound was multiplied by delivered prices in Gdansk. Because Gdansk was the most important port in the Baltic for goods brought from the West, its prices give a reliable indication of general price movements throughout the region.[25] These estimates served for making a preliminary test of connections between short-term movements of grain and herring. For all cases tested, prices were converted to quantities of pure silver. Included were all years between 1550 and 1780 for which a complete set of data exists. The dates cover the entire period of growth, stability, and decline in herring exports.

The Polish balance of payments for all goods and services swung into surplus sometime in the second half of the fifteenth century and remained there until the first half of the seventeenth century. At that point surpluses not only became the rule but rose dramatically with the rise in grain exports. Moreover, in the seventeenth century it appears that grain sales in part determined Polish demand for luxury goods; for example, cloth from Western Europe. Baltic grain producers did take out their increased earnings in cash, in silver and coin, but they also used their greater power to obtain goods, especially foods to diversify their diets.[26]

[25] The data on goods shipped through the Sound are from Bang and Korst, eds., *Tabeller over Skibsfart*. For Gdansk prices, see Julian Pelc, *Ceny W Gdánsku w XVI i XVII Wieku* (Lwów, 1937), pp. 64–65, and Tadeusz Furtak, *Ceny W Gdánsku W Latach 1701-1815* (Lwów, 1935), p. 152. On the importance of Gdansk, see Stanisław Hoszowski, "The Polish Baltic Trade in the 15th–18th Centuries," *Poland at the XIth International Congress of Historical Sciences in Stockholm* (Warsaw, 1960), pp. 139–40, 148. Rusiński, "Role of Polish Territories," pp. 120–21; and Unger, "De Sonttabellen," pp. 147–48.

[26] Marian Malowist, "The Economic and Social Development of the Baltic Countries from the Fifteenth to the Seventeenth Centuries," *Economic History Review*, 2nd ser., 12 (Dec. 1959), 184–87. Antoni Mączak, "The Balance of Polish Sea Trade with the West, 1565–1646," *Scandinavian Economic History Review*, 18, no. 2 (1970), 118–21. Henryk Samsonowicz, *Untersuchungen über das Danziger Bürgerkapital in der Zweiten Hälfte des 15. Jahrhunderts* (Weimar, 1969), pp. 31–35. Astrid Friis, "The Two Crises in the Netherlands in 1557," *Scandinavian Economic History Review*, 1, no. 2 (1953), 200–11. Maria Bogucka, "Die Bedeutung des Ostseehandels fur die Aussenhandelsbilanz Polens in der Er-

The results of comparing grain revenues with herring costs do not indicate a strong linear relationship between the two. While results are typically significant, the coefficients of determination are low.[27] This suggests that the success of Poles in finding grain markets was not the predominant factor in determining the volume of Dutch herring exports to the Baltic. Balance of payments figures derived from prices declared at the Sound show that the value of herring imported into the Baltic was not a consistent proportion of the value of wheat and rye exported from the two major ports of Gdansk and Elbląg.[28] That evidence supports the same conclusion: the sharp rise in Dutch herring sales to the Baltic late in the sixteenth century and the high level of production and exports in the first half of the seventeenth cannot be explained by the boom in grain sales in the same period. Further confirmation is offered by a comparison of the profits to be made in the Iberian grain trade with total expenditure on herring in Baltic ports.[29] Greater profits in the grain trade did not necessarily lead to greater herring purchases by the most likely buyers, the landowners, who appropriated the greater part of those grain profits.

The results can in part be explained by difficulties with the data. Smuggling renders the official returns of volume inaccurate, and probably inconsistently so. For the sake of simplicity the assumption is that all trade between Eastern and Western Europe went on through the Sound, which

---

sten Hälfte des 17. Jahrhunderts," *Der Aussenhandel Ostmitteleuropeas 1450–1650*, Ingomar Bog, ed. (Cologne, 1971), pp. 47–50. In the first half of the seventeenth century contemporaries complained about a flight of money from Poland to buy foreign luxuries. The conclusion made then was that there was a balance of payments deficit rather than a surplus, which seems the more likely explanation. The Netherlands ran an overall balance of payments deficit in the late sixteenth century. W. Brulez, "The Balance of Trade in the Netherlands in the Middle of the 16th Century," *Acta Historiae Neerlandica*, IV (1970), pp. 45–48. One writer claims that the Netherlands bought twice as much in value from the Baltic as it sold. Christensen, *Dutch Trade*, p. 367.

[27] V is volume and P is price. The commodities are herring (H), rye (Y), and wheat (W). D indicates shipments in Dutch bottoms, T total shipments. N indicates price in the Netherlands, G price in Gdansk. For $V_{HD} \cdot P_{HG} = f(V_{YD} \cdot P_{YG})$, N = 163 R @ .05 = .15 R^2 = .31. That is, for a sample = 163, the absolute value of R should exceed R @ .05 = .15 in order to reject the hypothesis at the 5 percent level of significance. The results are not as good using data for all herring shipped into the Baltic and all rye shipped out.

$$V_{HT} \cdot P_{HG} = f(V_{YT} \cdot P_{YG}) \qquad N = 166 \quad R\ @\ .05 = .15 \quad R^2 = .20.$$

The addition of wheat shipments also does nothing to improve results.

$$V_{HT} \cdot P_{HG} = f[(V_{YT} \cdot P_{YG}) + (V_{WT} \cdot P_{WG})] \quad N = 46 \quad R\ @\ .05 = .29 \quad R^2 = .01.$$

[28] Mączak, "Polish Sea Trade," pp. 135–36. Using his figures for prices of goods declared at the Sound and the volume of goods going in each direction the following ratios of the value of herring imports to the value of rye and wheat exports are derived:*1565* .040; *1575* .012; *1585* .072; *1595* .105; *1605* .195; *1615* .154; *1625* .233; *1635* .074; *1646* .185. The figures for those representative years show no consistency. The values for grain exports are for the two ports of Gdansk and Elbląg.

[29] N = 31 R @ .05 = .36 R^2 = .03. The data are for proifts in the Iberia wheat trade including those made from agio or disagio of the ducat generated by Maria Bogucka, "Merchants' Profits in Gdansk Foreign Trade in the First Half of the 17th Century," *Acta Poloniae Historica*, 23 (1971), 79–82. Idem, "Amsterdam and the Baltic in the First Half of the Seventeenth Century," *Economic History Review*, 2nd ser., 26 (Aug. 1973), 439–47.

of course was not true. Poland carried on an extensive trade through Moldavia with the Black Sea and Turkey; its deficit in that trade affected the power to buy goods, like herring, from the West. Consumption patterns changed in Poland over time. It may be true that up to about 1660 Baltic buyers had to pay cash, usually silver coins, for goods brought from the West, but by that date they certainly could use a more complex system. The ability to negotiate bills of exchange and to settle international debts multilaterally would have tended to undermine any direct relationship between exports and imports, and especially on any single trade route.[30]

The sale of grain surpluses by Polish landowners in the short run explains only a small proportion of the performance of Dutch exporters in finding a market in the Baltic for domestically produced herring. Additional explanation is provided by the volume rather than the value of Polish exports. The proportion of Dutch ships entering the Baltic in ballast fell from two thirds to less than one half by the end of the 1620s. Herring exports certainly made an important contribution to that shift. The correlation between the volume of herring shipped into the Baltic and the volume of grain shipped out is as strong as any of the relationships between revenues.[31] Not surprisingly, the relationship for all herring going into the Baltic and all rye being exported was not as highly correlated.[32] Figure 1 seems to confirm that some long-run connection did exist between the tonnage available going to the Baltic and herring exports from Western Europe. While the results leave a great deal unexplained, they nonetheless indicate that volume was as significant a determinant of Dutch herring exports as the value of Baltic grain sales.

IV

The price of herring or, more specifically, the profit which could be made in the herring trade seems to offer a fuller explanation for the vol-

[30] Antoni Mączak, "Der Polnische Getreideexport und das Problem der Handelsbilanz (1557–1647)," *Der Aussenhandel Ostmitteleuropas 1450-1650*, Ingomar Bog, ed. (Cologne, 1971), pp. 35–45. Bogucka, "Die Bedeutung," pp. 52–55. Hoszowski, "The Polish Baltic Trade," pp. 123–27. Jacob M. Price, "Multilateralism and/or Bilateralism: The Settlement of British Trade Balances with 'The North', c. 1700," *Economic History Review*, 2nd ser., 14 (Dec., 1961), 254–74. John Sperling, "The International Payments Mechanism in the Seventeenth and Eighteenth Centuries," *Economic History Review*, 2nd ser., 14 (April 1962), 446–68.

[31] $N = 207$ $R @ .05 = .14$ $R^2 = .31$ for the volume of rye shipped out of the Baltic in Dutch vessels with the volume of herring shipped into the Baltic in Dutch ships. Limited to the period 1600–1780, the result is slightly better. $N = 173$ $R @ .05 = .15$ $R^2 = .34$. It was the rye shipped out of the Baltic in any given year that was correlated with herring shipments into the Baltic, not rye exports for the previous year. Herring shipments depended more on tonnage available at a given point in time. For herring shipments compared with rye shipments out of the Baltic for the previous year, where data for both were available and using only 1600–1780, $N = 169$ $R @ .05 = .15$ $R^2 = .24$.

[32] For all cases, $N = 210$ $R @ .05 = .14$ $R^2 = .14$. For 1600–1780, $N = 175$ $R @ .05 = .15$ $R^2 = .18$. The addition of wheat to the quantity of grain sent out of the Baltic not only does not improve the correlation but it actually decreases the coefficient of determination. For all cases, that is 1550–1780, $V_{TH} = f(V_{TY} + V_{TW})$, $N = 210$ $R @ .02 = .16$ $R^2 = .16$.

ume of Dutch herring exports. Presumably, Dutch merchants sent herring overseas because they expected to make money. Profit figures are even more difficult to generate than those for revenue. One estimate for the early fifteenth century suggests that profits for Dutch merchants delivering herring in Prussia were about 60 percent of the sale price. A more sophisticated set of estimates for the first half of the seventeenth century shows consistent profits from selling Dutch-produced herring in Gdansk.[33]

TABLE 1
PRICES OF HERRING IN AMSTERDAM AND GDANSK AND PROFITS IN HERRING TRADE ON THAT ROUTE
(in grams of silver/last)

Year	Average price in Amsterdam	Freight	Other costs[a]	Average price in Gdansk	Profit in grams silver	Profit as % total cost
1631	1226	92.52	211	2025	495	+ 32
1632	1169	92.52	228	2268	778	+ 52
1634	1645	61.68	249	2293	337	+ 17
1638	1624	66.82	253	2349	405	+ 21
1640	1157	102.80	234	2349	855	+ 57
1641	1473	102.80	214	1944	219	+ 12
1643	1408	102.80	218	2025	296	+ 17
1645	1134	113.08	207	2025	571*	+ 39
1646	1097	113.08	167	1539	162	+ 12
1648	853	77.10	125	1134	79	+ 7
				Average:	420	+ 27

[a] Charges in Amsterdam about 1.5 percent; Sound toll about 2.5 percent; customs and charges at Gdansk about 8 percent; total of 4 percent of the purchase price + 8 percent of the selling price.
Source: Bogucka, "Merchants' Profits," pp. 84-85.

For other commodities the level of profits seems to have affected the volume of goods shipped. For example, there is some indication that the quantity of rye shipped out of the Baltic was related to the difference between the price paid by merchants for rye in Gdansk and what they could sell it for in Amsterdam. In whaling a similar situation prevailed, with the number of ships sent out being closely related to the price of whale oil.[34] Apparently the same was the case in the herring fishery, whose structure

[33] Bogucka found herring to be the most profitable commodity shipped from West to East, and she probably underestimated profitability. She used Amsterdam prices for full herring which is of higher quality and produced later in the year than matie *(matjesharing)*, for which prices are also available. The revised table here uses prices for matie on the Amsterdam bourse from Posthumus, *Inquiry*, vol. I, pp. 88–90. Duties and taxes are adjusted to reflect the different herring prices. Bogucka's estimates of freight charges are probably too high as well since she assumes that herring was charged the same as grain going in the other direction, which was not true. For problems with using Posthumus' Amsterdam herring prices see Kranenburg, *De Zeevisscherij*, pp. 193–95. He suggests that full herring prices are even suspect, and that they may not reflect the true cost of herring when it left Dutch ports. It is possible that the errors are consistent, however. Bohnke, "Der Binnenhandel," pp. 55–56.

[34] Mączak, "Polish Sea Trade," pp. 115–16. Jacobus R. Bruijn and C. A. Davids, "Jonas Vrij De Nederlandse Walvisvaart, in het Bijzonder de Amsterdamse, in de Jaren 1640–1664," *Economisch- en Sociaal-Historisch Jaarboek*, 38 (1975), 150–52.

resembled the whaling industry's. The difference between the price of herring in Gdansk and the price of matie on the Amsterdam bourse when compared with the shipments of herring in Dutch vessels into the Baltic shows a coefficient of determination of .43.[35] Charges that Dutch herring exporters were dumping, that is, using profits from the grain trade to offset losses in herring sales, do not appear to have been valid, and certainly not for the seventeenth century.[36] It seems that Dutch traders sent herring to the Baltic, and in the quantities selected, because of the profit to be made. Since the sample is small and there are data problems, one cannot place great confidence in the correlation; however, the trend in herring prices from the sixteenth through the eighteenth century is clear.

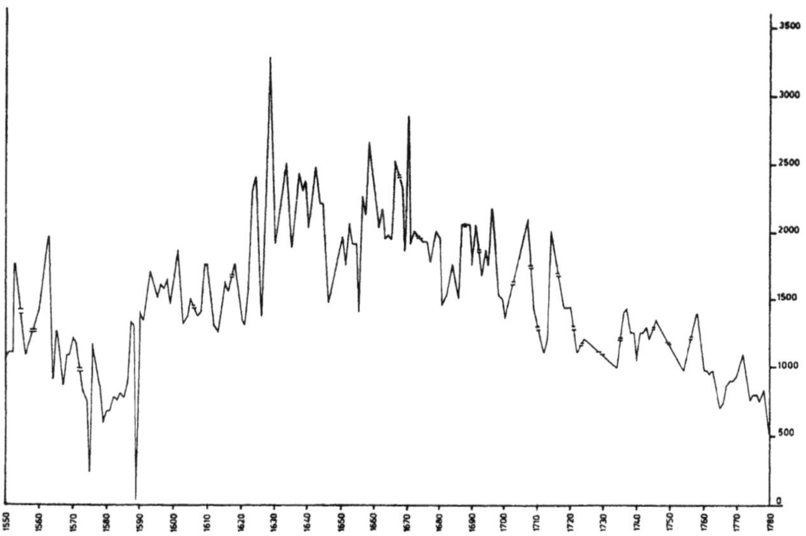

FIGURE 2
HERRING PRICES IN GDANSK IN GRAMS OF SILVER, 1550–1780

Source: Pelc, *Ceny W Gdańsku*, and Furtak, *Ceny W Gdańsku*.

The price of herring in Gdansk rose in the 1590s to well above the level for the previous two decades. This was followed by yet another rise in the 1620s and 1630s. Prices remained at the same relatively high level until the 1670s, when a long-term decline set in. Presumably herring prices in

[35] For $V_{HD} = f(P_{HG} - P_{HA})$ where the subscript A refers to the price in Amsterdam, N = 61 R @ .05 = .25. As expected, there is no significant correlation between the Gdansk-Amsterdam price differential and the total quantity of herring shipped into the Baltic; that is, $V_{TH} = f(P_{HG} - P_{HA})$, N = 63 R @ .05 = .25 $R^2 = .03$.

[36] Michell, "The European Fisheries," p. 177. Little correlation was found between the rye price difference in the Netherlands and Gdansk and the quantity of herring entering the Baltic in Dutch bottoms, which also contradicts the suggestion that Dutch traders were dumping herring; that is, $V_{HD} = f(P_{YN} - P_{YG})$, N = 203 R @ .05 = .14 $R^2 = .08$.

Amsterdam rose in the years around 1600 but, unfortunately, no data from the Bourse exist for those years. Amsterdam prices did fall in the late seventeenth and eighteenth centuries, though not as far or as fast as did Gdansk prices. That squeezed profits in the trade and moreover made selling herring at home more attractive. The success of Dutch exporters, it seems, depended on the absence of any alternate lower-cost suppliers. The rise in prices at the end of the sixteenth century coincides with a decline in the Swedish and Norwegian coastal fisheries and, in the third decade of the seventeenth century, with the final collapse of the Scania fishery. With no alternate supplies, Dutch traders seem to have been able to reap the benefits of higher prices, enjoying almost a monopoly. In the eighteenth century prices went down at the same time as the Norwegian and then the Swedish coastal fisheries revived.[37] The shipments of herring through the Sound reflect the change: while in the period 1661–1720 the Dutch share of herring imports to the Baltic was still about 60 percent, for the entire period 1661–1783 the figure was only 31 percent; by the 1760s Dutch ships carried less than 10 percent of herring going into the Baltic. Scots and Scandinavians replaced the Dutch as suppliers of herring, albeit in much reduced quantities. Coastal fisheries always enjoyed the advantage of lower capital costs than the deep-sea fishery, which used large seagoing vessels. It was the greater reliability of the deep-sea fishery combined with falling costs due to technical innovation which allowed Dutch producers to overtake Scandinavian competitors and then surpass them in the fifteenth and sixteenth centuries. The greater efficiency of the Dutch, for example, had been the cause for the shrinking of the Scania fishery from the mid-fifteenth century on. The expansion of the coastal fisheries after the 1670s was in part due to the high prices which they could get for their product, and in part to the fact that those fisheries had been neglected for so long that fish populations had been able to increase. Their competitive position was improved in the eighteenth century as Dutch herring busses became less productive, making two voyages annually instead of three, and bringing back fewer fish each year.[38]

The correlation of profits with herring sales in the Baltic is more apparent than real, however. First, the relationship between the two changed at the time of the wars of William III, that is, around 1700. To that date the price of fish delivered in Gdansk had always been greater—and by a large

[37] The fall in herring prices in Gdansk in the late seventeenth century coincided with wars and political crises in Poland which served to dampen demand. Hoszowski, "The Polish," pp. 119–22. Paul Simson, *Geschichte der Stadt Danzig bis 1626* (Danzig, 1918–1924), vol. II, p. 500. Schäfer, *Das Buch*, pp. xlii–ix.

[38] Willem S. Unger, "Trade through the Sound in the Seventeenth and Eighteenth Centuries," *Economic History Review*, 2nd ser., 12 (Dec. 1959), 208–09. Kranenburg, *De Zeevisscherij*, pp. 204–05. Kranenburg, "De Haringexport," pp. 252–57. At least not all Dutch ports suffered from falling herring production. Vlaardingen and to a lesser degree Maassluis noted significant growth even if the catch per buss was falling. Wätjen, "Zur Statistik," pp. 173–77. Unger, "The Netherlands Herring Fishery," pp. 350–53.

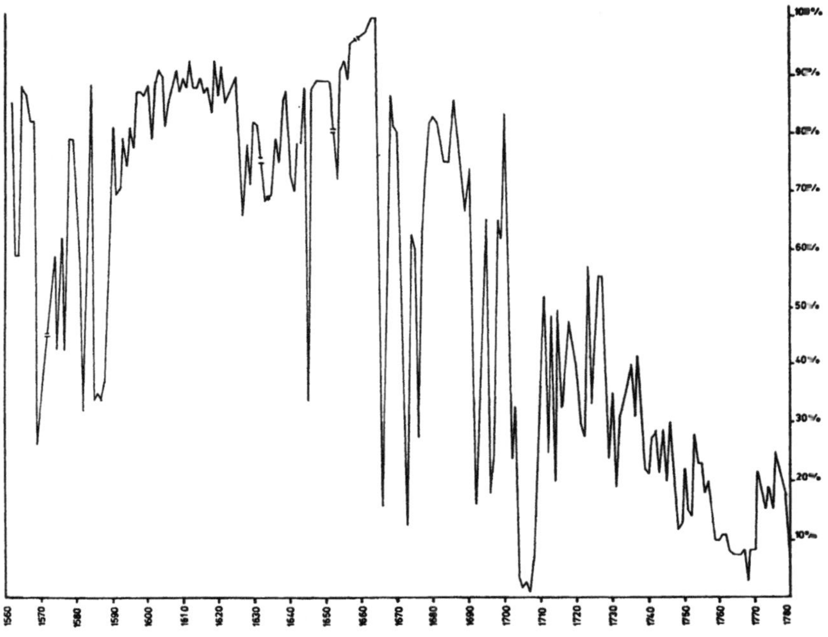

FIGURE 3
PERCENTAGE OF HERRING SHIPPED THROUGH THE SOUND IN DUTCH SHIPS, 1562–1780

Source: Bang and Korst, *Tabeller over Skibsfart.*

amount—than the price in Amsterdam. In the eighteenth century the Gdansk price was typically slightly lower than Amsterdam's. Second, the export of herring rose as Gdansk prices rose relative to Amsterdam prices, but once a great difference was reached, exports varied widely. The two patterns suggest not one but two relationships, one for the seventeenth and one for the eighteenth century. Tests revealed that in fact there is no significant correlation in those two time periods between the price differential and Dutch herring shipments.[39] Moreover, the distribution of observations contradicts the presumption of a linear relationship. Added to that, changes in herring prices from year to year showed little correlation with changes in the shipment of herring.[40] Herring prices were much lower in Gdansk than in Amsterdam in the 1760s and still herring went from the Netherlands to Baltic ports. In the 1670s, on the other hand, the price difference was the opposite and the volume of herring exported in

[39] For 1624–1694, N = 29 R @ .05 = .37 R^2 = .11, and for 1718–1780, N = 32 R @ .05 = .35 R^2 = .20.

[40] $(V_{TH})_{t+1} - (V_{TH})_t = f[(P_{HG})_{t+1} - (P_{HG})_t]$ where, for example, $(V_{TH})_t$ is the total volume of herring shipped out of the Baltic in year t. N = 147 R @ .05 = .16 R^2 = .01.

Dutch ships to the Baltic was also relatively low. The confused relationships tends to be confirmed when other, more complete Dutch price series are used.[41] Despite the evidence that there were other and more powerful factors at work, extant data show that the herring price differential between Gdansk and Amsterdam was positive up to 1690 and then often negative from 1691 to 1780. The former period coincides with years of a high level of exports, while the latter period coincides with years of depressed production and a low level of exports.[42] It cannot be shown conclusively that price differentials were important to output or exports, but it can be shown that both were affected by production history.

V

Unquestionably the most important factor in determining herring shipments to the Baltic in any given year was the quantity of herring shipped in the previous year. For herring carried into the Baltic in Dutch ships, the coefficient of determination was higher than for any other relationship examined.[43] The pattern is consistent and clear. Dutch traders had at their disposal extensive information about the state of the Polish market. The regular sailings, the large numbers of ships involved, the correspondence between merchants and their resident factors living in Baltic ports all created a massive supply of market intelligence for herring exporters.[44] It was not from ignorance that exporters chose to follow the pattern of the previous year. Rather, what fixed the scope of their trading was the limitations of productive capacity set by the state of the industry. Capacity was in turn fixed by the weather and the availability of fish, but especially by supplies of capital, both physical and human, and the state of the art, the knowledge of techniques for the best exploitation of that capital.

By the late sixteenth century Dutch fishermen could produce pickled herring, which was generally recognized to be of the highest quality, and deliver that product to Baltic markets, in the process turning a handsome profit. The long apprenticeship of the herring fishery, the process of learning-by-doing, yielded the high levels of output, the high level of exports, and the high prices and profits. Prices of inputs were under downward pressure as well, since the Dutch trading network—in part a result of the ability to produce high quality herring at competitive prices—supplied the salt, ships, cordage, nets, and so on at or below the prices paid by competitors. Moreover, Dutch fishermen knew how to get most from that

[41] Posthumus, *Inquiry*, vol. II, pp. 245–52, 274–82.

[42] For 1624–1690, N = 29. For 1691–1780, N = 34 and the differential was negative in 26 of the 34 cases.

[43] $(V_{HD})_t = f[(V_{HD})_{t-1}]$, N = 201 R @ .05 = .14 R^2 = .65. As expected, the results were not as good using all herring shipments. $(V_{HT})_t = f[(V_{HT})_{t-1}]$, N = 205 R @ .05 = .14 R^2 = 59.

[44] Dutch traders had factors in Baltic ports by the second half of the sixteenth century, by which time their correspondence was already extensive and regular. H. Enno van Gelder, "Zestiende-eeuwsche Koopmansbrieven," *Economisch-Historisch Jaarboek*, 5 (1919), 136–91. Christensen, *Dutch Trade*, pp. 215–16, 431–40.

equipment. Certainly the level of capital investment in the herring fishery, both fixed capital in ships and warehouses and working capital in barrels and salted and cured fish, contributed to the comparative advantage of Dutch producers. Capital came largely from the retained earnings of past voyages. It was true despite or perhaps because of the highly flexible financial arrangements in the fishery. Success in any one year was a significant determinant of production potential in the following year.

Capital investment, knowledge of markets, internal organization, and institutional arrangements all contributed to Dutch success. The competitive superiority of Dutch producers and exporters was a highly fragile one, however. Others could and indeed did copy Dutch curing methods and vessel types like the herring buss. Other producers could and did develop the necessary skills in manufacturing pickled herring of high quality. That took time, as did the accumulation of profits needed to capitalize a deep-sea herring fishery. Yet the slow process of accumulation could be sped up by temporary high prices, government protection of domestic markets for domestic producers, direct government subsidies, or by the short-term collapse of Dutch production.

In France, Colbert pursued a policy of supporting domestic production. He established a company to promote technical improvement in the fishery and used the company as a conduit for government subsidies. A high tariff was placed on herring imports in 1664 but that was changed to a general prohibition in 1687. The latter never fully worked, and after the Peace of Utrecht in 1713 the tariff of 1664 went back into effect. The result of those vain efforts to keep out Dutch herring was to raise French domestic prices and thus allow French fishermen higher earnings which facilitated a long-term rise in domestic production. By 1786 the value of the French herring catch was greater than that of the Dutch, and France imported only about 18 percent of herring consumed in the country.

England's policy was long devoted to protection and direct and indirect support of the fishing industry. The Elizabethans legislated extensively to aid the fisheries and the laws were, if anything, strengthened in the seventeenth century. English claims to sovereignty of the seas in the reign of James I appear to have had little effect on Dutch fishing. The Navigation Acts, the first dating from 1651, placed stricter limitations on the Dutch selling herring in England. From 1726 direct subsidies, or bounties, were paid by the British government to the herring fishery.

Elsewhere, Sweden offered protection of domestic markets and subsidies to its herring fishermen. A company was set up to produce herring in the Dutch manner in 1745, and Frederick the Great of Prussia did the same at Emden in 1769. In each case the goal was to promote a domestic herring fishery.[45] The potential for net gains in foreign exchange, amply

[45] Dardel, *La Pèche Harenguière*, pp. 138–40, 152–58. Lawrence A. Harper, *The English Navigation Laws* (New York, 1939), pp. 28–31. Adam Smith, *Wealth of Nations*, ed. Edwin Cannan (New York,

demonstrated by the Dutch for the better part of two centuries, were enough to convince Frederick and other monarchs of the value of a herring fishery. Different states had varied success with protectionist legislation. Overall the policies were effective because there was a technology which could be borrowed; general prohibitions of imports, though not totally enforceable, raised prices for new domestic producers; and governments were typically consistent in their policies toward the herring fishery. With all those expanding fisheries in the eighteenth century, just as with the Dutch earlier, past experience played a significant role in the growth, or decline, of total production.

In the seventeenth century Dutch herring fishermen held a virtual monopoly. They could and did maximize the rent to be earned from that unique position by limiting fishing effort, and therefore production, through extensive government regulation. As with all fisheries, as indeed with all common property resources, they faced declining average product with an increase in fishing effort, the latter denoted by the number of vessels (Figure 4). The relationship between the two is assumed to be linear, which does not distort the results. The curves AP and MP represent average product and marginal product respectively. Furthermore, it is assumed that the amount of fishing effort does not affect marginal and average cost, which are identical and constant. This curve also subsumes the opportunity cost, that is, the income which the fishermen could earn in alternate employment. With a level of costs of OC, production takes place at the point where economic yield is maximized, that is, where marginal cost equals marginal product. OV_d boats are then used, and a rent is earned equal to the rectangle ABCD. If the fishery is truly common property and open to all, new resources, boats, and fishermen will be attracted until the rent is eliminated. The number of boats would rise to OV_t with MC equated to AP.[46] Through regulation the Dutch government limited domestic producers' efforts, presumably trying to maximize rent or at least to keep the number of vessels in the fishery at or below OV_m. At that point marginal product would be zero and thus give what is called the maximum sustained physical yield.

Producers from other countries could enter the fishery, moving fishing effort closer to OV_t, but the Dutch industry could drive them out by a short-term reduction in price. The cost curve is the average for the entire herring fishery; so Dutch marginal and average costs were at or below

1937), pp. 484–88, 901–03. Smith was distressed about the net loss in gross income from the subsidy, but it seems to have been effective, directed as it was at increasing the production of herring to compete with the Dutch product. Staffan Högberg, *Utrikeshandel och sjöfart pä 1700-talet* (Stockholm, 1969), pp. 168–71. Reinhold Koser, *Geschichte Friedrich des Grossens* (Darmstadt, 1964), vol. III, p. 257.

[46] H. Scott Gordon, "The Economic Theory of a Common Property Resource: The Fishery," *Journal of Political Economy*, 62 (1954), 129–32. For a summary of the common property equilibrium applied to a fishery see Donald G. Paterson and James Wilen, "Depletion and Diplomacy: The North Pacific Seal Hunt, 1886–1910," *Research in Economic History*, 2 (1977), 83–86, 118–27.

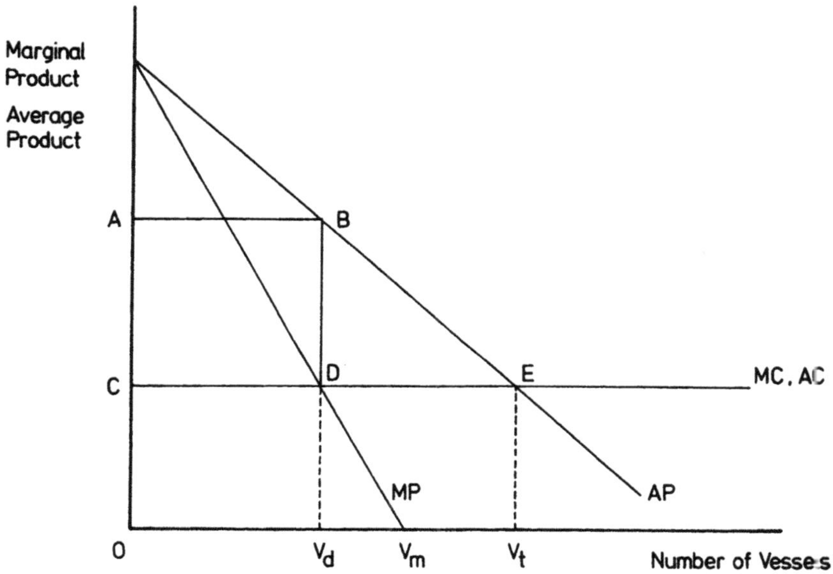

FIGURE 4

OC. The position of the Dutch, then, was secured not by ownership of the resource—common property is open to all—but rather by technical superiority, that is, lower costs, and by continuous and strict regulation of the fishery. So long as Dutch costs were lower than their competitors', they could maintain a virtual monopoly by careful marketing strategy. And so long as consumers understood the Dutch brand to indicate herring of consistently high quality, Dutch producers could and did command a rent.

While the potential Dutch supply curve, given their relatively lower costs, was S_d (Figure 5) they supplied herring along the curve S. Z, the quantity of herring supplied by the Dutch, corresponds to V_d (Figure 4). While all other herring suppliers suffered with a supply curve to the left of S, Dutch producers earned a rent, here equal to the rectangle VWXY. OW was the market price of the herring and OV the production cost to Dutch fishermen. But as other producers developed the necessary capital, equipment, and skills, their costs fell. Their supply curve shifted to the right, leading to a decrease in price, from OW to OU, and to an increase in output, from OZ to OT. The Dutch government and Dutch herring fishermen at that point chose not to change the old strategy of trying to maximize their rent. They might be perfectly willing to drop prices in the short run to drive competitors out of their markets, but Dutch producers were not willing to make a long-term commitment to greater fishing effort and larger volume sales at lower prices, that is, below OU. Instead they chose to maintain prices for their product at W, or even raise them above old levels. For example, in the 1670s there were complaints about the quality of Dutch herring delivered in Gdansk. In response the Dutch government made regulation of the domestic fishery even more restrictive. That move raised costs, and so the Dutch supply curve S_d tended to shift to the left. With prices going down, rents were squeezed and profits for Dutch fishermen along with their share of the market decreased. Oddly enough, continuing with a strategy that was initially designed to maximize rents not only served to decrease rents but also led to the long-run decline of the industry in the face of increasingly aggressive foreign competitors. A fall in the demand for herring in the eighteenth century—a shift in the demand curve D down and to the left—penalized Dutch producers more than others since they were trying to keep prices up. The changes in the cost position of competitors, the change in demand, and the market strategy of the Dutch herring fishery all contributed to the observed pattern of a sharp fall in absolute production and an even more dramatic drop in market share.

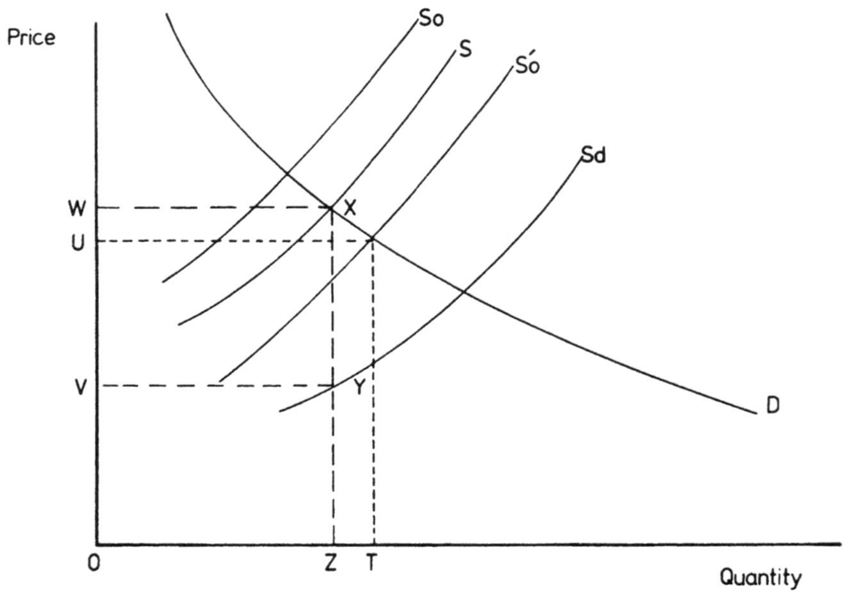

FIGURE 5

With the entry of other producers, the fishery, taken altogether, tended to move to equate the average costs of all producers with average product, shown by E (Figure 4). Average product fell. The average return from fishing effort, that is, the catch per boat, went down. That was true especially for Dutch fishermen, even though they were decreasing the number of their boats. While Holland had about 500 herring busses in 1630, there were only a little over 200 by 1730 and 150 by 1780.[47]

The advantage enjoyed by the Dutch herring fishery was destroyed by the simultaneous fall in costs for other producers and rise in costs for Dutch fishermen. The difficulties for Dutch producers in fact started long before the eighteenth century, and well before foreign governments began protecting their own herring fishermen. Wars fought in the North Sea meant an increased danger to the highly vulnerable busses. As early as 1625 Spanish forces began extensive and determined attacks on Dutch herring boats. From 1652 and the start of the First English War, Dutch fishermen had to face an almost constant threat of violence; at times, for their safety the government would not even let them leave port. In some years the province of Holland alone lost as many as 100 busses to pirates and privateers. The worst Dutch losses came during the wars of William III. In 1703 French warships burned the Enkhuizen herring fishing fleet in the Shetland Islands, a loss of about 100 busses in one day.[48] It was that disaster and other sizable if not comparable losses that appear to have destroyed the fragile Dutch competitive advantage. After the restoration of peace in 1713, Dutch production and exports did rise again, but the long-term competitive edge that technology and investment in that technology had given Dutch producers was gone. With the foundation built up through the sixteenth century having been badly damaged by war losses, other Western European producers found ways to supplant Dutch pickled herring in their own and in other markets. The decline, like the rise of Dutch herring exports and output, was self-reinforcing.

[47] Dardel, *La Pêche Harenguière*, pp. 138–58, on French attempts to replace Dutch imports with domestically produced herring. Johannes DeVries, *De Economische Achteruitgang der Republiek in de Achttiende Eeuw*, 2nd ed. (Leiden, 1968), pp. 137–41. Anthony Beaujon, *The History of Dutch Sea Fisheries: Their Progress, Decline and Revival*... (London, 1884), p. 286. Kranenburg, *De Zeevisscherij*, pp. 169–71, on the development of competing fisheries. On evidence for a shift in demand for herring in the eighteenth century, see Michell, "The European Fisheries," pp. 182–83.

[48] Wätjen, "Zur Statistik," pp. 142–51. Kranenburg, *De Zeevisscherij*, p. 204. Production per vessel was also subject to greater variation in war years (ibid., pp. 42–43). Jonathan I. Israel, "A Conflict of Empires: Spain and the Netherlands, 1618–1648," *Past and Present*, 76 (Aug. 1977), 44–48.

XIX

The Dutch Coal Trade in the Seventeenth and Eighteenth Centuries

Moving coal down the east coast of England from Newcastle to London was to contemporaries and historians alike an overwhelmingly important part of English shipping and critical as a basis for continued maritime expansion. The volume of the trade, the frequency of voyages and the ease of entrance into the trade all made it a pillar of English commercial and maritime strength. If coal was so important to English shipping then certainly there must be some question about the contribution of moving coal to Europe's largest merchant marine in the seventeenth century, that of the Dutch Republic. Peat was of course the greatest source of heat energy in the Netherlands in the seventeenth century. The peat was dug domestically and shipped along canals and rivers to nearby markets in the rapidly growing towns of Holland. Supplies lay close at hand and were very near or below the water level so it was easy to load the cut peat on boats. In the course of the seventeenth and eighteenth centuries a whole new network of canals was created to connect consumers and producer of peat.[1] Home heating was the original and principal use for peat. The expansion of Dutch industry in the fifteenth century put an end to that. As demand for fuel, especially from industrial users, grew rapidly, in the second half of the sixteenth century Dutch consumers began looking for other energy sources. Coal was used alongside peat in Holland long before the golden century. For home heating peat and wood were always preferred. On the other hand smiths always relied almost exclusively on coal. Brewers turned more and more to coal over time. Coal always had an advantage since a kilogram produced more heat than a kilogram of peat.[2] That made transportation and storage of a given quantity of energy in the form of coal easier. Storage was a problem for urban industry. Since coal had four times the calorific value of peat then not only space could be saved with coal but also a good deal of labour in stoking fires. There were other advantages. Peat had to be dried so in years when the spring was wet supplies of good quality peat were limited. There was a seasonal problem with peat. From 1 November to 1 February peat supplies from interior provinces like Overijssel and Groningen were not available. And the exhaustion of peat supplies in Holland had obviously even by the 1670's made imported coal more attractive as an energy source.

Resistance to the change to coal had many different reasons. While peat enjoyed the advantage of being easily shipped coal often had to be transferred from a seagoing vessel to another smaller inland one to reach Dutch urban markets. Both peat and coal were bulk goods. Shipping costs were a significant proportion of delivered cost, regularly more than 50% in the case of coal. It was difficult therefore for either good to sustain any increase in transport costs. Incidentally moving peat was important to domestic shipping. In the seventeenth century it took 3,000 ships or about 40% of the inland vessels in the Republic to move the annual peat output. The work employed 6,000 men and often members of their families. Small boats making frequent voyages moved the peat. The *aalman*, the *pont* and the *pleit* were the most commonly used boats for the job. They were all flat bottomed, had no posts and were poled rather than sailed. A 1450 Leiden law set the overall length of the *aalman* at a maximum of 10.8 meters. It had platforms at either end and could carry about 4.8 metric tons of peat. Boats got bigger over time but were limited because of the width of ditches and canals that ran along the peat bogs. The *turfpont* could carry from about 15 to about 50 metric tons of peat. Vessels that carried turf to Brabant and Flanders were larger, able to handle from 50 to even 100 tons.[3] Even though moving peat was important to navigation that in no way lessened the significance of moving coal to Dutch shipping in general. Holland and Zeeland had to major sources of coal. There was no domestic coal within the Dutch Republic so imports were the only possibility. Coal either came from modern Belgium and more precisely the prince-bishopric of Liége or it came from Britain, from England or Scotland. Coal is mentioned in the Maas region near Liége in 1195 and again in 1213. In all Europe only the English mines predate those of Liége. Dutch shippers were active in bringing coal from Newcastle to the Low Countries in the last years of the fourteenth century and probably before that. [4] Coal is not all the same. Three large categories were used in the seventeenth and eighteenth centuries. First there were strong or small coals which gave off an intense heat. They were much preferred by smiths. Second there were large coals which burned more slowly and had less sulphur so were used in industries such as brewing or brick- and lime-making. The third type was a mixture of dirt and coal which was very cheap and used for home heating or in some cases for salt boiling. It could not sustain the cost of transport and so was only sold close to mines. It was the first two types, forging coal and large coals which found their way into international commerce. In fact the two collectively were called 'ship coal'. English mines in the seventeenth century produced mainly for the London market where the coal was intended for home heating. They sent their large coals there and had to find a market for the other type of 'ship coal'. An easy outlet was the industries and especially the smithies of the Continental coast. During the First Anglo-Dutch War (1652–1654) there were complaints that there was no way for English mines to get rid of their second grade of small coals since the Dutch market was closed. Even in peace time by the mid-seventeenth century finding an outlet for small coals was a problem for shippers from England's major coal port, Newcastle.[5] No matter the type it was coal from those two sources – the southern Netherlands and the British Isles – which from the thirteenth century on competed for the market in the Rhine and Scheldt deltas. Coal from both could be sold there, in Holland and Zeeland, in Brabant and Flanders, at competitive prices because it could be moved to the region at low cost over water. The Dutch then always had access, at comparatively low cost, to the two greatest coal producing districts of Europe. It is not surprising then that coal found uses in Holland from the high middle ages on. Coal from Liége came on boats down the Maas so Dordrecht was the principal distribution point for that coal. It had to pass a number of tolls on the way. Even as early as 1272 toll collectors were having trouble with collecting the amount due them because coal was used to cover goods which had to pay a higher duty.[6] There is some indication that by the sixteenth century Maas coal reaching Dordrecht was being sent on to other markets such as Delft, a center of brewing, and to other towns in the Low Countries as well as in some cases to northern France, Spain and to England. Very little of the shipping was over long distances however.[7] As demand for Liége coal grew in Holland through 1600 Dordrecht became even more just simply a transshipment and distribution point. Dordrecht shipcarpenters even developed specific and unique types of vessels to carry coal. The Dordrecht *kool-haalder* was a long flat-bottomed boat, open and with no deck. There was a small rectangular cabin amidships where the single mast stood. The mast carried a square sail.[8] Traffic along the Maas was not always easy because of wars and government rules and tolls. Still for the sixteenth and seventeenth and through much of the eighteenth century Dordrecht was an important intermediary for moving coal from mines around Liége to consumers. Coal production around Liége dated from the high middle ages but it was probably not until the fifteenth century that there was significant production for export. In the sixteenth century production continued to grow. In 1545–'46 minimum output was 48,300 tons, rising to 90,000 tons in 1562–'63 and then dropping off to 37,000 tons in 1566–'67. Wars overseas and disturbances at home may well explain the low figure for the last date. Though there had long been a thriving market for coal in Liége and the level of consumption there was unique on the Continent there was also a sizeable surplus for export. In the sixteenth century about 40% or as much as 35,000 tons was sent out of Liége each year. The share was down to about one-third by the beginning of the eighteenth century but that still left some 50,000 tons for shipment to the Netherlands. There were other sources of coal in the southern Netherlands such as the Mons district so the exports of the Liége district to Holland are a minimum of imports. Around 1700 as many as 400 inland vessels carrying 25 to 30 tons each found work in that coal trade. There was also work for ships and men in the re-export trade. Liége coal enjoyed a reputation for high quality and it was much preferred to, for example, Scottish coal because it burned cleaner. In the early sixteenth century Liége coal competed with coal from Newcastle in French ports even though the latter was loaded directly on seagoing ships and the former had to travel some 250-300 kilometres by river to reach the sea. The re-export trade which began certainly by the sixteenth century continued through the eighteenth century. The relatively small river going vessels of the Maas could not make the trips to Rouen and other French ports. Presumably labourers on the Dordrecht docks transferred the Liége coal to boyers, a popular vessel for trade to northern France. The combination of a square sail and a spritsail on the mainmast and a lateen sail on the mizzen mast gave a high degree of manoeuvrability. Tonnages in the range of 50-100 made boyers small enough to negotiate rivers such as the Seine. Galjoots were also used for trade to France. They were faster and more narrowly built and thus not as well suited to the bulk trade in coal.[10] By the eighteenth century the principal coal port in South-Holland was no longer Dordrecht but Rotterdam. Trade through Dordrecht did not end but the better location of Rotterdam for international trade and the generally healthier economy of that town made it possible for it so supercede its upstream rival. Rotterdam specialized more in importing and re-exporting coal from the British Isles while Dordrecht, for reasons of geography, remained loyal to coal coming down the Maas. Amsterdam was in the sixteenth century the principal port for coal imports from the British Isles. The smiths of Amsterdam needed the small coals which England exported. In fact even in the second half of the seventeenth century Amsterdam along with Hamburg was the main destination for English coal exports. The two principal English exporting ports, Newcastle and Sunderlund, sent more than 30,000 tons to Amsterdam in 1675. Better than 20,000 tons was typical in the late seventeenth century from those two ports alone. At an average of 80 tons for ships going to northern England a minimum of 75 ships were needed to maintain that trade. All those ships were Dutch at least up to the 1660's. The low levels of exports of coal to the Baltic from Amsterdam strongly supports the conclusion that British coal was meant for use in local industry and for distribution to nearby towns and villages. Vessels from around the Zuiderzee and from the islands in the Waddenzee carried the coal from the northeast coast of England to Amsterdam. By the late eighteenth century some shipments were regularized, an industrialist for example chartering a ship to keep him supplied during one year or longer with coal from England.[11] There are no exact coal import figures for the port of Amsterdam before 1668. In that year 2,369 tons of smiths' coal entered the port from England, 924 tons of the same was shipped out. Figures become more abundant from Britain, but on the other hand she took over 19,500 tons of small coal from the same source. The total of all coal imports of all types to the town was 22,300 tons. Exports of small coals in the same year were about 170 tons so the re-export trade was negligible. In 1774 imports of all coal into Amsterdam totalled almost 22,000 tons and exports were just some 70 tons. At the end of the eighteenth century Amsterdam was taking even more coal from Britain. With the outbreak of war in 1795 imports fell off dramatically. But from 1789 through 1793 the average imports from Britain to Amsterdam were about 31,200 tons. The trade to Amsterdam alone must have offered full time employment for about 59 ships. The coal trade through Amsterdam faded relatively even from the mid-seventeenth century as Rotterdam came to take its place. As early as 1680 total imports into the Rotterdam admiralty district – which included Dordrecht – of British coal was up to 15,000 tons. Almost 6100 tons of that total was exported. Rotterdam importers over time turned their interest more to coal from Sunderlund rather than from Newcastle. The quality was higher. The relative rise of Sunderlund coal exports was in part due to the growth of the Rotterdam market. As with Amsterdam much of the coal went to local industry or to villages and towns in the district. By the late eighteenth century that consumption and the re-export trade had driven up imports to a level far above those of Amsterdam. One observer in 1777 put imports into the Zuiderzee at 26,320 tons from all sources. The quantity of coal reaching Maas ports at that time was over 110,000 tons and that was imports just from Sunderlund. There were additional and significant supplies from Newcastle and from Scotland.[12] 250 Ships would be the minimum number for coal carrying ships serving Rotterdam alone in the late eighteenth century. By that date, however, few if any of those ships were Dutch. In the late middle ages shipments came principally from Newcastle but also from ports like Lynn and Scarborough. From 1377 to 1403 occassional figures for six of those years show an annual average export from Newcastle alone by Hollanders and Zeelanders of 1505 tons. Overseas exports were volatile because of wars, shipping problems and variations in taxes. The share of exports in total English production fell over time. In the mid-sixteenth century it was still around 50% but sales overseas did not keep pace with the rapid growth in output. By the end of the seventeenth century foreign markets only took some 20%. Of those exports 80-95% went to ports in nothern France, the Low Countries or the North Sea coast of Germany. The markets fluctuated. For example Holland was the most important foreign destination in the 1620's but by the 1630's it had changed to Hamburg.[13] Coal exports to Holland alone required as much as 5,000 tons of English shipping in the second decade of the eighteenth century. By the last decade of the century that figure had risen to over 12,500. That was small compared to the total tonnage needed for carrying coal to the rest of northwestern Europe – 265,000 tons in the 1750's – or in the coasting trade – 125,000 tons in 1773. The carriage of English coal by then was almost exclusively in British ships. The total of all exports of coal rose over the eighteenth century. The total volume remained stable but the Dutch share of total exports tended to fall.[14] This made little impression on Dutch shipping since carriers from Holland had already been chased out of the trade. The tonnage of coal shipped was on the other hand very important to the English merchant marine. By the close of the eighteenth century clearly industrial uses dominated in Holland for coal. Rotterdam dominated imports and Britain dominated supply. All estimates from

English Coal Exports to Holland (metric tons)

Year (Christmas to Christmas)	Total of all exports	Tons to Holland
1710 – '11	66176	27296
1711 – '12	66907	24117
1712 – '13	79255	24013
1728 – '29	124127	31663
1729 – '30	87060	34613
1730 – '31	123197	34280
1731 – '32	128917	37925
1732 – '33	125018	33214
1733 – '34	133244	35243
1734 – '35	135350	36791

the eighteenth century and all official returns point to the importance of coal as a source of energy in Holland, to Holland's importance as a market for English coal exports and the importance of coal in international trade. Carrying coal down England's east coast was a simple task. Navigational problems were few. Winds were reliable. The trade was best served by small ships of inexpensive build and frequently old ones. They also had to be ships which were inexpensive to operate. It was in exactly that type of ship that Dutch shipbuilders excelled. They produced low cost bulk carriers which dominated the trades of nothern Europe. That was already true in the sixteenth century so Dutch ships carried the majority of English coal exports. English merchants preferred Dutch-built ships since they cost so much less to buy and to operate. They bought them directly from Dutch yards or captured them as prizes in one of the three Anglo-Dutch wars. In the 1660's and 1670's as much as one-third of English mercantile tonnage may have been of foreign build and almost every ton of that was for carrying bulk goods like coal. The English coal trade always employed a high proportion of tonnage not because of the distances, which were usually short, but because of the massive tonnage of coal which the ships moved. By the second half of the seventeenth century colliers were specialized vessels – not just any old ship pressed into service as had been the practice a century before. The change in design led to larger ships with greater carrying capacity with little or no increase in crew size. The average size of a cargo of coal into London in 1592 was just 51 tons but it was up to 66 tons by 1606, 75 tons by 1615, 126 tons by 1638 and it had jumped to 225 tons by 1701. In the same period the manpower needs for each ton fell by about 50%. Coal ships typically had small and efficient crews, usually only one man for each 20 tons or more of cargo. A collier of 1650 with a ratio of 21.5 tons/man had among the highest ratios of any sailing ship before the nineteenth century. The average size of colliers rose as the smaller ones disappeared and a few large ones – above 400 tons – entered the trade in the course of the seventeenth century. The ships in the export trade were typically larger, that is well above 100 tons. This applies to both English and Dutch ships. No matter the size it still took about two weeks to make the better than 600 kilometer trip from Newcastle to London.[15] The principal ship for the Newcstle coal trade was the hoy. The keel, a ship with pointed bow and stern, shallow draft and a wide beam – as much as one – half the typical length of 12 metres – was the vessel for handling coal in the middle ages. The design was similar to that of small Scandinavian vessels. The keel was used as a lighter along the River Tyne long after it had been supplanted in the longer distance carriage of coal. In the sixteenth century builders began making their coal carriers setting up the frame first and then adding the hull planks rather than starting with overlapping hull planks. The design change to skeleton-from shellbuilding was a major one. It made possible the construction of larger, lighter, and less costly colliers. The hoy, of skeleton design, owed a great deal to the *heu* or *heude,* a sea going ship of the Low Countries. There such vessels rarely exceeded 40 tons. They were large enough to have a full deck and a large trapezoidal wing transom. Originally there was a single low mast with a single spritsail. There was also a staysail on the forestay. The combination of pure fore-and-aft rig kept manning requirements very low. Netherlands shippers used such vessels on inland waterways but the sharp cut of the hull recommended the *heu* for use at sea.[16] Used for carrying coal in the seventeenth century the hoy was made larger and sometimes given a square sail as a topsail on the mainmast and even a jib. The spritsail was replaced by a gaffsail which was simpler to handle and equally effective on either tack since there was no big sprit running a 45° angle from the foot to the top of the sail as with the spritsail. It was still possible with a gaffsail to increase sail area by adding bonnets, small strips of canvas sewn to the bottom of the sail to make it longer and easily taken off when it was time to shorten sail. The gaffsail was easier to shorten too since all sailors had to do was lower the gaff as far down the mast as they thought necessary. These coal hoys enjoyed a good reputation for being nimble, highly mobile and surprisingly fast. The change in the seventeenth century to larger ships was also a change to a different type. Dutch *fluiten* proved highly effective in carrying all kinds of bulk cargoes, for example grain and salt, and so it was not long after the development of the type in the late sixteenth century that *fluiten* went into use in the coal trade. The hull was almost box-like, the bows bluff. Standard rig was three masts with square sails on the fore and mainmasts and a lateen sail on the mizzen mast. For the coal trade apparently there was in some cases a reduction to two masts and increasingly the use of a sprit- or gaffsail on one mast to decrease the already low manning requirements. Sterns were round and that feature was kept in colliers in the eighteenth century as ketchs, two-masted vessels, tended to dominate the trade. Many colliers were said to be cat-built and that may reflect descent from medieval Scandinavian design. Bows were bluff as with the *fluit* and almost square. The waist was deep and the bottom rather flat. Sometimes called coal cats they carried three but more often two masts. There could be topsails, but again those were rare, to keep down manning needs. The rig was similar to that on hoys but with the addition of a second and even a third mast. Hull design was very different from the more sharp hoy. To get more driving power builders included more square sails as a percentage of the total canvas. In the seventeenth century both Dutch and English shippers used the same types of vessels, principally *fluiten* or variations on the basic *fluit* design. By the eighteenth century English shipwrights especially those along the Yorkshire coast, had developed the skill of building specialized colliers, efficient, relatively large, rugged and durable. Captain James Cook on his first voyage of discovery in 1768 chose a 368 ton bark of collier design to make the trip to the South Pacific. Though she was a slow sailer she, like others of her design, was an excellent sea-boat.[17] The success of northeast shipwrights was part of the shift of the coal trade from East Anglia and from the Netherlands. East coast trade was long dominated by short voyages along that coast or to the Low Countries. In the fifteenth century surplus capacity in East Anglia turned to the carriage of coal from the Tyne to home ports for redistribution either locally or to London. Fishing boats in the off season could on occassion supplement earnings by moving coal. English vessels still dominated the carriage of coal in the 1540's. The expansion in total tonnage used along the east coast of England down to the end of the century brought new vessels from overseas into the trade. Shippers in the smaller East Anglian ports had to face not only competition from foreigners but also from London and Yarmouth shippers who in the course of the seventeenth century acquired cheap Dutch bulk carriers. Ultimately though decline in East Anglia was because of the rise of native shipping and shipbuilding in Yorkshire ports such as Newcastle, Scarborough and Whitby. It may well be that as the pace of growth in coal mining slowed in the late seventeenth century that mine owners in the North turned to shipping as a place to invest their capital.[18] Whatever the reason in the eighteenth century all competitors, English or Dutch, had been driven from the trade by captains, crews, ships and shipowners from Yorkshire. Coal did not just go to London or to Holland. There were other markets such as France, lands on the Baltic and especially Norway. In total volume those shipments were significantly less than shipments to Holland but they offered employment for ships and opportunities to combine coal carriage with moving other goods. Dutch ships travelled directly to Newcastle and after taking on coal might touch a Low Countries port before going on, for example, to Rouen. The Dutch ships would go out in ballast and then take the coal to ports such as Lisbon or Hamburg. Rotterdam ships from 1660 to 1790 carried coal to ports near and far away, from Cherbourg and Fécamp to Copenhagen and even Philadelphia.[19] Then they would have to return home in ballast or try to find some cargo, no matter how small, to carry home. Since coal was the bulky commodity it determined the quantity of tonnage used in the trade. Shippers tried a number of different schemes for getting around the problem of voyages with no paying cargo. The most successful solution was a threecornered trade with Dutch ships going to Norway in ballast to collect a cargo of timber and then carrying that to Newcastle where they took on coal for the Dutch market. The great advantage was that the timber trade was the only English trade which had a volume comparable to that of the coal trade. Moreover the same older capacious and slow ships were suited to both trades. A Rotterdam merchant chartered the Hopewell of Bridlington of 110 tons in 1732 to carry unspecified goods to Sunderlund and there to take on a load of coal for Bergen, Norway. In Norway the master was to deliver the coal and the goods from Rotterdam to the agent of the Rotterdam merchant and then within ten working days take on the goods specified by the agent and return to Rotterdam. She also carried a small quantity Rhenish brandy distilled in Germany on behalf of two Rotterdam wine merchants. In 1733 yet another charter sent James Willey, the master, to Sunderlund to load 96 tons of coal for Kristiansand, Norway. At the same time the same Rotterdam merchant was chartering another Hopewell, also of Bridlington, this one of 80 tons to make similar voyages for him, one in 1732 and one in 1733. There was a third Hopewell, this one of 150 tons, chartered out of Rotterdam in 1732 in the coal trade. She was to go to Sunderlund where the merchant's agent would load on as much coal as possible and then she was to carry that coal to Copenhagen. From there the destination was Kristiansand for a cargo of miscellaneous goods but probably largely wood.[20] It was not a coincidence. The similarities were caused by the intensity of the triangular trade – or rather triangular trades – created in the North Sea by the presence of massive English coal exports and demand for that coal throughout the region and especially in Holland. Foreign carriers dominated the English overseas coal trade from the second half of the sixteenth century. In 1593 seven out of every eight ships clearing Newcastle for a foreign port were not English. Foreign ships carried between 33% and 40% of coal coming to London from Newcastle in the 1590's, the equivalent of about 10,000 tons each year. The overwhelming majority of those foreign ships were Dutch. Domestic shippers were distressed at the loss to foreigners of employment for their ships while the English government saw potential foreign exchange slipping away along with a valuable training ground for seamen. The government embarked on a legislative programme to make the coal trade an English preserve. In 1563 all foreign shipping was prohibited from coastal trade. In 1615 the government issued a proclamation reminding Englishmen that acts of 1368, 1489 and 1541 were still in force which made it illegal to ship goods in or out of England in anything other than English vessels. In both cases the attack was on Dutch shipping and especially on Dutch carriage of English coal. There was a sharp rise in the proportion of English ships carrying coal to Dutch ports in the first two decades of the seventeenth century. The proportion increased through the century first because of the greater use by English shippers of Dutch-designed ships and second because of more and more restrictive legislation. In 1661 for example 158 of 190 cargoes from Newcastle went out in English bottoms. The tax structure worked to the disadvantage of foreigners as well. Non-English shippers had to pay a higher export duty than their English counterparts. In 1694 Dutch shippers paid 233% more than natives. At times the differential was even higher. Since taxes and duties were about half the cost of duties became more and more complex over time. Because of a tax on imports into London imposed in 1666 English coal delivered in Rotterdam in that year was cheaper than coal delivered in London. Again in 1739 there was a complaint that the Dutch could buy coal cheaper in England than could English manufacturers because of the tax structure. No matter the confusion and inconsistency of government the seventeenth century policy of excluding foreign shippers and taxing shipments by foreigners at a higher rate when and if they were permitted in the coal trade was effective. In the years 1711–1713 less than 5% of exports went out of England in foreign bottoms.[21] The English legislation was designed to retain the carriage of coal for natives. It succeeded and increasingly so from the 1660's. One major effect was to drive Dutch shippers for much of the seventeenth century to seek other less expensive sources of coal, places where they could use their relatively more efficient ships. They found a major source along the shores of the Firth of Forth. Scotland enjoyed a thriving coal export trade to Holland during the seventeenth century. The duties and restrictions common south of the Tweed did not apply in Scotland, that is until the Act of Union of 1707. Scottish great coals which looked something like slates were much preferred in Holland for home heating. They were cleaner in burning and also much cleaner to use, a major consideration for Dutch housewives. From the late sixteenth century coal production rose rapidly in Scotland and exports appear to have grown even faster. One reason was the rising duties on English coal exports. In 1614 Scotland sent out about 15,000 tons of coal. Dutch ships carried the overwhelming majority of that coal. In the 1620's it was said that Dutch ships were queueing in the Firth of Forth to get their coal cargoes. Though exports fell to about 18,000 tons per year in th 1680's they were back up to around 45,000 tons by the end of the century. Moving the coal took about 75 ships, mostly Dutch. The chief foreign market was unquestionable Holland.[23]

But in the last years of the seventeenth century sales there failed to grow despite the increase in total exports. In part it was because the English lowered their export duties so Dutch buyers tended to go to Newcastle and Sunderlund instead. Dutch shippers took some of the increase in Scottish output to customers in third countries. The preference for the Dutch market was a long standing one in Scotland, one not broken until late in the eighteenth century. The lack of a domestic merchant marine forced Scottish coal producers to rely on foreign shippers and the Dutch were the most efficient. The efficiency of the Dutch acted as a break on the development of the Scottish merchant marine in the seventeenth century, much more than the similar effect which it had in England. Many Scottish ships were in fact built in Holland. Ships were always smaller in the Scottish coal trade than the English simply because of the poorer quality of ports in the Forth. Investment in improving facilities by local landowners and by some Dutch merchants who were also mineowners did not overcome that handicap. Vessels from Rotterdam visited many small ports up and down the Forth shores. "Borrowstouness" for example turns up as a destination for Rotterdam ships much more frequently in the late seventeenth century than would be warrented by its size. The importance of such ports rose and then fell as miners exhausted coal seams close to the water. There was some re-export of Scottish coal, especially south to Brabant and Flanders. But high duties on Scottish coal in the southern Netherlands served to keep much of the coal in Holland. Ships carrying coal from Scotland also on occassion were involved in a triangular trade taking coal to Rotterdam and then going in ballast but with cash to Norway where they got a cargo of wood for Scotland.[24] The Scottish coal trade and indeed the coal trade in general was very simply organized. A bulky commodity that was neither highly differentiated nor subject to deterioration did not require an extensive system of factors and agents. In the English coastal trade it was the masters of the old leaky colliers who did the buying of the coal at Newcastle and the selling at London. That made possible stockholding and the regularization of sales. The simple commercial arrangements of the sixteenth and seventeenth centuries suited Dutch shippers perfectly. In the timber trade and the Baltic grain trade they bought goods on the spot for cash and thus kept prices low. It was a simple matter to transfer that practice to Newcastle and to all the small ports along the Firth of Forth. Dutch mercants from the second half of the seventeenth century had to gradually abandon direct dealing. They turned to chartering English or Scottish ships to bring cargoes of coal back for them for a fixed fee.[25] The need to use native English shippers was part of the reason for the increased use of chartering. The practice was also common in the three-corner trades where other goods were involved. Though there was a rise in chartering in the late seventeenth century the independent shipper did not disappear. In the eighteenth century that businessman was replaced in the Dutch coal trade however and for the same reasons as his scope was reduced in the English coasting trade. The growth in the scale of operations, the simple increase in the tonnage carried put the trade beyond the skills of such men. Coal supplied more than energy in the golden century. The coal trade gave employment to shipping both along the rivers of the Low Countries and across the North Sea to Yorkshire and the Firth of Forth. The approximately 6,000 tons of shipping used to bring coal back to Holland from the British Isles in the 1630's was small in a merchant fleet which totalled well over 300,000 tons. The proportion grew up to the 1660's and then fell as English protectionist legislation became effective. The figures understate the value of the trade to shippers however. It gave them a place to put older ships and a trade where it was always possible to put in an extra voyage while waiting for a cargo. For England the coal trade was more important to the total of shipping tonnage than for the Dutch merchant navy. By the 1770's for example the coal trade from Newcastle to London employed more shipping than all the rest of English trades.[26] The coal trade was important for England because it offered employment to a large number of seamen. That was true for the Dutch merchant navy but again it was not as significant in Holland. There were other advantages reaped by both Dutch and English shipping from the coal trade. The existence of the coal trade meant the opportunity to use ships for moving other goods. The emergence of three-cornered trades in the North Sea in the course of the seventeenth century was a direct result of the presence of bulky coal shipments. The coal trade fit into and contributed greatly to the creation of a trade network in the North Sea and the English Channel, a network which stretched in some cases to the Baltic and the Biscay coast of France. The coal trade contributed to the drive to develop and to build efficient bulk carriers which incorporated in their design lower labour requirements, even lower than for the grain trade which was the most important bulk trade for the Dutch. The Dutch coal trade did not have the glamour of the trade in spices to the East Indies. It did not compare in value to the wine trade to Bordeaux. It did not compare in volume to the grain trade to the Baltic. But it did with its frequent sailings create marginal employment for shipping and it did bring to the Dutch Republic an important source of energy, an alternative to local peat supplies and an invaluable input to domestic industries.

Notes

I am indebted to Simon Hart of the Amsterdam city archive and R.C. Smout of Edinburgh University for their help in completing the research for this paper.

1. J.W. DeZeeuw - "Peat and the Dutch Golden Age. The Historical Meaning of Energy-attanability," *A.A.G. Bijdragen*, 21 (1978), 3-31.
2. Z.W. Sneller - *Geschiedenis van den Steenkolenhandel van Rotterdam*, Groningen, 1946, 14-15.
3. J.W. DeZeeuw - "Peat and the Dutch Golden Age...," 17-21. Using his own figures the number of ships was closer to 3000 than his estimate of 4000.
   W.J. Diepeveen - *De vervening in Delfland en Schieland tot het einde der Zestiende Eeuw*, Leiden, 1950, 133-142.
   P. van Schaik - "De economische betekenis van de turfwinning in Nederland," *Economisch-Historisch Jaarboek*, 23 (1969), 191-193.
4. N.J.M. Kerling - *Commercial Relations of Holland and Zeeland with England from the late 13th Century to the Close of the Middle Ages*, Leiden, 1954, 121-122.
   J.A. van Houtte - *An Economic History of the Low Countries*, New York, 1977, 38.
   Z.W. Sneller - *Geschiedenis van den Steenkolenhandel...*, 10-11.
   P. van Schaik - "De economische betekenis van de turfwinning in Nederland," *Economisch-Historisch Jaarboek*, 33 (1970), 228.
5. J.U. Nef - *The Rise of the British Coal Industry*, London, 1932, I, 109-117.
6. J.F. Niermeyer, ed. - *Bronnen voor de economische geschiedenis van het Beneden-Maasgebied*, I, 1104-1399, The Hague, 1968, 78 (Sept., 1272).
7. Z.W. Sneller - *Geschiedenis van den Steenkolenhandel...*, 25-41, 84-85.
   T.L.M. Thurlings - *De Maashandel van Venlo en Roermond in de 16e Eeuw, 1473–1572*, Amsterdam, 1945, 72-75.
8. Nicolaes Witsen, *Architectura Navalis et Regimen Nauticum...*, Amsterdam, 1690, 189.
9. J.A. van Houtte - *An Economic History of the Low Countries*, 169-170.
   J.U. Nef - *The Rise...*, I, 128-129. All figures are reported in metric tons of 1000 kilograms.
10. Nicolaes Witsen - *Architectura Navalis...*, 181-184.
    Z.W. Sneller - *Geschiedenis van den Steenkolenhandel...*, 68.
11. Stedelijk Archief, Amsterdam, Notarieel Archief, 1129/64 (1659), 15623/830 (1784). J.U. Nef - *The Rise...*, I, 86. Z.W. Sneller - *Geschiedenis van den Steenkolenhandel...*, 71-75, 90.
12. H. Brugmans - "Statistiek van den In- en Uitvoer van Amsterdam, 1 October 1667–30 September 1668, *"Bijdragen en Mededelingen van het Historisch Genootschap*, 19 (1898), 140.
    J.G. Nanninga, ed. - *Bronnen tot de Geschiedenis van den Levantschen Handel*, III, 1727-1765, The Hague, 1952, 662-663, 688-689.
    Leonie van Nierop - "Uit de Bakermat der Amsterdamsche Handelsstatistiek," *Jaarboek Amstelodamum*, 15 (1917), 43-106. N.W. Posthumus - "Statistiek van den In- en uitvoer van Rotterdam en Dordrecht in het Jaar 1680," *Bijdragen en Mededelingen van het Historisch Genootschap*, 34 (1913), 529-537.
13. N.J.M. Kerling - *Commercial Relations...*, 122, Appendix F.P. van Schaik - "De economische betekenis...," 33 (1970), 227.
    Ralph Davis, *The Rise of the English Shipping Industry in the 17th and 18th Centuries*, London, 1962, 4, 8-10.
14. Ralph Davis - *The Rise...*, 184, 202-203.
    L.A. Harper - *The English Navigation Laws. A Seventeenth-Century Experiment in Social Engineering*, New York, 1939, 336, 339.
    Z.W. Sneller - *Geschiedenis van den Steenkolenhandel...*, 96.
15. Violet Barbour - "Dutch and English Merchant Shipping in the Seventeenth Century," *Economic History Review*, 2 (1930), 265-268, 289-290. Ralph Davis - *The Rise...*, 33-34, 60-62. J.U. Nef - *The Rise...*, I, 390-394.
    G.V. Scammell - "English Merchant Shipping at the End of the Middle Ages: Some East Coast

Evidence, "*Economic History Review*, Second series, 13 (1960), 332-334. The large size of eighteenth century colliers was even more impressive because of the small size of fifteenth and sixteenth century coal ships.

16. J. van Beylen - *Schepen van de Nederlanden*, Amsterdam, 1970, 173-174.
Roger Finch, *Coals from Newcastle. The Story of the North East Coal Trade in The Days of Sail*, Lavenham, Suffolk, 1973, 25-27, 36-38, 40.
G.P.B. Naish - "Ships and Shipbuilding," *A History of Technology*, C. Singer *et al.*, eds., III, London, 1957, 498-499.
W. Salisbury - "Early Tonnage Measurement in England, "*The Mariner's Mirror*, 54 (1968), 69-71.
17. R. Morton Nance - "British Ships through Dutch Spectacles, "*The Mariner's Mirror*, 5 (1919), 3-9. R.W. Unger - *Dutch Shipbuilding before 1800*, Assen, 1978, 36-38, 43-46.
J.C. Beaglehole - *The Discovery of New Zealand*, London, 1961, 24-25. The *Bark Endeavour* was 32 metres long, had a beam of 8.8 metres and drew 6 metres. She carried 3 masts with a full complement of sails typical of the largest of eighteenth century sailing ships.
18. Ralph Davis - *The Rise...*, 92-93. G.V. Scammell, "English Merchant Shipping...," 328-329, 336-339.
19. Stedelijk Archief Amsterdam, Notarieel Archief, 10277/592 (1746), 7290/277 (1709), 5916/767 (1710). Gemeente Archief Rotterdam, Scheepsverklaringen, 1660-1790, *Steenkolen.*
20. Gemeente Archief Rotterdam, Notarieel Archief, 2116/139, 206; 2117/243, 257, 258, 290, 314, 392; 2118/46, 135; 2326/158.
21. Ralph Davis - *The Rise...*, 227, 309. L.A. Harper - *The English Navigation Laws*, 14, 280-281, 335.
22. To Dutch ports the percentage was slightly higher – 6%.
Violet Barbour - "Dutch and English Merchant Shipping...," 283.
T.S. Ashton and Joseph Sykes - *The Coal Industry of the Eighteenth Century*, Manchester, 1929, 226. J.U. Nef - *The Rise...*, II, 233-238, 385.
23. R.C. Smout - *Scottish Trade on the Eve of Union 1660–1707*, Edinburgh and London, 1963, 190, 223-229.
24. Violet Barbour - "English Merchant Shipping...," 287.
R.C. Smout, *Scottish Trade...*, 54-44, 74, 155. English shipowners were worried during negotiations for the Act of Union about the more efficient Scottish ships which were largely Dutch built being allowed to trade freely to England. Many Dutchmen were part owners of Scottish ships as well as coalmines.
Gemeente Archief Rotterdam, Notarieel Archief, 511/219, 513/220, 514/618, 667/290, 940/933, 1480/235a. 1565/123. 'Allowa' was another popular port. 940/521, 1392/310, 1567/173.
25. J.U. Nef, *The Rise...*, I, 93-94. The highly stable Dutch *rijksdaalder* won quick acceptance in places such as Scotland where local coinage was unreliable. R.C. Smout - *Scottish Trade...*, 103, Scottish factors in Rotterdam rarely handled any coal business. Gemeente Archief Rotterdam, Notarieel Archief, 940/521 (1674), 1392/310 (1686), 1506/187 (1713).
26. J.R. Bruijn - "Scheepvaart in de Noordelijke Nederlanden, 1580-1650," *Nieuwe Algemene Geschiedenis der Nederlanden*, VII, Haarlem, 1980, 138. Ralph Davis - *The Rise...*, 176, 181. Adam Smith - *...the Wealth of Nations*, Edwin Cannan, ed., New York, 1937, 352.

XX

Grain, Beer and Shipping in the North and Baltic Seas[1]

The commodities carried by ships do not determine the design of those ships. Demand does not create technical change. There is no automatic or immediate connection between the needs of shippers and the actions of shipbuilders. Those technicians have other demands to face. The supplies of factors of production, for example, force them to make specific choices about the ways they build their ships and how they shape them. Equally the design of ships does not determine what goods will be traded. Supply of a good and demand for it must exist in two separate markets and the price at which the good is offered has to be the same or less than the buyer is willing to pay. If the difference in the offered and bid price is less than the cost of transporting the good then there will be no trade. It is and was the function of shippers to move goods between markets at a cost below the price differential between the two markets. Technical change in ship design could and did make trade possible by lowering transport costs and thus open new trades which did not exist before.

Beer was rarely if ever a traded good in the early Middle Ages. In the years from 1200 to 1500 it became a commodity of international exchange in northern Europe. Ships carried beer over the waters of the North and Baltic Seas to distant markets. The rise of the trade in beer was the result of a complex series of technical advances both in brewing and in shipping. There were changes in the good itself. The beer of 1500 was not the same drink as that produced and sold in 1200. Changes in product and production methods came at the same time as developments in the types of vessels shippers used in the North. It is difficult to describe precisely the progression of events but even more difficult to isolate the relationship between technical advances in shipping and in brewing. There are three principal questions raised by the history of trade in beer in the later Middle Ages. First, brewers were making beer which could be exported by the middle of the thirteenth century yet it was not until the second decade of the fourteenth that the trade deeply affected overseas markets. Second, the trade faced limitations and taxes in the Low Countries in the fourteenth century but despite those restrictions it continued to grow and along established routes not reaching its peak until early in the next century. Third, though less mysterious is the question of why the trade did finally decline after about 1420.

Production methods for beer throughout the Middle Ages were relatively simple at least for the great majority of producers, the peasants. Making beer took little more organization than baking bread so the home brewed beer could be and was made in almost any rural household. It was an "...activity deeply integrated in peasant society."[2] The beer was consumed very soon after it was produced. The process required little equipment and little skill though certainly some brewers and brewsters were better at it than others. Brewing in towns was technically the same as in the countryside at least up to 1200. It is possible that urban brewers could produce beer for less than their rural counterparts, if for no other reason than that they were specialists and could exploit economies of large scale production. They also typically had access to better copper work for their kettles and access to higher quality grains. They usually enjoyed protection by the town government, interested in protecting its own tax income.[3] Brewers in the countryside and in town used *gruit*, a mixture of herbs and spices, as an additive to flavour but also to help to preserve their beer. *Gruit* was mostly dried myrtle. It was reputed to have an intoxicating effect which may help to explain why brewers down into the nineteenth century resisted government efforts to stamp out its use. Myrtle grew wild in certain specific districts of the northern Netherlands so supplies there could be easily controlled. *Gruitrecht*, the right to dispense the mixture and thus tax it, was common in towns of the entire area of the lower Rhine.[4] Increasingly though another additive came into use, especially in the rising commercial breweries of the towns: hops.

Western Europeans used hops in brewing in the eighth century, that apparently largely in monasteries. Monks became consumers of large quantities of beer in northern Europe especially in districts where wine was not produced. In larger houses permanent brewing establishments had to be set up, almost invariably next to the bakery. Before the rise of urban brewing it was only in such monasteries that beer was produced on anything like a commercial scale. Romans had grown hops in vegetable gardens for use as a flavouring agent and in the ninth century monasteries were doing the same thing. By 800 it is almost certain that monastic brewers used hops to flavour and preserve their beer. It took some time before the plant came into use in towns and before use spread throughout Europe. For example only in the fourteenth century Scandinavians cultivated hops and when they did use hops in beer they mixed it with myrtle.[5]

The growth of towns increased the demand for commercially brewed beer. The confined quarters of towns often made it difficult for individuals to keep on producing their own beer. The larger numbers of townspeople expanded the market. By the thirteenth century urban centres boasted commercial breweries supplying local needs. It was in the thirteenth century if not before that those urban commercial brewers began to add hops to their product. This may well have been done first in towns in north Germany, especially those on the coast of the Baltic and North Seas. Hops were being traded in the early thirteenth century since the good entered duty free at Lübeck from 1218 to 1225. Hops came from a number of different places but were also exported by the towns. Wismar exported hops by sea in 1328. The towns developed regulations on trade in and use of hops, in Wismar for example in 1339. Other towns followed.[6] The hops not only gave a different and slightly bitter taste. The oil from the plant helped to preserve the beer. Since it could last longer it could survive being shipped.

The addition of hops created a new trade good. Though still called beer it was in fact a different commodity. The high quality made it possible for hopped beer to compete not only with *gruit* beer but also with wine. In Bruges for example by 1380 the holders of *gruitrecht*, the tax on *gruit*, complained about a fall in their income because of a shift to drinking hopped beer. The town accounts of Bruges from the fourteenth century show that the complaint was well founded. Income from the tax on local beer made with *gruit* fell in the period while income from a new tax on hopped beer rose rapidly. The tax records show that consumption of the new type of beer probably gained at the expense of wine as well the traditional *gruit* beer. Members of the court of the counts of Holland in the early fourteenth century drank imported hopped beer. They also drank locally produced beer.[7] The ability of hopped beer to compete with the old style of beer and also with wine depended on the quality of the product but also ultimately on its price. Hamburg beer commanded a higher price in the Low Countries in the 1320s than other beers because of its quality. In the early fifteenth century it was said to be as costly as good wine. A duty which raised the price by some 75% certainly contributed to that high price.[8] Competition among brewers and among towns existed in the export trade in beer and that forced down prices. Ultimately the price of the beer depended on the cost of raw materials and grain made up the overwhelming majority of that cost. The delivered price also depended on the cost of getting the beer from the brewery to the final consumer. That included the cost of packaging in barrels but above all the cost of shipping.

The north German towns of Lübeck, Rostock, Hamburg and Bremen made hopped beer certainly by the mid-thirteenth century. They soon started exporting the product. By the early fourteenth century they were sending it to England, the Low Countries and to Baltic lands. Beer from Bremen is mentioned as being imported into Amsterdam as early as 1274.[9] Bremen was first but was soon superceded by Hamburg. An old chronicle called Hamburg the *Brauhaus der Hansa*. Wismar followed and later so did towns inland. The ports had a comparative advantage in the trade since shipping beer by sea was always more economical than shipping overland. The Low Countries formed the logical market for exports from Hamburg. There the highest level of urbanization in northern Europe was combined with a large population with relatively high incomes and already used to consuming beer.

It was not until 1321 that the government of Holland acted to confine and limit the flow of trade in hopped beer. In that year the count, William III, forbade the import of Hamburg and eastern beer into the province. He was trying to protect his income from the tax on *gruit* which was falling in the face of the sharply increased consumption of imported hopped beer from north Germany. The resulting shortage of beer was so great and the complaints so numerous that just two years later in 1323 Count William lifted the ban but insisted that all beer for Holland would have to pass through the port of Amsterdam or through Medemblik. The tax soon became a lucrative one. Medemblik faded in comparison and so Amsterdam in the second half of the fourteenth century became the market for imported German beer and the transshipment point for sending the product via Flanders to the south. Stavoren in the province of Friesland just to the north played a secondary role. Though the regulations constricted the trade of the Hamburgers and though they had to pay a toll they did not object and certainly not effectively. They were forced to live with the new regulations. Difficulties with the Count of Flanders in the second half of the fourteenth century increased the willingness of Hamburg merchants to accept the rules of Holland. The volume of trade continued to grow throughout the century. The indication is strong that technical reasons explain why first it was not until after 1300 that German beer became a threat to Dutch producers and why second German shippers accepted and conformed to the Holland regulations governing the beer trade.[10]

From Hamburg there is evidence of the strength and importance of the brewing industry. A 1376 survey of the town showed that of the 1075 people mentioned with a trade 457(43%) were brewers. Of those 126 were brewers for the Amsterdam market and 55 for Stavoren. The output of all those Hamburg brewers must have been about 240,000 hectoliters(hl) per year. If all of them produced approximately equal amounts exports to those two ports in the Netherlands must have been about 100,000 hectolitres. Shippers would have had to supply transportation for about 10,000 tonnes of beer. The volume of production in Hamburg rose to some 300,000 hl per year in the early fifteenth century and 375,000 hl by about 1480. The volume of exports rose as well. The tax of the Counts of Holland gives an indication of how large the beer trade actually was. In 1352-4 105 ships brought at least 31,319 barrels of beer from Hamburg to Amsterdam, or an average of almost 300 for each vessel. In the tax year 1364 the total amount fell considerably, down to 17,514.[11] The export of beer from Hamburg to Amsterdam fluctuated widely in the 1340s and 1350s but rose strongly in the 1360s, averaging in the ten years for which data are available almost 32,000 barrels each year. It appears that exports to Holland fell in the next decade but returned to the former high levels in the 1390s. The barrel it appears held just a bit more than a hectolitre of beer and so was much smaller than a contemporary French wine barrel. If that was so then shippers only had to move about 3,500 tonnes of goods to Amsterdam and about half as much to Stavoren. The total figure for shipping required for the beer trade between Hamburg and the Low Countries then falls somewhere between 5,000 and 10,000 tonnes each year. Adding about 20% for the space taken up by the barrels then the capacity requirement for beer was equivalent to between about 21,000 and 42,000 registered tons.[12] Those shipments to Amsterdam formed close to half of all Hamburg exports by sea The shippers who carried the beer came not only from Hamburg but from the Netherlands as well. All of them probably used the same types of ships.

The trade in hopped beer from north Germany was always a coastal one. Even before 1300 Frisian traders travelled along the coast to Bremen and Hamburg supplying dairy products in exchange for, among other commodities, small quantities of beer. The Hamburg shippers who brought beer to Stavoren and the other ports of Friesland usually did not continue on in their own ships but transferred the goods to smaller vessels.[13] By the 1350s most of the beer trade was already in the hands of shippers from the Netherlands, especially from towns along the Zuider Zee. By 1364-66 Netherlands skippers did virtually all of the shipping of beer. They used vessels which could carry from 20 to 25 lasts. In 1380 a Hamburg ship could carry more than 1000 hl or better than 90 tonnes of beer in one trip though it is not sure how frequently that much beer went in one vessel at one time.[14] All evidence points to a design for such vessels similar to that of the boat of Celtic design found in Bruges at the end of the last century. That early cog seems to have been the forerunner of vessels used along the Frisian coast throughout the Middle Ages.[15] There is some discussion about when builders changed that type into a deep sea trader by the addition of a keel. Certainly by the thirteenth century though German traders used the new advanced cog on trade routes in the Baltic and North Seas. The greater carrying capacity per unit of length and the greater seaworthiness were certainly advantages to shippers. The new type may well have combined with the technical advance in brewing to create the trade between Hamburg and Holland. If that is true though it was still some time after the technical advances in brewing before the long distance trade in beer become a significant one.

Despite the presence of a sea going ship that could have sailed directly to Flanders shippers continued to bring their beer to Amsterdam for transshipment to smaller vessels so it could be carried on to markets just to the south. The enforced seasonality of the trade did not force shippers to maintain existing practice, that is to go through Amsterdam. Towns in northern Germany typically regulated the export of beer. The goal was to guarantee adequate domestic supplies of grain and to insure the quality of export beer. In the fifteenth century in Hamburg for example beer tasters waited at the entrance to the harbour so ships could not leave without first having the beer on board checked by those city inspectors. Essentially export was prohibited in the winter. Large export volume ran from April until mid-November.[16] The summer months would have been the best ones for sailing to Flanders directly. Brewers in Hamburg could not avoid regulations of their own town. But they could have avoided the tax levied on their imports by the Count of Holland by sending their ships directly to Flanders. They did not do it. They continued to ship their beer to Amsterdam. From there it went on to markets in Flanders, often in smaller inland vessels. The failure of producers or shippers to avoid the Dutch tax suggests that there was some technical limitation making it prohibitively expensive to sail directly to Bruges or its ports at Sluis and Damme. It is true that Hamburg brewers and merchants had established connections with Amsterdam. They had agents there called *liggers* and even before 1358 those men had formed themselves into a *hanze*. It is true that the Hanse and Hamburg in particular had periodic difficulties with the Count of Flanders. It is true that Amsterdam was well located for the distribution of beer around the province of Holland and to markets to the south. That was why William III had chosen it as the port for beer imports. It is also true that the counts of Holland in the thirteenth century did try to promote the *binnenvaart*, the shipping of goods from the East through Holland on her inland waterways. The locks at Spaarndam were 24 feet wide which would have made it possible for some cogs to make their way through the canals, lakes and rivers of Holland.[17] Though Hamburg beer did get through to Flanders there is evidence that suggests Holland was a more important final destination. It has been estimated that in 1369 only some 11% of Hamburg exports went to Flanders. From the 1370s however the volume appears to have risen considerably. Hamburg beer merchants set up their own *ligger* organization at Sluis by the end of the fourteenth century. That body was separate from the one maintained by other Hanse traders because of the large number of Hamburgers and their specialization in the shipping of beer.[18] Other beers from Germany, even from the Baltic, already made their way to Flanders by the mid-fourteenth century. It is often difficult to identify the exact origin of the beer.[19] Some ships carrying beer from Bremen, from Hamburg or even from Lübeck must have avoided Holland and made landfall on the Flemish coast. Many Hanse ships must have made the direct trip from Germany to supply the factory, the large Hanse depot in Bruges. Even so much if not the overwhelming majority of Hamburg beer exports still passed through Holland. The favourable treatment of the transit trade by the counts of Holland helped. The possibly greater per capita consumption of beer in Holland than in Flanders may have helped. But the design of ships was probably important too. Though the cog with a keel may date from the eleventh century it is in the fourteenth that a mature vessel appears in illustrations and in customs records. The presence of that more developed cog by the end of the thirteenth century certainly must have aided the export of hopped beer from north German ports. The need for vessels to carry beer may have led builders to refine the hull design and the rigging of the cog. By 1300 skippers presumably would have been able to go directly to Flanders in confidence.They would have hugged the shoreline in any case. That was standard navigational practice in northern Europe until the sixteenth century. In 1300 and even in 1350 skippers may have felt more comfortable and more certain of success if they travelled along the inland waterways of Holland. The greater security could only be had by limiting the size of cogs so that they could pass through the Dutch locks. The price was one skippers apparently were willing to pay.

When the Count of Holland prohibited the import of hopped beer in 1321 he also allowed the production of hopped beer in Holland, something prohibited up to that point for tax reasons. He was afraid that if Dutch brewers used hops he would lose income from the tax on *gruit*. The foreign competition drove him to legalize the new brewing method, making sure however that hops were taxed so that he would get just as much income as he would have had had brewers used *gruit*. That the count made abundantly clear in grants allowing the brewing of hopped beer in Delft (1326) and Haarlem (1327), for example.[20] The rule against importing hopped beer and the subsequent change to a tariff on imports of the good were imposed in part to protect a domestic industry. The duties combined with grants allowing the brewing of hopped beer proved highly successful.

Brewing developed rapidly in the towns of Holland. Already by 1326 the town of Delft had regulations on producing beer in the eastern style. It was called hop beer and also turf beer, presumably because larger quantities of peat were needed to produce it than traditional beer. Production of hopped beer in various towns had by about 1350 led to the cultivation of hops as a field crop. In the fourteenth century Haarlem already had about 50 breweries. By the end of the fifteenth century brewing was second only to the cloth manufacture among Holland's export industries.[21] The biggest foreign market and the one invaded first was Flanders. The taxes levied there made Dutch beer significantly less expensive than Hamburg beer and the tax on the latter was periodically raised. Dutch brewers also tried to imitate the German product, in some cases shipping their own product in Hamburg casks.[22] The shorter distance between producer and consumer helped the competitive position of Dutch beer in Flanders. It also helped in Friesland where in the first half of the fifteenth century the short distance trade along the coast to Hamburg declined. Frisians got their beer from Holland instead.[23] The shorter distances made it feasible to use smaller vessels, those designed for inland waterways. Such boats could make more frequent trips and more effectively distribute risk. There was little change since shippers had used such vessels to transship German beer before. The increased use of such inland craft came in the wake of significant improvement in their rig. It was some time in the thirteenth century that Dutch shipbuilders developed the simple fore-and-aft rig of two triangular sails in the same plane as the boat's keel. Such an arrangement, the forerunner of modern small boat rig, sharply decreased labour requirements.[24] The savings to shippers were probably being reaped just as the volume of beer produced and shipped in Holland was rising.

The increased production of beer in Holland meant that brewers needed more raw materials. The principal ingredient of beer was and is grain. The fall of population after the Black Death (1346-50) made it possible for farmers to concentrate their efforts on the better land and so increase grain output per hectare under cultivation and also per worker. Dutch agriculture was apparently already efficient compared to farming in other parts of Europe.

Despite those facts the sharp increase in beer output suggests that Dutch brewers had to import at least part of their grain from other sources. Hamburg had long been a source of grain. It was a market for rye and wheat brought down the Elbe and surplus grain went on by ship to markets in Friesland, Holland and even Flanders. On the other hand the relatively efficient Netherlands farms may have been a source of barley for Hamburg brewers, at least from the end of the fourteenth century.[25] Farmers throughout northern Europe apparently turned to growing more barley after the Black Death since its price did not fall as much or as fast as that of wheat. The reason for the buoyancy of barley prices was presumably caused by brewers' demand.

The ability of Netherlands farmers to turn to barley which would earn them a greater income depended in turn on there being a supply of food grains from other sources. At the end of the fourteenth century German shipbuilders developed a new type of cog joining the best features of that type with the hulk. The new combined type lowered the cost of moving surplus grain from the Baltic to western Europe. The trade had started by the end of the thirteenth century. Shrinking markets in the Baltic after 1350 led producers to look west for possible buyers. Danzig even exported grain already malted and ready for brewing, that by the end of the fifteenth century if not earlier. The volume shipped climbed all the faster as the cost of transport dropped. Holland ports were able to accomodate the new larger type. About half of the cost of Baltic grain delivered in Bruges was the shipping cost so improvements in ship design could and did pay handsome dividends for shippers.[26] They increased the supply of all grains in Holland and so in turn made it easier and less expensive for brewers to produce their product. The combined cog-hulk was well-suited to carrying a bulk commodity like grain. Beer was a bulky commodity too, made even more bulky by the barrels in which it was shipped which took up space as well. It is doubtful, incidentally, that consumers in the Netherlands returned the empty barrels in which Hamburg beer was imported. Moving grain was always easier than moving beer. Grain did not need special treatment. It did not have to be packed in barrels. There was no danger of damage to the container. There was no fear that moving the good around would hurt it. Disturbing beer in the cask could speed up fermentation and certainly cause the beer to be cloudy. In general grain put less of a demand on shippers and on ships. If that was not enough it was always cheaper to move grain than beer. Beer is much heavier than the grain from which it is made since it is, after all, more than 90% water.

The share of the Flemish market held by beers from Hanse towns rose from 1388 to 1411, that at the expense of Holland beers. The German success was temporary though. By the end of the fifteenth century such exports had shrunk dramatically. A reduction of the excise tax charged on their beer in Holland had helped Hamburgers in that market from 1403 to 1418. That too proved temporary. When the import duty was reinstituted the income from it was small reflecting the sharply lower volume of beer imports. The society of Hamburg *liggers* in Amsterdam still existed in 1407 but that appears to have been the last mention of them. The end of the fourteenth century proved to the first high point of north German beer exports.[27]

North German brewers, faced with the loss of their largest export market in the Low Countries did not abandon brewing nor brewing for export. They continued to produce for their own towns and for nearby villages. They also produced high quality beers such as Jopen or Danziger beer for the wealthy of the Low Countries. Danzig proved to be both an exporter and an importer of beer, in the latter case from Hamburg and Lübeck. German brewers sought out and successfully penetrated new markets especially in Scandinavia. The loss of the Netherlands market probably led Hanse merchants to concentrate more on northern countries as an outlet. Hamburg beer did not fare well there but breweries of the Wendish towns, especially Wismar, Stralsund and Rostock were still prominent exporters to Sweden and Norway in the sixteenth century. In the region around Bergen consumers developed a taste for strongly hopped beer because of the flow of German imports. Sweden supplied hops, especially to Danzig. In 1491 14% by value of total Swedish exports were hops sent to Danzig. Even Amsterdam by the late fifteenth century had become a source of hops for Danzig though the volume was small. Hops did not flow in the other direction. Though hops had been exported from Wismar as early as 1328 sales to Denmark were prohibited by the Hanse on political grounds in 1363, 1367, 1368 and 1422. Wismar issued similar restrictions repeatedly in the fifteenth century.[28] Wismar had been exporting beer since about 1327 and continued to do so. In 1464/5 there were 182 brewers in that town alone. North German towns, most notably Lübeck, tended to prohibit the import of beer from other towns along the coast. Swedish regulations on the import of Wismar and other beers were not as strict as in the Hanse towns which led shippers to look north for a new market.[29] The success of German brewers in selling their product in Scandinavia was based on the greater availability of grain in their home towns, on the quality of the product and also on the political strength of the Hanse, on its ability to pressure local governments not to protect domestic brewing. The ability of the Hanseatic League to impose favourable regulations on others declined in general in the fifteenth century though it did not decline as much in Scandinavia.[30]

The typical carrier for that exported beer was presumably the cog. From Hamburg in 1404 and 1405 three ships left with cargoes of 653, 375 and 308 barrels of beer, each barrel containing more than one hectolitre. In the fifteenth century beer ships left the Elbe with 300 to 700 barrels on board.[31] Cogs were large enough to handle those volumes easily. The combined form of cog and hulk had no trouble taking the cargoes over the open sea to ports in the Low Countries or Scandinavia. But transporting beer over the open sea appears to have always been the exception. The trip from Wismar to Denmark or north to Stockholm and even the voyage to Bergen in Norway could be and typically was a coastal one.

Beer was never able to sustain high transport costs. The volume compared to its value made transport costs a significant share of the delivered price. Governments always found beer an item worth taxing. The resilient demand in the face of rising prices and even more resilient in periods of rising real incomes made it ideal for the tax collector. Imported beer was an excellent candidate since it had to pass through fixed points and so could not avoid the customs collector. Beer could be produced virtually anywhere so there was always some competition from local sources of supply. With beer available from domestic brewers and tax collectors searching for ways to raise their incomes the international trade in beer was all the more sensitive to changes in transport costs. Typically only better beer was traded. When threatened with higher transport costs, higher taxes or greater local production the standard solution was to brew and trade in beers of higher quality and therefore of higher value. That is what Hansards did in the fifteenth century. The original boom in the beer trade which started at the end of the thirteenth century was caused by the introduction of a new product to international commerce: hopped beer. The slow development in the trade before that may well have been caused by slow developments in the design of ships. The boom in the beer trade lasted for a little more than a century. The relatively short period of high and rising volumes in the trade ended because of the taxation and protectionist policies of governments, especially of the County of Holland, because the cost of transport did not fall appreciably and because, with the improvements in ship design around 1400, any fall in the cost of moving beer was equalled or exceeded by a fall in the cost of moving grain. If shippers could move the raw material, which incidentally had many more uses, at savings similar to those in moving the final product it made more sense to ship the grain. That is precisely what Hanse merchants did in the fifteenth century.

The cog was not the only vessel used in northern Europe in the

Middle Ages. Even Hanse shippers appear to have used other ship types in both the Baltic and the North Seas. But it seems the cog was the dominant type for moving beer. The trade in beer was originally coastal and remained so up to and through the end of the thirteenth century. Even after the boom in the shipping of hopped beer began traders typically carried most of the product along the Frisian coast from Hamburg to Stavoren and Amsterdam. There is no evidence to suggest that Hansards used anything other than cogs to take beer to Scandinavian ports in the fifteenth century. It may be that cogs were still in use in inland trade in the Netherlands carrying beer in the seventeenth century.[32] In the twelfth and much of the thirteenth century it appears that traders had problems moving beer from Bremen and Hamburg to west Friesland and to Holland. By 1300 any problems had been solved since volumes rose significantly and continued to do so. There may have been technical improvements in cogs at that time, in carrying capacity, in seaworthiness which offered advantages to the beer producers. The sharp and relatively sudden rise in demand for shipping services to carry hopped beer to the Low Countries may have pressured shipbuilders into modifying their designs or into more fully exploiting existing and known design features. It is doubtful that the keel was added to the flat-bottomed cog only in the late thirteenth century. Archaeological evidence is apparently not precise but a date after 1250 seems too late. On the other hand it is possible that builders, especially in Hamburg, may have gone over completely to the new and improved design modifying and improving it because of the sudden need for more vessels in the beer trade. Even though the cogs were better than before they apparently still were not good enough or efficient enough to convince shippers to give up their traditional coastal route or give up going through the port of Amsterdam. The structure and organization of the trade remained. That included paying the taxes levied by the counts of Holland. Technical change was not great enough to make it advisable or feasible to abandon what was known. The improvements made in cog design at the end of the fourteenth century came at a time when the beer trade was reaching its medieval peak. Oddly enough those improvements may have played a role in the decline of the beer trade. Certainly the new type could carry beer for less than before and in larger volumes. But it could do the same for grain. The greater availability of grain, in part because of better ships, combined with changes in the economy and in government policies to decrease the need for carrying beer over long distance.

The history of brewing in northern Europe in the Middle Ages cannot be separated from the history of ship design. That statement is true for any industry be it medieval or modern. That ability to move goods economically is critical to growth, development and technical advance in any industry. In turn the presence of goods at competitive prices was and is critical for technical advance in the design and building of the ships which carry the goods. For brewers in northern Europe in the thirteenth, fourteenth and fifteenth centuries there was certainly a great deal of interest in how to make beer and how to make it better. Governments shared that interest, albeit for purely fiscal reasons. The concern for improvements in techniques did not generate the birth of technological society. But the interest in technical advance on the part of brewers and of shipbuilders and the mutual relationship between the two, created by the international market in beer, did promote continuing improvements in ways of meeting demand. It may be that the beginnings of the concentration on how to do things, on how to improve technology which was to be and is still a hallmark of European society, can be found in the creative tension between advances in two technologies such as that generated in the fourteenth century between brewing and shipbuilding.

NOTES

1. The John Simon Guggenheim Foundation and the Social Sciences and Humanities Research Council of Canada supported the research for this paper.
2. Odd Nordlund, *Brewing and Beer Traditions in Norway. The Social Anthropological Background of the Brewing Industry*, Oslo: Universitetsforlaget, 1969, 283.
3. Wolf Bing, "Hamburgs Bierbrauerei vom 14. bis zum 18. Jahrhundert," *Zeitschrift des Vereins für Hamburgischen Geschichte*, 14(1909), 214.
4. G. Doorman, *De Middeleeuwse Browerij en de Gruit*, The Hague: Martinus Nijhoff, 1955, 24, 27-34. O. Nordlund, *Brewing and Beer* , 126, 132-4, 216-22.
5. Walter Horn and Ernest Born, *The Plan of St. Gall. A Study of the Architecture and Economy of, and Life in the Paradigmatic Carolingian Monastery*, Berkeley: University of California Press, 1979, vol. II, 249-264. G. Doorman, *De Middeleeuwse Brouwerij...*, 39-40. O. Nordlund, *Brewing and Beer...*, 57-58, 203. Harald Thunæus, *Ölets historia i Sverige*, Stockholm: Almqvist & Wicksell, 1968-69, vol. I, 70.
6. Friedrich Techen, "Das Brauwerk in Wismar," *Hansisches Geschichtsblätter*, XXI(1915), 316-326.
7. Wolf Bing, "Hamburgs Bierbrauerei...," 220-221. Raymond DeRoover, "Les comptes communaux et la comptabilité communale à Bruges au XIVe siècle," *Finances et Compatabilité Urbaines Du XIIIe au XVIe Siècle*, Colloque International Blakenberge, 6-9-IX-1962, Blakenberge: Pro Civitate, 1964, 94, 100-101. Gerald Stefke, *Ein Städtisches Exportgewerbe Des Spätmittelalters in seiner Entfaltung und ersten Blüte*, Hamburg, 1979, xxx-xxxvii, 55-58.
8. Gerald Stefke, *Ein Städtisches Exportgewerbe...*, xliv-xlv. Hans Hunteman, *Das deutsche Braugewerbe vom Ausgang des Mittelalters bis zum Beginn der*

Industrialisierung, Nürnberg: Verlag Hans Carl, 1971, 16.
9. G. Doorman, *De Middeleeuwse Brouwerij*..., 18.
10. H.J. Smit, "De Registers van den Biertol te Amsterdam," *Bijdragen en Mededeelingen van het Historisch Genootschap*, XXXVIII(1917), 1. G. Doorman, *De Middeleeuwse Brouwerij*..., 18. Wolf Bing, "Hamburgs Bierbrauerei...," 219.
11. H.J. Smit, "De Registers van den Biertol...," 31-34, 37-40, 45-49.
12. Wolf Bing, "Hamburgs Bierbrauerei...," 243. Hans Hunteman, *Das deutsche Braugewerbe*..., 11-12, 16. F.C. Lane, "Tonnages, Medieval and Modern," *The Economic History Review*, Second series, XVII(1964), 214-220. Gerald Stefke, *Ein Städtisches Exportgewerbe*..., xlix-liv, 63-78, 118-122. Friedrich Techen, "Das Brauwerk in Wismar,", 165-167.
13. F.C. Berkenvelder, "Frieslands handel in de late Middeleeuwen," *Economisch-Historisch Jaarboek*, XXIX(1963), 139-140, 166-167.
14. F. Ketner, *Handel en Scheepvaart van Amssterdam in de Vijftiende Eeeuw*, Leiden, E.J. Brill, 1946, 5-6. H.J. Smit, "De Registers van den Biertol...," 6. Hans Hunteman, *Das deutsche Braugewerbe*..., 14-15.
15. Ole Crumlin-Pedersen, "Cog-Kogge-Kaag Træk æf en frisisk skibstypes historie," *Handels- of Sofartsmuseets Årborg*(1965), 81-144.
16. Gerald Stefke, *Ein Städtisches Exportgewerbe*..., 34, 67-68. Wolf Bing, "Hamburgs Bierbrauerei...," 244-245, 253.
17. H. Brugmans, "De Binnenvaart door Holland in De Deertiende Eeuw," *Mededelingen der Koninklijke Akademie van Wetenschappen*, Afdeling Letterkunde, 54, B, 5(Amsterdam, 1922), 86-88, 91-95, 97-98.
18. Wolf Bing, "Hamburgs Bierbrauerei...," 212, 217-222.
19. Gerald Stefke, *Ein Städtisches Exportgewerbe*..., 123, 127-131. Friedrich Techen, "Das Brauwerk in Wismar," 197-202.
20. G. Doorman, *De Middeleeuwse Brouwerij*..., vi, 18, 50-51.
21. G. Doorman, *De Middeleeuwse Brouwerij*..., 48-50, 59-60. H.P.H. Jansen, "Holland's Advance," *Acta Historiae Neerlandicae*, X(1978), 12.
22. Wolf Bing, "Hamburgs Bierbrauerei...," 227-230.
23. F.C. Berkenvelder, "Frieslands Handel...," 168-169, 176, 179.
24. Richard W. Unger, *Dutch Shipbuilding Before 1800 Ships and Guilds*, Assen, Van Gorcum, 1978, 54-56.
25. F.C. Berkenvelder, "Frieslands Handel...," 156. Hans Hunteman, *Das deutsche Braugewerbe*..., 30-31. Gerald Stefke, *Ein Städtisches Exportgewerbe*..., 47-49.
26. Victor Lauffer, "Danzigs Schiffs- und Waarenverkehr am Ende des XV. Jahrhunderts," *Zeitschrift des Westpreussischen Geschichtsvereins*, XXXIII (1894), 39-41. Jorma Ahvenainen, *Der Getreidehandel Livlands im Mittelalter*, Helsinki: Keskuskirjapaino, 1963, 67, 71, 76-77. Richard W. Unger, *The Ship in the Medieval Economy, 600-1600*, London: Croom-Helm, 1980, 169-171.
27. Hans Hunteman, *Das deutsche Braugewerbe*..., 14, 18. Wolf Bing, "Hamburgs Bierbrauerei...," 226, 235-36. F. Ketner, *Handel en Scheepvaart van Amsterdam*..., 16, 124. Victor Lauffer, "Danzigs Schiffs- und Waarenverkehr...," 28. H.J. Smit, "De Registers van den Biertol...," 5-6.
28. Harald Thunœus, *Ölets historia i Sverige*, I, 71-72. Wolf Bing, "Hamburgs Bierbrauerei...," 225. Friedkrich Techen, "Das Brauwerk in Wismar," 324-326. Odd Nordlund, *Brewing and Beer*..., 225. F. Ketner, *Handel en Scheepvaart van Amsterdam*..., 166-167.
29. Friedrich Techen, "Das Brauwerk in Wismar," 266, 269, 176-178, 182, 194-195.
30. Hans Hunteman, *Das deutsche Braugewerbe*..., 19.
31. Wolf Bing, *Hamburgs Bierbrauerei*..., 250-51.

32. The cog remained in use in Holland as an inland trader into the nineteenth century. It seems to have been popular with traders to the port of Brouwershaven, Zeeland, and the name of the town suggests a connection with moving beer. The design of the seventeenth century cog was apparently little altered from that of its fourteenth century predecessor. Richard W. Unger, *Dutch Shipbuilding...*, 53. Nicolœs Witsen, *Architectura Navalis*, Amsterdam, 1690, 192.

INDEX

For Product Safety Concerns and Information please contact our EU
representative GPSR@taylorandfrancis.com
Taylor & Francis Verlag GmbH, Kaufingerstraße 24, 80331 München, Germany

www.ingramcontent.com/pod-product-compliance
Ingram Content Group UK Ltd.
Pitfield, Milton Keynes, MK11 3LW, UK
UKHW021503240926
475702UK00035B/222
* 9 7 8 1 1 3 8 3 8 6 2 4 2 *